P9-DFF-216

METHODS IN MOLECULAR BIOLOGY™

Series Editor
John M. Walker
School of Life Sciences
University of Hertfordshire
Hatfield, Hertfordshire, AL10 9AB, UK

For other titles published in this series, go to
www.springer.com/series/7651

Cardiovascular Genomics

Methods and Protocols

Edited by

Keith DiPetrillo

Novartis Institutes for BioMedical Research, East Hanover, NJ, USA

Humana Press

Editor
Keith DiPetrillo
Novartis Institutes for BioMedical Research
1 Health Plaza, Bldg. 437
East Hanover, NJ 07936
USA
keith.dipetrillo@novartis.com

ISSN 1064-3745 e-ISSN 1940-6029
ISBN 978-1-60761-246-9 e-ISBN 978-1-60761-247-6
DOI 10.1007/978-1-60761-247-6

Library of Congress Control Number: 2009927009

Printed on acid-free paper

springer.com

Preface

Heart disease and stroke are the leading causes of death worldwide *(1)*. Atherosclerosis and hypertension are major risk factors for the development of heart disease and stroke, and kidney damage is an important risk factor for overall cardiovascular mortality. Despite the availability of medicines to ameliorate these illnesses, there remains a clear need for more effective therapies to reduce the global burden of cardiovascular disease.

One strategy for delivering improved therapies is deciphering the genes and cellular pathways that underlie cardiovascular disease. The last decade has brought tremendous advancement in technology for understanding the genome, including whole genome sequences of multiple species, platforms for high-quality measurement of genome-wide gene expression, high-throughput genotyping of single-nucleotide polymorphisms and copy number variations, as well as statistical packages for integrating phenotype information with genome-wide genotype and expression data. While these advances have led to better understanding of genome structure and function, the goal of identifying genes that cause cardiovascular disease remains distant. This is partly because linking genes with physiological function frequently requires combining genetic, physiological, bioinformatics, and statistical methods, a difficult task for most laboratories to achieve without collaboration. Therefore, the objective of this book is to provide methods for cardiovascular phenotyping of rodent models and for statistical and bioinformatic integration of phenotype data with genome-wide genotype and expression data. Understanding these diverse methods can allow an individual laboratory to utilize these genomic methods independently or to be better prepared to collaborate with scientists having expertise in other disciplines to uncover genes affecting cardiovascular disease.

Because mice and rats are the primary experimental species for genomic studies, Chapters 1–9 of the book focus on methods for evaluating cardiovascular phenotypes in these rodent models. Since some phenotyping methods, such as non-invasive and invasive blood pressure measurement, assessment of stroke, or echocardiographic evaluation of heart failure, are useful for evaluating both mice and rats, the protocols for both species are presented in the same chapter. Other methods, such as evaluating renal function and surgical induction of heart failure, have been commonly used in rat models, so chapters discussing these methods are focused on applying the protocol to mice. In the case of atherosclerosis, one chapter presents an experimental protocol for examining atherosclerosis in rats and a second chapter discusses how best to employ the various protocols that have been published for evaluating atherosclerosis in mice. Overall, the protocols presented in Chapters 1–9 should help investigators to reliably assess several important cardiovascular phenotypes in mouse and rat models.

Chapters 10–18 focus on statistical and bioinformatics methods for integrating phenotype data with genome-wide genotype and gene expression data. The first six chapters discuss statistical and computational methods for linkage mapping and association studies, including single gene, genome-wide, and haplotype association studies, in experimental species or humans. These methods are important for mapping genes that underlie cardiovascular phenotypes. The subsequent chapters present bioinformatics methods for analyzing gene expression data from microarray studies and for linkage mapping of genome-wide gene expression in humans and rodents. The final chapter discusses a bioinformatics strategy for intergrating phenotype, genotype, and gene expression data to identify the genes most likely to affect the phenotype of interest.

Combining the phenotyping protocols presented in Chapters 1–9 with the statistical and bioinformatics methods outlined in Chapters 10–18 will facilitate cardiovascular genomics studies in many laboratories. However, the statistical and bioinformatics methods described in Chapters 10–18 are not disease-specific and can be used for genomic analysis of most phenotypes. Hopefully, these protocols will enable researchers to identify causal genes and novel molecular targets that lead to new treatments for cardiovascular disease.

Keith DiPetrillo

Reference

(1) Lopez, AD, Mathers, CD, Ezzati, M, et al. (2006) Global and regional burden of disease and risk factors, 2001: systematic analysis of population health data. Lancet **367**(9524):1747–1757.

Contents

Contributors

Arthur Berg • *Department of Statistics, University of Florida, Gainesville, FL, USA*
Heather Hogie • *Data Sciences International, Saint Paul, MN, USA*
Andrew Brass • *School of Computer Science and Faculty of Life Science, University of Manchester, Manchester, UK*
Yun Cheng • *Department of Statistics, University of Florida, Gainesville, FL, USA*
Gary Churchill • *The Jackson Laboratory, Bar Harbor, ME, USA*
Alan Daugherty • *Cardiovascular Research Center, University of Kentucky, Lexington, KY, USA*
Keith DiPetrillo • *Novartis Institutes for BioMedical Research and Novartis Pharmaceutical Corporation, East Hanover, NJ, USA*
Aysan Durukan • *Department of Neurology, Helsinki University Central Hospital, Helsinki, Finland*
Minjie Feng • *Novartis Institutes for BioMedical Research and Novartis Pharmaceutical Corporation, East Hanover, NJ, USA*
Paul Fisher • *School of Computer Science, University of Manchester, Manchester, UK*
Lude Franke • *Groningen Bioinformatics Center, Groningen Biomolecular Sciences and Biotechnology Institute, University of Groningen, Haren, The Netherlands; Institute of Cell and Molecular Science, Barts and The London School of Medicine and Dentistry, London, UK; Genetics Department, University Medical Centre Groningen and University of Groningen, Groningen, The Netherlands*
Deborah A. Howatt • *Cardiovascular Research Center, University of Kentucky, Lexington, KY, USA*
Daniel A. Huetteman • *Data Sciences International, Saint Paul, MN, USA*
Ritsert C. Jansen • *Groningen Bioinformatics Center, Groningen Biomolecular Sciences and Biotechnology Institute, University of Groningen, Haren, The Netherlands; Genetics Department, University Medical Centre Groningen and University of Groningen, Groningen, The Netherlands*
Stephen Kemp • *School of Biological Sciences, University of Liverpool, Liverpool, UK*
Ron Korstanje • *The Jackson Laboratory, Bar Harbor, ME, USA*
Yao Li • *Department of Statistics, University of Florida, Gainesville, FL, USA*
Jing Liu • *Novartis Institutes for BioMedical Research and Novartis Pharmaceutical Corporation, East Hanover, NJ, USA*
Hong Lu • *Cardiovascular Research Center, University of Kentucky, Lexington, KY, USA*
David L. Mattson • *Department of Physiology, Medical College of Wisconsin, Milwaukee, WI, USA*
Christopher Newton-Cheh • *Center for Human Genetic Research, Massachusetts General Hospital, Boston, MA, USA*
Harry Noyes • *School of Biological Sciences, University of Liverpool, Liverpool, UK*

DEBRA L. RATERI • *Cardiovascular Research Center, University of Kentucky, Lexington, KY, USA*
DEAN F. RIGEL • *Novartis Institutes for BioMedical Research and Novartis Pharmaceutical Corporation, East Hanover, NJ, USA*
JAMES C. RUSSELL • *Alberta Institute for Human Nutrition, University of Alberta, Edmonton, AB, Canada*
MARTINA SCHINKE • *Novartis Institutes for BioMedical Research, Cambridge, MA, USA*
KEITH SHEPPARD • *The Jackson Laboratory, Bar Harbor, ME, USA*
JONATHAN B. SINGER • *Novartis Institutes for BioMedical Research, Cambridge, MA, USA*
J. GUSTAV SMITH • *Program in Medical and Population Genetics, Broad Institute of Harvard and MIT, Cambridge*
RANDY SMITH • *The Jackson Laboratory, Bar Harbor, ME, USA*
ROBERT STEVENS • *School of Computer Science, University of Manchester, Manchester, UK*
JOSEPH SZUSTAKOWSKI • *Novartis Institutes for BioMedical Research, Cambridge, MA, USA*
OLEG TARNAVSKI • *Novartis Institutes for BioMedical Research, Cambridge, MA, USA*
TURGUT TATLISUMAK • *Department of Neurology, Helsinki University Central Hospital, Helsinki, Finland*
BRUNO M. TESSON • *Groningen Bioinformatics Center, Groningen Biomolecular Sciences and Biotechnology Institute, University of Groningen, Haren, The Netherlands*
SHIRNG-WERN TSAIH • *The Jackson Laboratory, Bar Harbor, ME, USA*
RONGLING WU • *Department of Statistics, University of Florida, Gainesville, FL, USA; Center for Statistical Genetics, Department of Public Health Sciences, Pennsylvania State College of Medicine, Hershey, PA, USA; Department of Statistics, Pennsylvania State University, University Park, PA, USA*
SONG WU • *Department of Statistics, University of Florida, Gainesville, FL, USA*
YUNYU ZHANG • *Novartis Institutes for BioMedical Research, Cambridge, MA, USA*
FEI ZOU • *Department of Biostatistics, University of North Carolina, Chapel Hill, NC, USA*

Chapter 1

Modes of Defining Atherosclerosis in Mouse Models: Relative Merits and Evolving Standards

Alan Daugherty, Hong Lu, Deborah A. Howatt, and Debra L. Rateri

Abstract

Mouse models have become the most common model for defining mechanisms of atherosclerotic disease. Many genetic manipulations have enabled the development of atherosclerosis in mice due to either endogenous or diet-induced hypercholesterolemia. This availability of lesion-susceptible mice has facilitated many studies using pharmacological and genetic approaches. Unfortunately, this expansive literature on mouse atherosclerosis has generated many contradictions on the role of specific pathways. A contributor to these inconsistencies may be the multiple modes in which atherosclerosis is evaluated. Also, for each specific technique, there are no consistent standards applied to the measurements. This chapter will discuss the imaging, biochemical, and compositional modes of evaluating atherosclerosis with suggestions for standard execution of these techniques.

Key words: Atherosclerosis, mouse, imaging, cholesterol.

1. Introduction

Atherosclerotic diseases, such as ischemic heart disease and stroke, are the most common causes of morbidity and mortality in developed countries. There are some well-validated pharmacological approaches for attenuating atherosclerotic diseases, of which the most uniformly successful focus on dyslipidemias and the renin–angiotensin system (1). However, there are still considerable unmet needs for improved therapies in patients with atherosclerotic diseases. Therefore, the identification and validation of new targets to prevent atherosclerosis remain a high priority. A common step in identifying and validating new therapeutic targets and strategies involves the use of animal models of atherosclerosis. The

K. DiPetrillo (ed.), *Cardiovascular Genomics*, Methods in Molecular Biology 573,
DOI 10.1007/978-1-60761-247-6_1, © Humana Press, a part of Springer Science+Business Media, LLC 2009

most common species used in contemporary studies is mouse, in which atherosclerosis can be induced by dietary and genetic manipulations (2).

Human atherosclerosis is a chronic disease that progresses over decades in a clinically silent process. The evolution of lesions over these decades involves complex compositional changes in the relative presence of different cell types. Atherosclerotic diseases most frequently become a clinical imperative when lesions evolve to structures that promote acute thrombosis resulting in restricted blood flow to vital organs, such as the heart and brain. Many mouse models of atherosclerosis recapitulate facets of the formative phase of human disease, although there is less agreement that lesions in mouse models develop to the stage of generating atherosclerosis-associated thrombosis (3, 4).

The most common mode of assessing the clinical presence of atherosclerosis is by procedures such as angiography and ultrasound (external and intravascular) in which the primary endpoint is lesion size. Lesion size is also the major endpoint for most mouse atherosclerosis studies. While size continues to be an important parameter for assessing atherosclerosis severity, it is becoming increasingly clear that the common clinical manifestation of atherosclerosis occurs due to acute thrombosis, which is determined by lesion composition (5). As a consequence, there has been an increased emphasis on compositional analysis of lesions from mouse models of atherosclerosis.

There is now a vast literature that has used mouse models to determine atherogenic mechanisms and the effects of interventions that promote or inhibit the development of disease. As can be seen in **Fig. 1.1**, there was a small number of studies using mouse models of atherosclerosis prior to the generation of

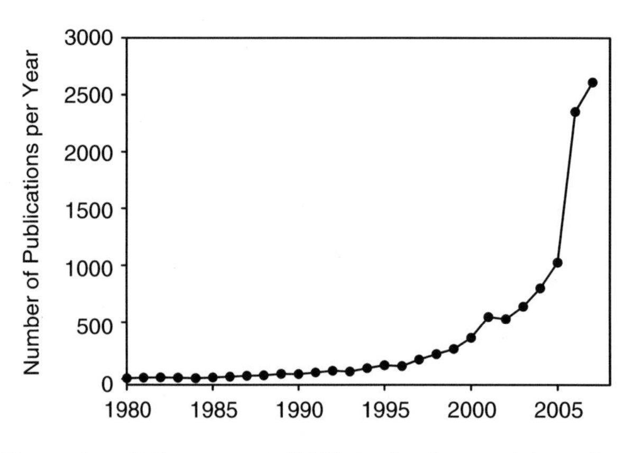

Fig. 1.1. The number of hits per year in PubMed using the search terms "mouse" and "atherosclerosis."

genetically engineered mice to enhance lesion susceptibility. Since the availability of these mice, primarily apolipoprotein (apoE) −/− (6, 7) and low-density lipoprotein (LDL) receptor −/− (8) mice, in the early 1990s, there has been a precipitous increase in publications. Unfortunately, this large literature is replete with contradictory studies, particularly in studies using compound-deficient mice in which a gene of interest is manipulated in an atherosclerosis-susceptible background. For example, genetic manipulations of class A scavenger receptors have shown this protein to promote, retard, or have no effect on atherosclerosis (9–12). There are also many examples in which the same genetic manipulation has generated different results (13, 14).

There are several potential reasons for the contradictory literature. One is likely to be the different methods of quantifying atherosclerosis. **Table 1.1** provides selected references that have detailed protocols for measuring mouse atherosclerosis. Even within a specific method of lesion analysis, there are several approaches in its execution that could lead to different results. The rigor in the application of standards will differ between forums. For example, the data needed to assist a pharmaceutical company in deciding to advance a potential anti-atherosclerotic compound to a human study may be far beyond the reasonable scope of data needed to justify a more incremental advance that is common in publications. The major emphasis of this chapter is to provide criteria for critically evaluating results from mouse models of atherosclerosis.

Table 1.1
References that provide detailed protocols for evaluation of mouse atherosclerosis experimental designs and measurements

	Reference
Experimental design	Daugherty and Rateri (20) Whitman (30)
Aortic root analysis	Paigan et al. (15) Purcell-Huynh et al. (18) Daugherty and Whitman (16) Baglione and Smith (17)
En face analysis	Tangirala et al. (19) Daugherty and Whitman (16)
Biochemical analysis	Daugherty and Whitman (16)
Compositional analysis	Lu et al. (26)

2. Measurements of Atherosclerosis

2.1. Ex Vivo Imaging Methods

The most common mode of assessing lesion size in mice is by removing aortas (*see* **Note 1**) and using imaging software to define the size of lesions in either tissue sections of aortic roots or the intimal surface of filleted aortas. Fortunately, both of these measurements may be performed on the same mouse, although this has not been common practice.

2.1.1. Aortic Root

The analysis of atherosclerosis in the aortic root was the first described process of measuring lesions in mice and remains as one of the most commonly used (15). The aortic root is the site of the greatest proclivity to developing lesions in mice and aortic root analysis is commonly the only method amenable to studies in which profound hyperlipidemia was not achieved to promote robust atherogenic responses, such as studies in which modified diets are fed to C57BL/6 mice that are wild type for genes such as LDL receptors and apoE. Detailed protocols for tissue sectioning and staining have been described previously (16, 17). However, there are many issues regarding the execution of these methods that may have profound impact on the interpretation of results.

1. *What region of the aortic root should lesions be measured?*

   The original description of aortic root analysis was restricted to a specific region (15). Lesion size was quantified in the aortic root region that was proximal to the valve cusps where the aorta has a rounded appearance. No lesions were measured in the aortic sinus, as defined by the presence of aortic valves. In many subsequent publications measuring atherosclerosis in mice, it became common to include the aortic sinus in the measurement of atherosclerosis. An informal assessment of the literature shows that most publications quantify lesions in the aortic sinus only.

   Although the original description of measuring lesions in the aortic root did not include the sinus, it is not clear whether there is any adverse consequence of measuring atherosclerosis in different regions of the aortic root. However, the size of lesions, as defined by the area on tissue sections, may vary greatly along the length of the aorta (18). Therefore, the selection of the region for acquiring tissue sections will have a major impact on the measurements and requires a reliable landmark aortic structure. The most common landmark is the appearance of aortic valves from distal sectioning of the myocardium. In the description of the original method and in our own experience, this tends to be a variable landmark. Instead, the disappearance of the valve cusps provides a more reproducible landmark.

To overcome the concern of obtaining aortic sections from different regions, serial sections should be obtained throughout lesions in aortic roots. The length of this can vary depending on the severity of the disease. In addition to overcoming the potential error of selecting different regions, this also permits the analysis of whether lesions have changes in thickness or length.

Overall, tissue sectioning throughout lesions in the aortic root overcomes potential errors due to sampling bias. It also provides an indication of whether the lesions are expanding in thickness and/or laterally. The most time-consuming task in aortic root analysis is usually the mounting of the tissue and cutting until the valves become visible. The visualization of the valves is the indication that subsequent sections need to be acquired for atherosclerosis analysis. Relative to mounting and initial cutting, it is not an onerous task to acquire sections throughout lesions and measure their area.

2. *How many tissue sections should be used to measure atherosclerosis?*

As noted above, there may be considerable differences in lesion areas in tissue sections from different aortic root regions. The combination of this variance (*see* **Note 2**) and the imprecision of defining the specific location for acquiring a tissue section highlights the importance of measuring lesion size in multiple tissue sections. However, it is not advisable to perform measurements on a randomly selected number of tissue sections. Instead, it is advisable to measure lesions in aortic root tissue sections that are spaced 80–100 μm apart, with the number of sections used to measure lesion size dictated by the distance they extend throughout the aortic root.

3. *What parameter should be used to define lesion "area"?*

Oil Red O staining of neutral lipids is frequently used to facilitate identification of lesions. This staining is particularly helpful when there is a preponderance of small lesions adjacent to valve cusps. In this region, the angle of the elastin fibers may hamper the clear definition of the intimal–medial boundary. However, the area of Oil Red O staining seldom covers the entire lesion area in a section. The major reasons for this include the presence of non-lipid-engorged smooth muscle cells, extracellular matrix, and non-neutral lipids, such as unesterified cholesterol. Consequently, neutral lipid staining routinely underestimates the size of lesions. Of greater concern is that change in composition can affect neutral lipid staining independent of size. Therefore, the analysis of lesion size usually requires the manual manipulation of image software to outline lesions defined by the internal elastic lamina and the luminal boundary. In advanced lesions in mice, there have been a limited number of published examples of the internal elastic lamina being fragmented and atherosclerotic lesions extending into the media. This scenario would call for some arbitrary decision on defining the lesion boundary. However, encroachment of lesions into the media is a relatively unusual phenomenon in mice.

4. *Representation of lesion size*

Image analysis software permits the use of calibrations to generate the absolute area of atherosclerotic lesions. These are represented either as mean lesion area per region or as a mean of multiple regions. Lesion size has also been represented as a percent of the lumen area. However, since lumen area may change during the atherosclerotic process, the representation of lesion areas in absolute units is preferable.

2.1.2. En Face *Analysis of Aortic Intima*

En face analysis of mouse atherosclerosis is a scaled-down version of an approach that has been used in many large animal species. Since the initial description of the *en face* analysis of lesions in aortic intima, its use has gradually increased (19). A detailed protocol for the analysis has been described previously (16, 20). Briefly, the aorta is dissected free, cut to expose the intimal surface, pinned, and digital images are acquired. The analysis of these images usually involves tracing the outline of the aorta to determine intimal area followed by determination of surface area of the lesion. Data are expressed either as lesion area or as a percent of intimal area.

1. *Standardization of area of analysis*

The original description of *en face* analysis of atherosclerosis in mice used the entire aortic intimal area, as defined by the site where the aorta emerges from the ventral aspect of the ventricle to the iliac bifurcation (19). In this study, lesions were measured in mice fed with high-fat diets for prolonged intervals that had extensive lesions throughout the aorta. Although prolonged high-fat feeding or protracted aging can lead to lesions throughout the aorta, lesions are most prevalent in the arch in the earlier stages of the disease (21). As a consequence, many studies only measure lesions in restricted areas of the aortic intima. Therefore, the area of lesion analysis must be strictly defined. Fortunately, unlike analysis of lesions in the aortic root, the ability to reference landmarks to identify and quantify the region of analysis is much easier in aortic intima that has been opened and pinned.

However, there are facets of area for standardization that need to be considered when analyzing lesions in the aortic arch. Common to all forms of *en face* analysis, there needs to be a clear definition of the proximal and distal ends of the tissue. In our own studies, the arch area is defined by the ascending aorta immediately after it emerges from the ventral aspect of the heart to 3 mm distal to the branch of the subclavian artery. There is also a need to standardize the inclusion of the three arterial branches off the aortic arch, particularly for the innominate artery, since it routinely contains large lesions for which inconsistent inclusion may influence data. In our own studies, we analyze the most proximal 1 mm of the innominate and left carotid arteries.

Another issue is standardizing the dissection of the aortic arch. The mode used in the original description of Tangirala et al. (19) made a longitudinal cut through the inner curvature of the aorta. The second cut was a longitudinal cut through the outer curvature, followed by mid-line cuts through the three branches off the aorta. This mode of cutting the tissue permits it to be pinned flat and avoids distortion of areas that can occur if the tissue is only cut longitudinally through the inner curvature.

2. *Use of neutral lipid stains*

Lesions have been visualized and quantified by the *en face* approach with or without lipid staining. The combination of the transparency of mouse aortas and the opacity of lesions negate the need to use stain to visualize lesions when they are sufficiently advanced. However, neutral lipid stains assist in visualizing small lesions, especially if they are in mice with more opaque aortas, such as occurs during angiotensin II infusion (22). As mentioned for analysis of aortic root, neutral lipid is not present throughout all forms of atherosclerosis, so there could be a discrepancy between area of stained and unstained lesions. Lipid staining also has the potential confounding issue that any residual adventitial adipose tissue will be intensely stained and may be confused with intimal staining of lesions by inexperienced operators. Finally, lipid staining requires the use of organic solvents, which negates the ability to perform tissue sterol analysis as a complimentary approach to lesion size evaluation.

3. *Multiple operators to analyze lesions*

The definition of lesion boundaries may frequently be an arbitrary decision, particularly in lesions that are small. Therefore, it is advisable to have at least two operators to determine the image area of lesions with an adjudication process for the operators to compare their data. It is best to do this while viewing the original aortic tissue under a microscope, since camera images may generate artifacts, such as glare, that may be mistaken for a lesion.

4. *Description of lesions to consider thickness*

As noted above, atherosclerotic lesions are formed initially in the aortic arch, with the inner curvature being a preferred site. However, continued lesion formation may be disproportionately represented by increased thickness compared to increased area. Therefore, operators should consider apparent thickness during their visual inspection of lesions. If the analysis shows no change in lesion area, but visual analysis is consistent with different thickness, secondary analysis for lesion quantification would be justified.

For most mouse studies, the measurement of aortic sterols may be considered as a complimentary approach (discussed in **Section 2.2**).

2.1.3. Atherosclerosis Analysis in Other Vascular Beds

1. *Innominate artery*

   The original detailed description of atherosclerotic lesions in the innominate arteries of hypercholesterolemic mice demonstrated more advanced compositional characteristics compared to other regions (23). Like analysis in the aortic root, tissue sectioning in this region has the advantage that it may be orientated so that tissue sections may be acquired perpendicular to the plane of the media. This feature reduces artifactual distortion of lesion size that may be generated by cutting lesions in different planes. Relatively few publications have measured lesion size in this region (24). The challenges to using this region include difficulty in establishing landmarks and unusually small length of lesions. The latter concern may be overcome by acquiring and measuring a sufficient number of sections along the length of lesions. Unlike the extensive literature on lesions in aortic roots, there is insufficient literature to judge whether 80–100-μm interval for tissue sections is appropriate for the innominate region.

2. *Longitudinal sections of aortic arch*

   Although there are a few publications that have analyzed lesions in longitudinal sections of the aortic arch, the primary concern is the ability to obtain tissue sections that do not distort the size of lesions due to the plane in which they are cut. For example, in the aortic arch, only sections that are acquired at the center of the arch will provide authentic lesion thickness and area measurements. To satisfy this requirement, the tissue region being cut would have to be flat. Also, only the sections in the mid-point of the arch may be used for lesion measurements. Given the difficulty of satisfying these criteria, other approaches to atherosclerotic lesion size analyses are preferred.

2.2. Biochemical Methods

Although not used in many publications, measurement of aortic sterol content has been one approach used to quantify mouse atherosclerotic lesions (25). This approach is only valid for lesions that are predominantly filled with unesterified and esterified cholesterol, which is the case for most mouse atherosclerotic lesions. The detailed approach to the analysis has been presented in a previous edition of *Methods in Molecular Medicine* (16).

1. *What is the optimal mode of normalizing the sterol content of lesions?*

   The main issue is the mode used to normalize the data. The only factor that is likely to cause confusion is tissue wet weight. Since the tissue is so small (only 10 mg if the total aorta is used), the wet weight of the tissue may be influenced greatly by either residual fluid or excessive removal of fluid. Although sterol content may be normalized to parameters such as protein and dry weight, it is preferable to use intimal area of the identical aortic areas that are used to measure lesion size by *en face* ex vivo imaging.

2.3. Compositional Analysis

Compositional analysis in the context of this chapter considers the use of histological and immunostaining techniques in tissue sections of atherosclerosis. The potential issues of acquiring immunostaining that is restricted to the development of chromogen only at loci of authentic antibody–antigen interaction have been discussed in a previous issue of *Methods in Molecular Medicine* (26).

The determination of mouse atherosclerotic lesion composition is relatively new and is frequently not described in much detail in publications. As an overview, it generally involves the demarcation of a lesion as an area of interest followed by some form of color segmentation analysis in which a threshold is set for a specific color or hue to determine the relative area of staining.

1. *How many sections should be quantified?*

As described for measurement of lesion area in the aortic root, the reliability of the measurement would be enhanced by the use of sufficient number of sections to prevent acquisition of data that are subject to sampling error. For example, in the case of a classic fibrolipid lesion, compositional analysis would detect a relatively high smooth muscle cell content at lesion edges with a progressive increase in macrophages and lipid toward the center of lesions. The pragmatic issue is the number of sections needed to provide authentic compositional data that encompass the heterogeneity of lesions versus the time needed to perform this analysis. No studies have been published to provide an indication of the variance of composition in mouse lesions, as performed in this region to determine lesion size (18). In the absence of data, it is difficult to suggest standards. However, it is clear that lesion composition frequently differs along the length of a lesion, which necessitates that compositional analysis be performed on multiple tissue sections.

2. *How to define area of staining?*

All histological and immunostaining techniques result in a range of color intensities that usually require an arbitrary decision to determine the border of the staining. Since there is seldom a situation in which a definite threshold exists for defining staining for a specific component, it is advisable to have at least two operators agree on the designation of the threshold.

3. *What is the meaning of area of a specific cell defined?*

Many cell markers are used to define the area of a specific cell that occupies a lesion in a tissue section. However, this interpretation will only be accurate if the antigen being immunostained is present throughout the cell in the plane viewed by the microscope. For example, CD68 is a lysosomal protein that is commonly used to determine "macrophage area." Immunostaining of very thin tissue sections would restrict the positively reacted area to just this intracellular organelle. However, much of the compositional analysis is performed on frozen tissues that are cut on a cryostat with a thickness of 8–10 μm. Since macrophages are ~7 μm in diameter, the two-dimensional view of the tissue section from the microscope could provide the impression that CD68 positivity decorates the entire cells. This impression would be enhanced by the use of many chromogens that smear slightly beyond the region of primary antibody interactions. However, macrophages within atherosclerotic lesions commonly become lipid engorged and may hypertrophy to 50 μm or more. This hypertrophy is the result of the intracellular deposition of lipids in droplets. Therefore, much of the area of a macrophage will be devoid of CD68. Consequently, the area of CD68 immunostaining will differ greatly from the actual macrophage area. Since it is unlikely that cell-specific markers will be found that cover the entire area of specific cell types, especially in hypertrophied states, the interpretation should be restricted to the area of chromogen development by the specific protein being detected rather than inferring an area covered by a cell type.

4. *Inferential interpretation of compositional analysis*

Data from compositional analyses are frequently used as a basis for using terms such as vulnerable and unstable in lesion analysis. These terms are highly controversial (3, 4). Part of this controversy stems from the disagreement on whether mouse atherosclerotic lesions progress to the stage of plaque rupture. Another aspect of this dispute is that vulnerable lesions in humans generally refer to a specific tissue configuration of a macrophage lipid-laden core encapsulated by smooth muscle cells and extracellular matrix that exhibit thinning at the shoulder regions (27). However, mouse compositional analysis infrequently considers the spatial distribution of components. Instead, it more commonly determines a ratio of staining of components such as neutral lipid, extracellular matrix components, macrophages, and smooth muscle cells, without regard to spatial arrangements. In the absence of a widely reproducible mouse atherosclerosis model that exhibits plaque rupture, it would be advisable to restrict the interpretation to a statement of the data rather than to include inferential suggestions of plaque vulnerability based on the data.

3. Issues for Establishing Standards Using Combined Approaches

The preference for each study to require multiple modes of quantifying atherosclerosis with compositional analysis in multiple vascular beds has to be balanced by the pragmatic aspects of time consumed to perform these techniques in a high-quality manner. This section provides some suggestions for guidelines to provide practical restrictions for evaluating atherosclerosis in mice. The following suggestions are based on our experiences in designing studies and peer reviewing numerous manuscripts on experimental atherosclerosis.

1. *Should more than one vascular bed be measured by ex vivo imaging?*

The most critical human diseases that arise from atherosclerosis are ischemic heart disease and stroke, due to atherosclerosis in the coronary and carotid arteries, respectively. It is not clear whether atherosclerotic lesions develop by identical processes in these two regions. Although there is some controversy (3, 4), development of atherosclerosis in mice does not generally lead to disease consequences that are similar to humans. Certainly, the regions in which atherosclerosis is usually quantified in mouse studies (aortic root, aortic intimal, and brachiocephalic artery) do not have direct correlations to the consequences of the disease in humans.

Given the potential for atherosclerotic lesion development by differing mechanisms in different vascular beds, a uniform change in lesion sizes in multiple regions in response to a specific manipulation provides an indication of the potential widespread applicability of the manipulation. However, as described at the outset, atherosclerosis studies are time-consuming in the generation and analysis of the lesions. Therefore, it is not clear whether every study requires measurement of atherosclerosis in more than one vascular bed. Rather, a rigorously performed study using a single vascular bed should provide sufficient information to provide insight into atherosclerotic mechanisms.

2. *Should ex vivo imaging be used in conjunction with another parameter to indicate atherosclerotic lesion size?*

En face analysis of lesions in the aortic intima is a common mode of analysis for determining the area of atherosclerotic lesions. A major deficiency of this approach is that it provides a two-dimensional measurement of a three-dimensional atherosclerotic lesion. Therefore, a potentially critical element of lesion thickness or volume is missed in this assay. If an intervention promotes a large difference in the area of lesions, as measured by *en face* analysis, it is hard to conceive a scenario in which the volume of the lesions would not also be greatly reduced. However, in the case of lesion areas being the same, but visual observation indicating a difference in lesion thickness, some measurement of lesion thickness or volume is warranted. Lesion thickness may potentially be defined on histological sections. While this approach may provide authentic information, it is also prone to the problems of sampling error as described for the aortic root. An alternative approach is the use of aortic sterol content. This has been used in past studies in which *en face* lesion measurements failed to discriminate between groups, but differences were readily apparent using sterol analysis (28). There are a few studies in which ex vivo image analysis demonstrated no changes, but visual inspection indicated thickness and volume changes that would warrant a combination of *en face* analysis and tissue sterol content.

3. *Should lesion size measurements be combined with compositional analysis?*

The answer to this question largely relies on the purpose of the study. If the emphasis of a study is that the mechanism reduces only size, then the need to perform compositional analysis is reduced. Compositional analysis is frequently applied to define a cause of changes in the dimensions of lesions. Unfortunately, it is not possible to define whether size changes are the cause or consequence of compositional changes. Therefore, unless the specific conclusion of a study requires the determination of lesion composition, it seems unreasonable for this analysis to be performed without justification. Based on these considerations, we provide **Table 1.2** as suggestions for standards to evaluate studies describing effects on atherogenic mechanisms.

Table 1.2
Summary of suggested standards for mouse atherosclerosis analysis

Ex vivo imaging	Aortic root	1. Section up to 1,000 μm of aortic root depending on lesion size. 2. Measure lesion area on sections spaced 80–100 μm throughout lesions in root. 3. Measure lesion areas by morphometric criteria of boundaries defined by internal elastin lamina and lumen. 4. Represent data as absolute areas. 5. At least two operators agree on lesion designation in images.
	En face	1. Standardize intimal area to the same area used for ex vivo imaging. 2. At least two operators agree on lesion designation in images.
Biochemical	Sterol measurement	1. Perform appropriate normalization process.

(continued)

Table 1.2 (continued)

Compositional	Histological and immunological staining	1. Multiple sections need to be analyzed to account for potential lesion heterogeneity. 2. At least two operators perform color segmentation analysis. 3. Strict interpretation of stained area to specific entity.
Analysis of multiple vascular beds		1. Preferable for defining the uniformity of the response, but not required.
Lesion size and composition		1. The combination is only warranted if required to substantiate the conclusions of the study. 2. If performed, given the potential for difference effects of lesion size in different vascular beds, composition analysis needs to be performed in the same region as size measurements.

4. Notes

1. Atherosclerotic lesions in the intima of arteries in mice are relatively easily dislodged from the media. This fragility of attachment is enhanced in mice that have undergone irradiation and repopulation with bone marrow-derived stem cells to determine atherogenic mechanisms (29). The analytical considerations described in this chapter are prejudiced on the assumption that lesions are not lost during processing.
2. Despite the uniformity of genetic background and environment in mouse atherosclerosis studies, it is common that there is a large variance in these studies. Consequently, there needs to be sufficient number of mice to obtain statistically robust results. Also, the variance and distribution of data are often not compatible with parametric analysis. Therefore, data frequently need to be analyzed using non-parametric tests that are usually less sensitive for detecting a statistical difference between groups.

Acknowledgments

The authors' laboratories are supported by the National Institutes of Health (HL08100 and HL62846).

References

1. Rader, DJ, Daugherty, A. (2008) Translating molecular discoveries into new therapies for atherosclerosis. *Nature* **451**, 904–913.
2. Daugherty, A. (2002) Mouse models of atherosclerosis. *Am J Med Sci* **323**, 3–10.
3. Schwartz, SM, Galis, ZS, Rosenfeld, ME, et al. (2007) Plaque rupture in humans and mice. *Arterioscler Thromb Vasc Biol* **27**, 705–713.
4. Jackson, CL, Bennett, MR, Biessen, EA, et al. (2007) Assessment of unstable atherosclerosis in mice. *Arterioscler Thromb Vasc Biol* **27**, 714–720.
5. Falk, E. (1999) Stable versus unstable atherosclerosis: clinical aspects. *Am Heart J* **138**, S421–S425.
6. Piedrahita, JA, Zhang, SH, Hagaman, JR, et al. (1992) Generation of mice carrying a mutant apolipoprotein-E gene inactivated by gene targeting in embryonic stem cells. *Proc Natl Acad Sci USA* **89**, 4471–4475.
7. Plump, AS, Smith, JD, Hayek, T, et al. (1992) Severe hypercholesterolemia and atherosclerosis in apolipoprotein-E-deficient mice created by homologous recombination in ES cells. *Cell* **71**, 343–353.
8. Ishibashi, S, Goldstein, JL, Brown, MS, et al. (1994) Massive xanthomatosis and atherosclerosis in cholesterol-fed low density lipoprotein receptor-negative mice. *J Clin Invest* **93**, 1885–1893.
9. Suzuki, H, Kurihara, Y, Takeya, M, et al. (1997) A role for macrophage scavenger receptors in atherosclerosis and susceptibility to infection. *Nature* **386**, 292–296.
10. Whitman, SC, Rateri, DL, Szilvassy, SJ, et al. (2002) Macrophage-specific expression of class A scavenger receptors in LDL receptor(–/–) mice decreases atherosclerosis and changes spleen morphology. *J Lipid Res* **43**, 1201–1208.
11. Herijgers, N, de Winther, MP, Van Eck, M, et al. (2000) Effect of human scavenger receptor class A overexpression in bone marrow-derived cells on lipoprotein metabolism and atherosclerosis in low density lipoprotein receptor knockout mice. *J Lipid Res* **41**, 1402–1409.
12. Daugherty, A, Whitman, SC, Block, AE, et al. (2000) Polymorphism of class A scavenger receptors in C57BL/6 mice. *J Lipid Res* **41**, 1568–1577.
13. Witztum, JL. (2005) You are right too! *J Clin Invest* **115**, 2072–2075.
14. Curtiss, LK. (2006) Is two out of three enough for ABCG1? *Arterioscler Thromb Vasc Biol* **26**, 2175–2177.
15. Paigen, B, Morrow, A, Holmes, P, et al. (1987) Quantitative assessment of atherosclerotic lesions in mice. *Atherosclerosis* **68**, 231–240.
16. Daugherty, A, Whitman, SC. (2003) Quantification of atherosclerosis in mice. *Methods Mol Biol* **209**, 293–309.
17. Baglione, J, Smith, JD. (2006) Quantitative assay for mouse atherosclerosis in the aortic root. *Methods Mol Med* **129**, 83–95.
18. Purcell-Huynh, DA, Farese, RV, Johnson, DF, et al. (1995) Transgenic mice expressing high levels of human apolipoprotein B develop severe atherosclerotic lesions in response to a high-fat diet. *J Clin Invest* **95**, 2246–2257.
19. Tangirala, RK, Rubin, EM, Palinski, W. (1995) Quantitation of atherosclerosis in murine models: correlation between lesions in the aortic origin and in the entire aorta, and differences in the extent of lesions between sexes in LDL receptor-deficient and apolipoprotein E-deficient mice. *J Lipid Res* **36**, 2320–2328.
20. Daugherty, A, Rateri, DL. (2005) Development of experimental designs for atherosclerosis studies in mice. *Methods* **36**, 129–138.
21. Nakashima, Y, Plump, AS, Raines, EW, et al. (1994) ApoE-deficient mice develop lesions of all phases of atherosclerosis throughout the arterial tree. *Arterioscler Thromb* **14**, 133–140.
22. Daugherty, A, Manning, MW, Cassis, LA. (2000) Angiotensin II promotes atherosclerotic lesions and aneurysms in apolipoprotein E-deficient mice. *J Clin Invest* **105**, 1605–1612.
23. Rosenfeld, ME, Polinsky, P, Virmani, R, et al. (2000) Advanced atherosclerotic lesions in the innominate artery of the ApoE knockout mouse. *Arterioscler Thromb Vasc Biol* **20**, 2587–2592.
24. Reardon, CA, Blachowicz, L, Lukens, J, et al. (2003) Genetic background selectively influences innominate artery atherosclerosis – Immune system deficiency as a probe. *Arterioscler Thromb Vasc Biol* **23**, 1449–1454.
25. Daugherty, A, Pure, E, Delfel-Butteiger, D, et al. (1997) The effects of total lymphocyte deficiency on the extent of atherosclerosis in

apolipoprotein E −/− mice. *J Clin Invest* **100**, 1575–1580.

26. Lu, H, Rateri, DL, Daugherty, A. (2007) Immunostaining of mouse atherosclerotic lesions. *Methods Mol Med* **139**, 77–94.
27. Falk, E. (2006) Pathogenesis of atherosclerosis. J Am Coll Cardiol **47**, C7–C12.
28. Daugherty, A, Zweifel, BS, Schonfeld, G. (1991) The effects of probucol on the progression of atherosclerosis in mature Watanabe heritable hyperlipidaemic rabbits. *Br J Pharmacol* **103**, 1013–1018.
29. Fazio, S, Linton, MF. (1996) Murine bone marrow transplantation as a novel approach to studying the role of macrophages in lipoprotein metabolism and atherogenesis. *Trends Cardiovasc Med* **6**, 58–65.
30. Whitman, SC. (2004) A practical approach to using mice in atherosclerosis research. *Clin Biochem Rev* **25**, 81–93.

Chapter 2

Evaluating Micro- and Macro-vascular Disease, the End Stage of Atherosclerosis, in Rat Models

James C. Russell

Abstract

Development of effective treatment or, more critically, preventative measures against atherosclerosis and cardiovascular disease will require animal models that mimic the disease processes seen in humans and permit identification of the genetic and physiological factors. The Rat is normally resistant to cardiovascular disease, but a number of genetic mutations make affected strains of rats highly susceptible to atherosclerosis and micro- and macro-vascular disease that is highly analogous to human disease. These models of obesity develop the metabolic syndrome and type 2 diabetes, hyperinsulinemia, hyperlipidemia, vascular and myocardial dysfunction, and end-stage lesions in the heart and kidney. The models offer the prospect of both genetic and molecular biology studies that are linked directly to spontaneous cardiovascular disease and exploration of putative preventative or treatment approaches, including pharmaceutical agents.

Use of small animal models of cardiovascular disease is dependent on appropriate experimental design and techniques that take account of the complex nature of the disease processes. Detailed experimental procedures for the use of rat models, including handling and treatment of animals, choice of experimental variables and endpoints, assay methods, and histological and electron microscopy techniques are covered in this chapter.

Key words: Atherosclerosis, rat, vascular function, vascular lesions, myocardial function, ischemic lesions, dyslipidemia, metabolic syndrome, histology, electron microscopy.

1. Introduction

Atherosclerosis has been recognized as an inflammatory process since the studies of Virchow and Rokitansky in the nineteenth century (1–3). It remains the major determinant of cardiovascular disease (CVD), the most important cause of death in developed societies and, increasingly, worldwide (4). CVD is not a single defined entity, but an end-stage pathophysiological state induced by a number of

K. DiPetrillo (ed.), *Cardiovascular Genomics*, Methods in Molecular Biology 573,
DOI 10.1007/978-1-60761-247-6_2, © Humana Press, a part of Springer Science+Business Media, LLC 2009

poorly understood disease processes. CVD itself is a misnomer as it affects both micro- and macro-vascular vessels ranging from the aorta, major abdominal arteries, coronary, carotid, and intracerebral arteries to renal glomerular and retinal capillaries (5). The resultant, largely intimal, lesions lead to occlusion or rupture with ischemia, infarct (myocardial or other), or hemorrhage. At a micro-vascular level the results are glomerular sclerosis in the kidney, transient ischemic attacks in the brain, and retinal damage.

In recent years, a number of theories have been advanced to explain the development of atherosclerosis and CVD. The most prominent is based on the high lipid content of atherosclerotic lesions and the strong association between high-plasma low-density lipoprotein (LDL) cholesterol and CVD. This hypothesis has yielded extensive understanding of lipid and lipoprotein metabolism and the role of the genetic mutations that result in defects in the LDL receptor, familial hyperlipidemia, and premature CVD (6). While hypercholesterolemia is probably responsible for 35–40% of CVD, the 1995 forecast that imminent control of hypercholesterolemia would result in virtual elimination of CVD (7) proved to be premature. There is now a growing appreciation of the contribution (probably also in the range of 40%) of the prediabetic metabolic syndrome and type 2 diabetes to atherosclerosis and CVD (8). The metabolic syndrome is characterized by abdominal obesity, insulin resistance and hyperinsulinemia, and elevated plasma levels of the triglyceride-rich lipoprotein fractions VLDL (very low density lipoprotein) and chylomicrons. Both type 2 diabetes and metabolic syndrome are related to a strong environment–genome interaction, with some populations at particularly high risk (9). Population studies have begun to identify specific polymorphisms that play a strong role in the development of CVD (10), but are yet to define the mechanisms. These metabolic dysfunctional states are associated with striking micro-vascular damage contributing to a major fraction of renal disease, as well as myocardial and cerebral vascular disease. In addition to these factors, it has long been recognized that hypertension plays a role in exacerbating end-stage CVD.

Effective treatment protocols for atherosclerosis and CVD, and even more critically, prevention of the disease burden they impose, will require detailed understanding of the multiple contributing mechanisms. While human studies are essential to identify the issues, the mechanistic knowledge will have to come from studies in animal models. Such models must accurately mimic aspects of the human disease, and no one model can cover all dimensions of the multi-factorial origins of CVD. Relatively few animal models spontaneously develop CVD and many commonly used models depend upon highly abnormal conditions, for example, feeding high-cholesterol diets to species, such as the rabbit, that are herbivores (11). The critical issues involving the use of animal models in the study of CVD have been reviewed recently (5). In summary, the best models are non-human primates, which are normally impractical for reasons of cost, longevity, and ethics, and swine, which again are relatively large and expensive, particularly the "mini pigs." Effective small animal models are essentially confined to rodents, including rats, mice, and more exotic species such as hamsters and sand rats. Choice of an appropriate model is critical and dependent on the definition of the experimental question to be addressed. Use of an inappropriate model will vitiate the study. For instance, cholesterol-fed rabbits develop cholesterol-laden intimal lesions, but not if specific pathogen free; development of lesions is dependent on the chronic inflammatory state induced by bacterial pulmonary infections by agents such as *Pasteurella multocida* (12).

1.1. Selection of Animal Model

In balance, rats and mice are the optimal models for mechanistic studies of CVD, being small, relatively cheap, and amenable to genetic manipulation. None of the mouse models, including the various gene knockout strains, have been shown to spontaneously develop advanced atherosclerosis and CVD comparable to that seen in humans. These models generally require high-lipid diets or other environmental manipulation to induce vascular lesions and, thus, are best used to study the effects of hyperlipidemia (bearing in mind the significant differences in lipid metabolism compared to humans). Similarly, the genetically normal rat is highly resistant to atherosclerosis and requires both a high-cholesterol diet and treatment with thyrotoxic compounds to show significant hypercholesterolemia and vascular disease. However, there are a number of rat strains carrying mutations that lead to obesity, insulin resistance, hyperinsulinemia, and hypertriglyceridemia. One strain progresses to overt type 2 diabetes on a high-fat diet. These strains are suitable for study of early and full-blown diabetes and show both macro- and micro-vascular disease. The BB rat is a spontaneous model for type 1 diabetes, but has not been extensively studied for vascular disease. Streptozotocin has been widely used to induce pancreatic β-cell death and failure of pancreatic insulin secretion and, thus, a cytotoxin-produced quasi type 1 diabetes (13). This model is not really analogous to the autoimmunity-induced type 1 diabetes of the BB rat (14) and humans and may not accurately model vascular disease associated with type 1 diabetes.

1.2. The Fatty or fa Rat Strains

1. *The Fatty Zucker*

The *fa* rat mutation was initially described by Zucker and Zucker in 1961 (15) and consists of a glycine to proline substitution at position 269 of the leptin receptor (ObR), resulting in a 10-fold reduction in binding affinity for leptin (16). Rats homozygous for the *fa* mutation (*fa/fa*) develop a variant of the metabolic syndrome, becoming obese, moderately insulin resistant, and hypertriglyceridemic, but with no progression to diabetes or cardiovascular complications (17). Heterozygous animals or those homozygous wild-type (Fa/*fa* or Fa/Fa) are lean and metabolically normal.

2. *The ZDF Rat*

Zucker Diabetic Fatty (ZDF) rats were developed from the original Zucker strain (18). When fed a high-fat chow, the *fa/fa* ZDF rats, particularly the females, become insulin resistant and hyperinsulinemic and convert to a frank type 2 diabetes with very high plasma glucose levels as young adults. Diabetic ZDF rats exhibit many of the complications of the hyperglycemic diabetic state, particularly micro-vascular damage leading to glomerulosclerosis and retinopathy (19). In contrast, there are no reports of atherosclerosis or macro-vascular disease in these animals, although there is evidence of vascular dysfunction in aorta, coronary, and mesenteric arteries in adult to middle-aged obese ZDF rats, probably related to the overt diabetes (20). There is cardiac dysfunction that appears to be related only to hydronephrosis and not atherosclerosis (21).

1.3. The Corpulent (cp) Strains

In 1969, Koletsky isolated a mutation in a rat line originating in a cross between Sprague Dawley and SHR (spontaneously hypertensive) rats (22). The mutation, later designated *cp*, is a T2349A transversion, resulting in a Tyr763Stop nonsense codon leading to absence of the transmembrane portion of the ObR leptin receptor and, thus, of activity of all of the isoforms of the ObR (23). The original obese rats developed a fulminant atherosclerosis with advanced lesions, including dissecting aortic aneurisms (23). The original strain, now designated as SHROB, has been maintained for 60 plus generations in various laboratories. The atherosclerosis-prone character has been lost due to improper breeding and the strain is now commercially supported by Charles River Laboratories (Wilmington, MA). Over recent years, most of the *cp* rat strains have been established on a commercial basis, including those designated as: Crl:JCR(LA)-*Lepr*^{cp}, SHHF/MccCrl*Lepr*^{cp}, and SHR/OBKOLCrl-*Lepr*^{cp} (details available at www.criver.com). *See* reference (5) for detailed background.

1. *The JCR:LA-cp Strain*

This strain has maintained the highly insulin-resistant and atherosclerosis-prone traits of the original Koletsky colony. Insulin resistance and hyperlipidemia develop rapidly in young male *cp/cp* rats of the JCR:LA-*cp* strain and are highly correlated with the development of CVD, both macro-vascular and micro-vascular (24–29). Most importantly, the *cp/cp* males spontaneously develop ischemic lesions of the heart (27) and are prone to stress-induced myocardial infarcts that can be fatal. Studies by scanning and transmission electron microscopy show extensive atherosclerotic lesions in major vessels throughout the arterial system of middle-aged *cp/cp* male rats (30). The lesions closely resemble the intimal atherosclerosis seen in human aorta and coronary arteries.

2. *The SHHF/Mcc-cp Strain*

The SHHF/Mcc-*cp* strain, developed from the SHR/N-*cp*, exhibits a cardiomyopathic/congestive heart failure trait, with a very high incidence of fatal congestive heart failure (31). This strain is a unique rat model of an important clinical problem and has been used to address issues related to cardiac dysfunction and ameliorative treatments (32, 33).

1.4. Other Rat Strains

Other rat strains that develop obesity and/or variations of type 2 diabetes have been described. These strains remain relatively poorly characterized and have not been demonstrated to develop atherosclerotic disease. The strains include the Goto-Kakizaki (GK) rat, a lean insulin-resistant strain (34); Zucker Diabetic Sprague Dawley (ZDSD), a strain with diet-sensitive development of hyperinsulinemic type 2 diabetes (*see* http://www.preclinomics.com/); the Otsuka Long-Evans Tokushima Fatty (OLETF) rat characterized by a mild type 2 diabetes and vascular dysfunction (35); and the SHR/NDmcr-*cp* rat, a strain derived from the original SHR/N-*cp* strain that exhibits hypertension, obesity, and vasculopathy (36).

1.5. Experimental Endpoints

Quantitative endpoints are an integral part of experimental design and allow detection and comparison of the severity of disease, as well as measurement of the efficacy of putative treatments. Choice of endpoint(s) is a function of the experimental aims. The principal CVD endpoints fall into metabolic, functional/physiological, and pathological categories. They are as follows:

1. *Metabolic Dysfunction*

Hyperlipidemias, the pre-diabetic state, and diabetes (type 1 and type 2) are the most important metabolic precursors to atherosclerosis, although there is significant overlap between these. While there are no established rat models of LDL receptor-mediated hypercholesterolemia, strains that develop obesity and/or the metabolic syndrome exhibit VLDL hyperlipidemia and chylomicronemia with delayed lipoprotein clearance (26, 37, 38). Measurement of fasting plasma lipids, especially of individual lipoprotein fractions, and postprandial kinetics provide indices of the pre-atherosclerotic status. Similarly, in strains that exhibit insulin resistance with hyperinsulinemia, fasting, and even more strongly postprandial, insulin levels are strongly correlated with vascular dysfunction and pathology.

2 . *Vascular Function*

The earliest overt dysfunction in atherosclerotic disease is vasculopathy evident in hypercontractility responses to noradrenergic agonists and impaired endothelium-dependent relaxation. Classical isolated organ bath techniques, using rings of aorta or mesenteric resistance vessels, allow quantitative measures of vascular dysfunction in different kinds of arteries.

3. *Myocardial Function*
Isolated heart perfusion techniques allow direct assessment not only of myocardial function, but also of the coronary artery flow regulation and relaxation. Recent in vivo echo-cardiographic studies in rats have shown the possibility for chronic monitoring of impaired cardiac function in atherosclerosis-prone strains.

4. *Thrombosis*
Early atherosclerotic vascular dysfunction also leads to a prothrombotic status. This is characterized by platelet activation and impaired endothelial cell NO release, together with elevated plasma levels of plasminogen activator inhibitor-1 (PAI-1) and reduced lysis and persistence of intravascular thrombi. Changes in plasma PAI-1 and platelet function provide indices of vascular status.

5. *Renal Micro-vascular Dysfunction*
Glomerular capillaries are micro-arteries and prone to damage induced by hypertension and the underlying pathophysiological factors leading to atherosclerosis in larger vessels. The early-stage damage leads to increased glomerular permeability and leakage of albumin into the urine. Further development of the pathological processes leads to fibrosis and scarring, rendering the sclerosed glomeruli non-functional. The early stages can be quantified by the urinary albumin concentration and later stages by the measurement of glomerular filtration rate or by histopathological analysis of individual glomeruli.

6. *Vascular and Myocardial Lesions*
Atherosclerosis, as such, can be quantified in rat arteries, although this can only be done postmortem and in larger arteries. The lesions are relatively small, given the small size of the rat, and the most effective techniques are transmission electron microscopy (TEM) and scanning electron microscopy (SEM). TEM or light microscopy of thick sections from TEM specimens gives a cross-sectional view of a limited area of the arterial wall and allows identification of the internal composition of the lesion. In contrast, SEM gives a surface view of a large area of the intimal surface of the artery and reveals a range of lesions from raised lesions, to de-endothelialization, adherent thrombi, and luminal macrophage adherence and infiltration.

Myocardial lesions, the real end-stage of CVD, can be identified by light microscopy of conventional sections of the heart and quantified. Old, scarred, ischemic lesions constitute a preserved lifetime record of myocardial events that were large enough to be visible after the infarcted tissue had scarred and contracted.

7. *Micro-vascular Lesions – Glomerulosclerosis*
The fraction of the renal glomeruli that are sclerotic is a measure of accumulated micro-vascular damage in the rat. It is readily quantified by conventional histology of kidney sections.

2. Materials

2.1. Solutions

1. Krebs buffer (for aortic ring vascular function): 116 mM NaCl, 5.4 mM KCl, 1.2 mM $CaCl_2$, 2 mM $MgCl_2$, 1.2 mM Na_2PO_4, 10 mM glucose, and 19 mM $NaHCO_3$.
2. HEPES-buffered physiological saline: 142 mM NaCl, 4.7 mM KCl, 1.17 mM KH_2PO_4, 1.2 mM $CaCl_2$, 10 mM HEPES, 5 mM glucose.
3. Krebs–Henseleit solution (for heart perfusion and coronary artery function): 118 mM NaCl, 4.7 mM KCl, 1.2 mM $CaCl_2$, 2 mM $MgCl_2$, 1.2 mM Na_2HPO_4, 10 mM glucose, and 19 mM $NaHCO_3$.
4. 10% Neutral-buffered formalin (primary fixative for conventional histology, containing 3.7% formaldehyde): 37% formaldehyde solution diluted 1:10 with phosphate buffer, pH 7, also commercially available and ready to use.
5. Tyrode's solution (balanced salt solution to flush circulatory system prior to perfusion fixation): 137 mM NaCl, 12 mM $NaHCO_3$, 0.9 mM NaH_2PO_4, 4 mM KCl, 0.5 mM $MgSO_4$, 2.5 mM $CaCl_2$, and 5.0 mM glucose (pH 7.2).
6. Tyrode's solution with 2.5% glutaraldehyde (fixative solution for arterial system): basic balanced salt solution with addition of 2.5% EM grade glutaraldehyde.
7. Tyrode's solution with 1.25% glutaraldehyde and 1.85% formaldehyde (fixative for perfusion): fixation containing half effective concentrations of both EM and conventional fixatives, allowing subsequent further fixation for either EM light microscopy.
8. 4% Paraformaldehyde in 0.1 M sodium cacodylate (non-denaturing fixative allowing staining of fixed sections with specific antibodies): Polyoxymethylene 4% w/v in cacodylate buffer (*see* www.abdn.ac.uk/emunit/emunit/recipes.htm).
9. 1% Aqueous Osmium Tetroxide (OsO_4) (postfixative for TEM): *see* www.nrims.harvard.edu/protocols/Osmium_ Tetroxide_Fix.pdf.
10. 2% Aqueous uranyl acetate (a heavy metal stain applied to increase contrast in TEM thin sections): *see* web.path.ox.ac.uk/~bioimaging/bitm/instructions_and_information/EM/neg_**stain**.pdf.

2.2. Equipment

1. Graz Vertical Tissue Bath System (Harvard Apparatus, Holliston, MA: www.harvardapparatus.com) or equivalent.
2. Isometric myograph system (Kent Scientific Corp., Torrington, CT: www.kentscientific.com/products).

3. Microliter syringe (Hamilton Company, Reno, Nevada: www.hamiltoncompany.com/syringes).
4. Isolated Heart Apparatus (size 3; Hugo Sachs Elektronik, March-Hugstetten, Germany), equipped with a Type 700 transit time flowmeter, *see* www.harvardapparatus.com.
5. Calibrated roller pump (i.e., Varioperpex II Pump 2120; LKB-Produkter, Bromma, Sweden, or equivalent).
6. Universal Servo Controlled Perfusion System (Harvard Apparatus, Holliston, MA: www.harvardapparatus.com).

2.3. Software

1. ALLFIT (a public domain program, *see* reference (42), program is available from this author or Dr. Peter J. Munson, e-mail: munson@mail.nih.gov).
2. Isoheart^{7}W program V 1.2 (Hugo Sachs Elektronik, March-Hugstetten, Germany).
3. Photoshop (Adobe Systems Inc., San Jose, CA).

3. Methods

3.1. Animal Treatment and Handling

Due to the dependence of atherogenesis on a complex environment–genome interaction, the housing, diet, and handling of experimental animals must be carefully controlled. Stress, varying from that induced by shipping to the effects of unknown humans entering the animal room, can significantly alter metabolism and CVD progression. Any microbiological infection, from viruses to parasites such as pinworm, is contraindicated. Rats should be demonstrated virus antibody free (VAF) and monitored for parasites and virus antibodies on an ongoing basis. This may be arranged through commercial organizations such as Charles River Laboratories (www.criver.com/), Research Animal Diagnostic Laboratory, University of Missouri (www.radil.missouri.edu), or by the veterinary staff of the animal facility.

1. *Diurnal Cycle*

   Rats are nocturnal and sleep during the light phase of the diurnal cycle. Thus, the animals' metabolism is reversed from that of humans and it is indicated that metabolic studies be conducted during the dark period of the diurnal cycle. Rats that are to be subjected to such studies are best maintained on a reversed light cycle, with lights off from 0600 to 1800 h, or equivalent timing adjusted for convenience in performing procedures. All manipulations of the rats are performed under a dim light in a dedicated procedure room.

2. *Housing*

Rats are traditionally maintained on aspen wood chip bedding in polycarbonate cages. However, we have converted our housing to a controlled isolated caging system (Techniplast, Slim Line^{TM}, Techniplast S.p.a., Buguggiate, Italy), in which all incoming and exhaust air is HEPA filtered. Our laboratory uses Lab Diet 5001 (PMI Nutrition International Inc., Brentwood, MO) as a standard food. The chow is available as pellets or as non-pelleted powder used for diets incorporating pharmaceutical or nutritional additives. The pelleted diet is also available in an autoclavable formulation which is important for maintaining animals in a secure barrier facility. Different diets are not interchangeable and are a significant variable that should never be changed during an experimental protocol.

Upon weaning, young rats are normally group housed for 1 week to reduce stress. Older rats are normally housed in pairs, as they are social animals and this can reduce stress levels. However, it is not good practice to house lean and obese male rats together beyond 6 weeks of age, as the lean rat will be aggressive toward the obese animal. Careful humidity control (50% relative humidity is indicated) is essential in the animal rooms. This is a significant problem in areas, such as Western Canada, where humidity levels can be very low, leading to ring tail disease in young animals.

Breeding colonies of genetic models are a core part of the research and should be carefully maintained. Staff must wear only facility clothing and footwear in the animal unit, as well as hats, masks, and gloves. Food and other items should be autoclaved before being brought into the unit and caging subjected to high temperature wash. These barrier precautions have been successful in preventing infection with viral or bacterial pathogens in genetic model colonies over many years. Because of the significant metabolic and physiological responses of rats to infectious processes, it is most important that experimental animals be maintained using similar protocols. Failure to do so will compromise the quality of the results.

3. *Breeding*

Breeding genetic models of a polygenetic disorder, such as atherosclerosis and CVD, requires a clear plan and allowance for inevitable genetic drift, which may result in the loss of some of the unique phenotypic characteristics of a colony. In the case of some strains with the *cp* gene, this has led to reduced severity of metabolic dysfunction. Most importantly, the inbreeding needed to create congenic strains also bred out the propensity for the development of the severe cardiovascular disease that is seen in the original Koletsky Obese SHR rat (22). To maintain the JCR:LA-*cp* rat, we ensured genetic stability of the colony through a formal assortive breeding program with ten separate lines, using a technique described by Poiley (39).

3.2. Drug Administration

In both mechanistic and pharmaceutical studies, there are often requirements to administer agents to rats in a defined manner, but with a caveat that the procedure should not be noticeable or stressful to the animals (*see* **Note 1**). Our standard approach is to incorporate agents (pharmaceutical or nutritional) in the rats' chow diet, measure food consumption, and control treatments on a per kilogram body weight basis. The method is as follows:

1. Agents to be incorporated, usually at doses in the range of 1–50 mg/kg BW/d, are intimately ground dry in a mortar and pestle with 5–10 g of powdered rat chow. The resultant powder is then mixed thoroughly with powdered chow to make a final mass of 1 or 2 kg in a small commercial pasta machine (Bottene, Marano, Italy (www.bottene.net)). For calculation of concentrations, based on rat body weights and food intake, *see* **Note 2**.
2. The mixed chow is moistened with approximately 200 mL of water, the volume being adjusted to give an appropriate consistency and extruded through a brass die machined to form 1 × 2 cm rounded pellets. The pellets of variable lengths are dried on a rack in a forced draft oven at 35°C.
3. Rats are weighed and food consumption determined twice weekly. New food is made up weekly based on the last week's food consumption data.

3.3. Anesthesia

Anesthesia is essential for surgical procedures and euthanasia, with blood and tissue sampling, while avoiding stress that will alter metabolic parameters (*see* **Note 3** for comments on appropriate and inappropriate methods). The method described is appropriate using either isofluorane or halothane. With either agent, the induction area and operating table should have an air scavenging system to ensure that personnel are not exposed to the anesthetic vapors.

1. Purge a bell jar chamber with isofluorane, 3.0–3.5% in oxygen at 1–1.5 L/min, delivered from a calibrated vaporizer.
2. Place rat in the chamber and continue isofluorane flow.
3. Once the rat loses consciousness (animal supine and non-responsive), transfer the rat to the operating/dissection area with heating pad and attach an anesthetic face mask over the nasal/mandible area. Maintain isofluorane flow at 200 mL/min and reduce concentration in increments to 2±0.5% for maintenance.
4. Test depth of anesthesia (e.g. no eye movement on contact with sterile Q-tip, no pedal response to foot pinch, visual observation of breathing depth and rate). These should be monitored continually throughout the procedure.

5. *Euthanasia procedures* (*see* **Section 3.8**) may be initiated at this point.
6. *For recovery procedures*, apply a sterile lubricating ophthalmic ointment to eyes and shave the surgical area using electric clippers.
7. Apply povidone–iodine (Betadine 10%) solution with sterile gauze pads to surgical sites.
8. Wash the surgical sites with 100% ethanol using sterile gauze pads.
9. Place sterile surgical drapes around the surgical sites.

3.4. Blood Sampling

A method permitting repeated blood sampling is necessary in many studies and should be conducted on unrestrained animals to avoid stress. This also obviates the need for anesthesia, which can significantly affect the metabolic endpoints and lead to myocardial infarction in susceptible rats. The method described, sampling from a tail snip, depends upon calming the rat by quiet handling and minimal restraint.

1. The procedure should be conducted during the dark, active, part of the diurnal cycle in subdued light and in an isolated room free from noise and other activities. All personnel should be known to the rats and work quietly to avoid stress.
2. In either the animal or surgery room, a warming plate is prewarmed to ~35°C and should be warm to the touch, but not hot.
3. The rat is placed on the heating plate, covered with a wire basket for approximately 10 min to warm, and the tail to vasodilate.
4. The rat is gently removed from the heating plate, held on the forearm, and transferred to the procedure area. The rats tend to attempt to hide in the dark space under the upper arm and will remain quiet if handled gently.
5. The rat is held against the body of the operator by gentle pressure and the tail held in the fingers of the left hand. Using the other hand, the conical portion of the tail (0.5 mm tip) is removed with sharp sterile scissors, with care being taken not to cut the distal bone.
6. The tail is "milked" by gently stroking from the base to near the tip using the free hand and the blood drops that form on the tip are collected in a Microtainer® blood collection tube (BD Medical, Franklin Lakes, NJ) containing anticoagulant, as indicated.

7. When blood sample is complete, pressure is applied to the snipped tail with a sterile gauze pad until clotting occurs. The animal is then returned to its home cage to recover and is monitored.
8. The tail bleeding procedure may be repeated for up to five times in one day, with total volume not exceeding 15% of the blood volume of the rat (~75–80 mL/kg based on lean body mass (40)) or 2.5–3.0 mL for a 250–300-g rat). A recovery period of 1 week should be allowed before any additional blood sampling.
9. Rats are accustomed to the procedure 1 week in advance, through an initial sham procedure using a tail pinch with the fingers to simulate the tail cut.

3.5. Insulin/Glucose Metabolism

Insulin resistance, resultant hyperinsulinemia, and type 2 diabetes are major factors in CVD and need to be assessed quantitatively in many studies. A standardized meal tolerance test (MTT) procedure uses a tail bleed to obtain blood samples without restraint and with minimal disturbance (24).

1. The rat is deprived of food over the light period (normally overnight if animals are kept on a reversed light/dark cycle).
2. Two hours into the dark period, the initial (0 time) fasted blood sample is taken (*see* **Section 3.4**). A 0.8-mL blood sample is taken during each repeated sampling, and up to 1.5 mL can be collected in one sample without distress.
3. The rat is returned to its home cage and given a 5-g pellet of rat chow. The pellet is normally eaten within 15 min, and the timing is started when it is half consumed.
4. Further tail bleed samples are taken at intervals up to 150 min. Bleeding is stopped after each sample.
5. Samples are subsequently assayed for insulin and glucose.
6. Insulin-resistant rats maintain euglycemia under this protocol, but at the expense of a very large postprandial insulin response that can reach 1,000 mU/L (*see* **Note 4** for interpretation).

3.6. Plasma Lipid Measurements

There is an established practice in metabolic studies to measure plasma lipid concentrations in the fasted state (in the rat, after a 16-h period of food deprivation). This results in relatively consistent values, but does not reflect the metabolic status of an animal in its normal day-to-day situation. In contrast, measurements on blood samples taken 2–3 h into the light period, when the rat has eaten its first meal of the day, are more representative of the animal's ongoing metabolism (*see* **Note 5**).

3.7. Fat Challenge

Abnormal metabolism of apoB-containing, triglyceride-rich lipoproteins is a prominent component of the proatherogenic status accompanying the metabolic syndrome and type 2 diabetes. A fat challenge test allows quantitation of the dysfunction as an experimental endpoint, analogous to and conducted in a similar way to the insulin/glucose MMT.

1. Fat-supplemented food is pre-made using "English Double Devon Cream" (47% milk fat w/w). Mix 170 g of cream with 170 g of Lab Diet 5001 (PMI Nutrition International Inc., Brentwood, MO) powdered rat chow with gloved hands to form 5-g pellets (24% milk fat w/w).
2. Rats are fasted overnight as for MTT.
3. A fasted blood sample (0.4 mL) is taken 1 h into the dark period (0 time).
4. The rat is placed into its own cage with water ad libitum. A 5-g (pre-weighed) fat-enriched pellet is placed in the feed hopper.
5. When half of the pellet is eaten, the time is recorded as time 0, the reference point for the rest of the experiment.
6. Additional blood samples (0.4 mL each) are taken at time points 2, 4, 5, 6, 8, and 10 h.
7. Blood samples are collected in heparinized Microtainers.
8. Samples are spun in a centrifuge and plasma is pipetted into vials and stored at –80°C.
9. When all blood samples have been collected, the animal is returned to its normal feeding schedule.
10. Animals are allowed to return to their normal physiological state for 7 days before any other procedures or euthanasia is performed.

3.8. Euthanasia with Blood and Tissue Sampling

This technique enables fresh tissue samples to be taken, while providing an acceptable method of euthanasia.

1. Overnight fasting, prior to tissue collection, may be required depending on experimental endpoints.
2. Tissues and samples that may be taken include urine, plasma/serum, heart, aorta, kidney, liver, individual fat pads, and skeletal muscle and should be taken in this order. The tissues required should be clearly identified prior to the procedure and prepared for accordingly.
3. The rat is anesthetized as described in **Section 3.3** and placed in dorsal recumbency on the operating table.
4. The depth of anesthesia (surgical plane) is confirmed by the absence of pedal response, blink reflex, and breathing rate and depth. Breathing should be slower and deeper than normal.
5. If sterile samples are required, the incision site(s) should be shaved and prepared aseptically, using povidone–iodine and alcohol.

6. Using scissors, a 5–8-cm incision is made down the midline of the skin and then on the abdominal wall (linea alba) to expose the length of the peritoneal cavity.
7. If urine is required, it is collected first (from the bladder) with an 18G needle and syringe and placed in a tube on ice.
8. The chest is opened up the sternum using a substantial pair of scissors and the diaphragm cut to fully expose the heart. Cardiac puncture (left ventricle) is performed immediately with an 18G needle and 10-mL syringe. The blood is collected, placed in a heparinized tube, and stored on ice (*see* **Note 6**).
9. Rapid death of the animal is ensured by resection of the heart or cutting of the right atrium with resultant blood loss.
10. Remaining tissues or organs are removed, death confirmed, and the carcass bagged and labeled before disposal.

3.9. Aortic Vascular Function

Vascular contractile and relaxant dysfunction, both in resistance and conductance arterial vessels, is a major contributor to the pathological sequelae and a marker for atherosclerosis. Measurement of vascular function provides a quantitative index of vascular disease severity, based on isolated arterial segments.

1. Rats are euthanized as per **Section 3.8**.
2. After cardiac puncture and blood sampling, the heart is dissected and reflected forward allowing the aortic arch and thoracic aorta to be dissected and excised en bloc.
3. The heart and attached thoracic aorta are isolated, placed in ice-cold Krebs solution (*see* **Note 7**) in a Petri dish, and adhering fat and connective tissue trimmed off. This is done by slightly lifting the fat up and sliding the scissors, cutting away all extraneous tissue and fat while not hitting, nicking, or stretching the aorta.
4. The aorta is placed on a wet filter paper with clean buffer and kept moist. Using a sharp scalpel blade, slice into sections or rings no more than 3 mm long. Ensure that there is no "crushing" of the tubular tissue (this rubs the endothelial cells) by using a slight rolling motion. The paper and tissue rings are placed in a Petri dish with cold Krebs, carried to the apparatus, and the rings mounted within 10 min (*see* **Note 8**).
5. The aortic rings are mounted on stainless-steel hooks under a 1.5-g resting tension in 10 or 20-mL organ baths in a tissue bath system, bathed at 37°C in Krebs solution, and gassed with 95% O_2 and 5% CO_2.
6. The tissues are allowed to equilibrate for 45 min before measurements are started. During this time, the resting tension is readjusted to 1.5 g as required and the tissues washed every 15 min. Flushes and manual adjusting are kept to a minimum as they cause stress to the tissue.

7. Stock solutions (0.1 M) are made of the noradrenergic agonist phenylephrine (PE); the endothelium-dependent NO-releasing agent acetylcholine (ACh); the direct NO donor *S*-nitroso-*n*-acetyl-penicillamine (SNP); and the inhibitor of nitric oxide synthase, L-NAME (*see* **Note 9**).
8. PE solutions are added to the tissue baths with a microliter syringe (Hamilton Company, Reno, NV), cumulatively, to give concentrations of 10^{-9}–10^{-4} M (*see* **Note 9**) and the contractile force produced measured and recorded as a concentration–response curve.
9. Rings are washed and equilibrated, then pre-contracted with PE to 80% of maximal contraction based on the dose–response curve.
10. The relaxant dose–response of the aortic rings to ACh and SNP is determined through cumulative additions of solutions to yield 10^{-9}–10^{-4} M; normally half of the rings are treated with each agent.
11. L-NAME may be added to one or more baths at 10^{-4} M to inhibit NOS activity and to provide a direct measure of NO-mediated effects *(41)*.

3.10. Mesenteric Resistance Artery Function

1. After euthanasia and removal of heart and/or aorta as above, the mesenteric arcade, 5–10 cm distal to the pylorus, is excised and placed immediately in ice-cold HEPES-buffered physiological saline.
2. Arterial rings (approximately 300 μm in diameter) are cut to 2 mm lengths, threaded onto two stainless steel wires of 25 μm in diameter, and mounted in an isometric myograph system (*see* **Note 10**).
3. Norepinephrine (NE) solutions are added to the tissue baths with a microliter syringe, cumulatively, to give concentrations of 10^{-8}–10^{-5} M and the contractile force produced measured and recorded as a concentration–response curve.
4. Arteries are washed, equilibrated and pre-constricted to 50% of the maximal contractile response to NE, and cumulative concentrations of SNP (10^{-9}–10^{-4} M) or ACh (10^{-9}–10^{-6} M) are added. Reduction in the contractile force is recorded and expressed as % relaxation.

3.11. Interpretation of Vascular Function Data

Hypercontractile response to noradrenergic agonists, endogenous (such as NE) or synthetic (such as PE), is due to increased sensitivity of the vascular smooth muscle cells (VSMC) and leads to a propensity to vasospasm. Reduced relaxant response to SNP indicates that the VSMC have impaired sensitivity to NO and thus the hypercontractility originates in the medial layer of the arteries.

Reduced relaxant response to ACh indicates that the endothelial cells have reduced NO production and/or release and that the hypercontractility originates in the intima and endothelial cells.

The dose–response curves are best described by the Logistic Equation, with the four parameters being the minimum, or zero, response; the maximum response; the concentration of the agonist giving 50% of the maximum response (EC_{50}); and the slope of the curve at the EC_{50}. Changes in the parameters are important quantitative end points and need to be tested using appropriate statistical techniques. There are programs that calculate the parameters through curve fitting algorithms and plot both the raw data and best fit lines, for example, Prism (GraphPad Software Inc., La Jolla, CA). Statistical analysis based on the Prism-calculated values is not optimal and direct ANOVA using the raw data at individual agonist concentrations is misleading. The most powerful approach is the program ALLFIT (an open source program (42)) that treats the entire data sets as a whole, determines the best fit parameter values for each condition, and tests for significance of intergroup differences.

3.12. Heart Perfusion and Coronary Artery Function

The coronary arteries run intramurally (within the ventricular wall) in rats, making isolation of coronary arteries for myographic study difficult. However, cardiac vascular function can be studied effectively in the isolated perfused heart and yield data on coronary artery function.

1. Following the protocol in **Section 3.9**, the heart is excised rapidly, with ~1 cm length of the aorta intact, and placed in a Petri dish containing ice-cold Krebs–Henseleit solution (*see* **Note 11**).
2. The aortic root is cannulated using a polyethylene cannula, secured with fine cotton suture ligature, and mounted on an Isolated Heart Apparatus (*see* **Note 11**). The heart is perfused through the aortic root and coronary arteries at 37°C in the Langendorf (constant pressure) mode at 100 mmHg pressure with Krebs–Henseleit solution gassed with 95% O_2 and 5% CO_2.
3. Data are acquired using the Isoheart[7]W program V 1.2, including heart rate, perfusion pressure (maintained at 100 mmHg), coronary flow, LVP systolic, LVEDP, and dLVP/d*t* maxima and minima.
4. Reactive hyperemia is measured by stopping perfusate flow for 60 s. When perfusion is restored, there is a reactive increase in coronary artery flow. The peak postischemia flow is expressed as percent of the baseline flow.
5. Coronary endothelial and medial function is assessed by addition of bradykinin or SNP to the perfusate to give concentrations over the ranges of 10^{-9}–10^{-5} M and 10^{-10}–10^{-4} M, respectively (*see* **Note 11**). Dose–response curves are calculated and analyzed as described in **Section 3.11**.

6. Between perfusion studies the liquid handling system must be thoroughly flushed and cleaned overnight on a regular basis with a non-toxic detergent solution, as supplied by Hugo Sachs Elektronik, to prevent bacterial contamination.

3.13. Cardiac Histology

1. At euthanasia, the heart is rapidly excised without the aortic arch and placed in 10% neutral-buffered formalin solution (3.7% formaldehyde) for 4 h to fix.
2. The fixed heart tissue is cut transversely into four segments, base (aortic root) through to apex, using a sharp scalpel. The segments are placed into a tissue processing cassette, cut side up together in a standard array, and then dehydrated and embedded in paraffin using conventional techniques.
3. The paraffin blocks are sectioned and two adjacent sections stained with hematoxylin and eosin (H&E) and Masson's trichrome, respectively. The slides are examined blindly by an experienced observer and the incidence of the four stages of ischemic lesions recorded (*see* **Note 12**). Stages 1–3 are most readily visualized in the H&E stained slides, with the Masson's trichrome used to confirm Stage 4 (old, scarred) lesions. The number of lesions of each stage found in the individual hearts is averaged and statistical comparisons made using non-parametric statistical methods, such as the Rank Sum Test.

3.14. Renal Histology – Glomerulosclerosis

1. At euthanasia, one kidney is dissected and excised (always the same side, preferably left), cut using a sharp scalpel through the cortex to the pelvis on the long axis, and the two halves placed in neutral-buffered formalin to fix.
2. The fixed halves are placed in a tissue cassette, cut side up, dehydrated, and paraffin embedded using standard procedures. The blocs are sectioned and the sections stained with H&E or periodic acid-Schiff stain (PAS), which gives better definition, but is less commonly used.
3. Light microscopic images of four fields of view of each kidney are taken of the slides at × 2 magnification using a digital camera system. The images are taken of the cortical area in a consistent pattern and with the operator blinded to the experimental group.
4. Images are visualized using Photoshop, examined blind, and all glomeruli in each field (minimum of 40 per kidney) rated as normal or sclerotic (*see* **Note 13**). Results are expressed as the percent of glomeruli that exhibit sclerosis.

3.15. Transmission Electron Microscopy

Electron microscopy is the definitive technique to detect and quantify atherosclerotic lesions in arteries of the size seen in the rat. The widely used alternatives, such as Oil Red O staining of the *en face* preparation of the aorta, give no information on the nature of the lesions, only that lipid is present. Quantitative measurement of lesions also requires perfusion fixation as the artery contracts on loss of blood pressure. Common practice is to simply dissect the thoracic and abdominal aorta, which is pinned flat and stained with Oil Red O or Sudan Black. This protocol leads to variable changes in area and serious artifacts. There are two different electron microscopy techniques, transmission and scanning. Transmission electron microscopy (TEM) provides a cross-sectional view of the arterial wall at high magnification, permitting identification of the types of cells (macrophages, foam cells, fibroblasts, smooth muscle cells) and materials (lipid, proteoglycan, collagen) in the intimal space that underlies the endothelial layer. The integrity of the endothelial layer and intimal thickness can also be established.

1. Euthanasia is conducted as in **Section 3.8** up to Step 8, the stage of blood removal through the needle and syringe.
2. The syringe is removed leaving the 18G needle in place in the left ventricle. A cannula terminated in a male Luer-Loc fitting is attached to the needle in situ.
3. The right atrium is cut with a pair of fine scissors, reaching behind the heart on the operator's left, thus opening up a drainage site on the venous side of the circulation.
4. The cannula is perfused with Tyrode's solution at 100 mmHg pressure until ~100 mL has been passed using a fixed pressure device and continuous monitoring of pressure. The needle in the ventricle may shift against the ventricular wall and occlude, thereby reducing the flow rate, as can happen during blood sampling (*see* **Note 14**).
5. When the arterial system has been flushed, the perfusate is changed to Tyrode's solution containing 2.5% glutaraldehyde as fixative and the perfusion continued at 100 mmHg for a further 100 mL. If conventional histology is to be performed, this stage of the fixation procedure is conducted by perfusion with Tyrode's solution containing 1.25% glutaraldehyde and 1.85% formaldehyde (effectively half the concentration of glutaraldehyde for EM fixation and half the concentration of formaldehyde of the milder fixation for light microscopy). If immunochemical staining is to be used, the perfusion fixation is conducted with Tyrode's solution containing 4% paraformaldehyde in 0.1 M sodium cacodylate.
6. The fixed (stiff) internal organs are dissected out by cutting with fine scissors above the aortic arch and folding the block of tissue forward and out. The heart and aorta are dissected carefully with adherent fat and fibrous tissue. The aortic arch and/or thoracic/

abdominal segments are placed in 2.5% glutaraldehyde in Tyrode's solution and the heart in 10% neutral-buffered formalin. Both are fixed for several days, and the aorta will be kept indefinitely in the 2.5% glutaraldehyde solution.

7. The aortic arch is dissected free of all adherent fat and connective tissue using a dissecting microscope and micro-scissors. This is essential to prevent tissue damage resulting from excessive thickness and presence of water in the sample during processing.
8. The branches on the arch are cut short and the arch cut with fine scissors along the greater and lesser curves, yielding two curved mirror image halves of the arch. Samples (full thickness 1–2 mm in size) are cut from one half, focused on the lesser curve of the arch and the ostia of the branches.
9. Samples are postfixed in 1% aqueous OsO_4 for 1 h at 4°C and stained en bloc with 2% aqueous uranyl acetate for 1 h at room temperature. They are dehydrated through graded ethanol and embedded in Spurr's resin.
10. Toluidine blue-stained 1-μm sections from each block are mounted on glass slides for light microscopy. Ultrathin sections (50 nm) are cut on an ultramicrotome, mounted on hexagonal 200-mesh copper grids, and stained with uranyl acetate and lead citrate.
11. Toludine blue-stained sections give a cross-sectional view of the intimal lesions and thickness can be quantified by image analysis. The ultrathin sections are examined in a transmission electron microscope and give information on cellular and subcellular elements of the lesions.

3.16. Scanning Electron Microscopy

Scanning electron microscopy (SEM) provides an image of the luminal surface at magnifications ranging from $\times 10$ to $\times 10{,}000$. The entire surface of a sample, such as the whole aortic arch, can be examined for the presence of endothelial damage, raised intimal lesions, adherent macrophages, and thrombi (at both large and cellular scales). Images can be recorded at a resolution of 4,000+ lines and quantified.

1. Follow Steps 1–7 from **Section 3.15.**
2. The aortic arch halves used for SEM are postfixed in 1% aqueous OsO_4 for 1 h at 4°C, dehydrated through graded ethanol, critical point-dried from propylene oxide in a critical point dryer, mounted on aluminum stubs, and sputter coated with gold.
3. The samples are examined overall in a scanning electron microscope, at magnifications ranging from $\times 4$ to $\times 4{,}000$. Particular attention is paid to the lesser curve of the arch and the regions adjacent to the ostia of the branches, which are atherosclerosis prone.

4. All lesions are recorded as a digital image at 4,000 lines resolution and each type of lesion is assigned a severity score, based on areal extent and character of the lesion, for each animal (*see* **Note 15**).

4. Notes

1. We have established that obese or metabolically abnormal rats are hyperresponsive to the stress of physical restraint and simple handling with significant and long-lasting metabolic changes (43, 44). These changes can seriously compromise experimental data. For example, a common approach to administration of agents to rats has been to inject the agent on a daily basis (subcutaneously, intramuscularly, intraperitoneally, or through the tail vein). All routes involve repeated restraint and some pain and are thus stressful. The technique of gastric gavage similarly imposes a repetitive stress and additionally creates a necessity to dissolve or suspend the agent in an aqueous medium, which can often be difficult.
2. The concentration of food additives may be calculated as follows:

$$C(\text{mg/kg}) = BW(\text{kg}) \times D(\text{mg/kg})/FC(\text{g/d})$$

   where C is the final concentration of additive in the prepared food, BW is the body weight determined 2 × per week, D is the desired dose/day, and FC is the measured daily food intake of the group of animals. This procedure has been demonstrated to give dosages within ±3% of calculated values (45).
3. Rats have conventionally been anesthetized with pentobarbital which is injected intraperitoneally. More recently, other agents, such as ketamine and xylazine, have been used (46). Obese and metabolically dysfunction rat strains, in particular, do not tolerate these injectable agents well. Typically rats with large fat depots do not reach a surgical plane of anesthesia using a normal dosage and a further increment must be given. Following this, the anesthesia becomes too deep and irreversible respiratory depression develops with rapid death.
4. The conventional method of assessing insulin and glucose metabolism in the rat has been the intravenous glucose tolerance test (IVGTT), with the euglycemic insulin clamp being the ultimate approach. Both methods present serious problems in obese insulin-resistant rats. Because the IVGTT shows relatively small differences in the rate of clearance of an injected glucose load (typically 0.5 g/kg) as a function of insulin sensitivity/resistance (47), it is difficult to detect experimental changes in the severity

of insulin resistance using this test. A euglycemic insulin clamp can give detailed information on the glucose clearance and hepatic output, effectively as a concentration response curve against plasma insulin, and is practical in obese rats (48). However, the procedure is complex, lengthy, and requires either that the rat be anesthetized throughout or that indwelling cannulae be surgically implanted in advance. The relatively lengthy anesthesia required for both the IVGTT and the euglycemic insulin clamp leads to very significant stress in these rats. Obese rats respond to stress with significant variations in insulin and glucose metabolism, as well as central nervous system changes (43, 44), all of which can obscure experimental effects.

In contrast to the IVGTT, which bypasses the complex gut hormone responses to food intake, a meal tolerance test (MTT) is much more sensitive to changes in insulin and glucose metabolism. The standardized MTT does not require anesthesia, minimizes stress on the rats, and can be administered repeatedly to the same animal (24). The insulin response is short-lived and the plasma concentrations at 30 min provide an excellent index of insulin sensitivity, which is essentially the plasma insulin level required to maintain euglycemia. Treatments that lower insulin resistance result in a reduced 30 min insulin concentration and can, in some cases, completely prevent the postprandial insulin response (45, 49).

5. Fasting plasma triglyceride and insulin levels are significantly higher in obese insulin-resistant rats and are higher again in the fed state, but glucose concentrations are in the normal range. The development of hyperinsulinemia and hypertriglyceridemia as the juvenile rat matures can be followed sensitively in the fed state (50), but is obscured in the fasted state.

6. Cardiac puncture requires access via the diaphragm, which in turn collapses the lungs, and initiates hypoxia, painless, irreversible unconsciousness, and death. It should be kept in mind that once the chest is opened, no further anesthetic agent reaches the circulation and that the brain levels will drop rapidly due to uptake by fatty tissues. Thus, it is essential that the cardiac puncture and blood sampling be initiated immediately before the level of anesthesia falls.

   A volume of 8–11 mL of blood is readily collected. Good collection depends on the needle puncture of the ventricle being made about $^1/_3$ of the way up the heart from the apex at an angle of about 30° and the bevel of the needle facing up. If the blood does not flow readily from the beating heart, the bevel of the needle is obstructed against the ventricular wall and the angle of the needle should be reduced with gentle pressure applied toward the back of the heart.

7. Krebs buffer solution for aortic ring vascular function is made from two stock solutions that are made up in RO water at 10 × and 20 × final concentrations, aliquoted into 200-mL screw-capped tubes, and stored at –20°C.

Solution 1 g/L		Solution 2 g/L	
NaCl	6.87	$NaHCO_3$	2.09
KCl	0.35	Glucose	1.81
$MgSO_4$	0.24	Na_2HPO_4	0.131
$CaCl_2$	0.176		

1. Keep the two solutions separate to prevent precipitation.
2. Prepare Krebs by mixing 200 mL of solution 1 and 200 mL of solution 2 into 2 L of RO water.
3. Check pH after solution has been bubbled for ½ h with 95% oxygen/5% CO_2 in the baths at 37°C to saturate. It may be necessary to adjust pH to ~7.4 with 0.1 M HCl.
4. 0.1 M HCl can be made by adding 8.6 mL of concentrated HCl to 1 L of RO water. (Will usually need to add ~17 mL to 1 L of Krebs to bring pH down.)

8. Aortic tissue is sensitive and must be harvested immediately after exsanguination, with care not to injure the endothelial layer. When all external fat and clotted blood from inside aorta are removed, place the aorta on a wet filter paper with clean buffer and keep moist. Using a sharp scalpel blade, slice into sections no more than 3 mm long. Carefully ensure that there is no "crushing" of the tubular tissue (this rubs the endothelial cells) using a slight rolling motion. Damaged endothelial cells will not secrete NO, either spontaneously/physiologically or in response to ACh, leading to an absence of NO-mediated relaxation and hypercontractilty.
9. Protocol for preparation of agonist/antagonist solutions.

Agent	Mol weight	g/mL	Stock M	Further dilutions
PE	203.7	0.0204 g	10^{-1} (10^{-4} in bath)	*
ACh	181.7	0.01817	10^{-1} (10^{-4} in bath)	*

(continued)

Agent	Mol weight	g/mL	Stock M	Further dilutions
L-NAME	269.7	0.02697	10^{-1} (10^{-4} in bath)	No further
SNP	222	0.0222 §	10^{-1} (10^{-4} in bath)	§

- Use RO water for all solutions.
- Baths – 10 mL; * – 0.1 mL of stock conc in 1 mL RO water = 10^{-2} (or 10^{-5} in bath); 10^{-6} mol/L would be made with 0.1 mL of 10^{-5} mol/L made up to 1 mL with RO water and so forth.
- § – Will not dissolve in 1 mL, use 10 mL (10^{-5} mol/L in bath will give full effect).
- SNP – *S*-nitroso-*n*-acetyl-penicillamine; ACh – acetylcholine chloride.
- L-NAME – *N*-nitro-L-arginine methyl ester; PE – phenylephrine hydrochloride.

All equipment must be cleaned daily to prevent buildup of salts and bacterial contamination.

10. Mesenteric arteries of 300 μm diameter are resistance arteries, in contrast to the aorta, which is a conductance artery. These different vessels have different physiological responses and give results that provide a range of information on vascular dysfunction. The agonists and equipment used are different to suit the type of artery being studied.
11. Rat hearts are perfused on an Isolated Heart Apparatus (size 3) (Hugo Sachs Elektronik, March-Hugstetten, Germany), equipped with a Type 700 transit time flowmeter, and data acquisition with the Isoheart[7]W program V 1.2 (Hugo Sachs). Perfusion is performed using Krebs–Henseleit solution gassed with 95% O_2 and 5% CO_2. Bradykinin (a non-cholinergic, endothelium-dependent vasodilator acting on its own BK receptor) and SNP (a direct NO donor and vasodilator) are added to the perfusate through a "T" fitting upstream to the debubbling chamber using a calibrated roller pump. The following variables are measured; heart rate, perfusion pressure (maintained at 100 mmHg), coronary flow, LVP systolic, LVEDP, and dLVP/d*t* maxima and minima. Reactive hyperemia is measured following 60 s of stopped perfusion, with the peak postischemia flow expressed as percent of the baseline flow. ACh is not used to assess endothelial cell function, as the coronary artery does not express cholinergic receptors, but does have BK receptors.

12. The myocardial lesions are a record of end stage of CVD, over the life time of the rat. They are identified as follows:

    Stage 1: areas of necrosis. These are recently infarcted areas of an age from ~6 to 24 h, before chronic inflammatory cell (CIC) infiltration has been established, and characterized by a darkened "glassy" appearance in the histologic section.

    Stage 2: areas of cell lysis with CIC infiltration. These are lesions aged 24 h to 3 week. The most recent lesions show only numerous chronic inflammatory cells infiltrating the necrotic zone, evident by dark prominent staining of the nuclei. Advanced Stage 2 lesions are characterized by cell "drop out" (empty or unstained areas where myocytes have been lysed and scavenged by the CIC).

    Stage 3: nodules of chronic inflammatory cell infiltration. These are small isolated nodules of CIC without evidence of lysis of myocytes, appear in normal rats, and are probably not significant.

    Stage 4: old, scarred lesions. Stage 4 lesions are the most important, as they reflect the cumulative record of earlier stage lesions that were large enough to remain identifiable after the scarring and contraction of the repair process.

Micrographs of typical myocardial lesions are shown in **Fig. 2.1**. The mean incidence of each stage yields a consistent index of the ischemic damage to the myocardium (51).

13. Characterization of glomeruli as sclerotic has been summarized, with excellent representative micrographs (52). Sclerotic glomeruli show enlarged Bowman's space, often fissures in the glomerular tuft and fibrosis. The use of Photoshop, with high resolution recorded microscopic images, allows both large-scale, low-magnification assessment and higher magnification (electronically) examination of individual glomeruli.
14. Good perfusion fixation is dependent on maintaining the desired 100 mmHg pressure in the ventricle. If the needle occludes, partially or completely, flow will drop and pressure upstream from the heart at the pressure transducer will rise, due to the normal drop in pressure within the length of tubing leading to the heart. The needle should be gently moved away from the ventricular wall by trial and error until the measured pressure drops slightly and the flow resumes. The flow can also be monitored through the outflow from the cut right atrium. An effective fixation is usually accompanied by sudden and widespread muscle spasm throughout the body. Perfusion should be continued until the entire 100 mL of fluid has been infused.

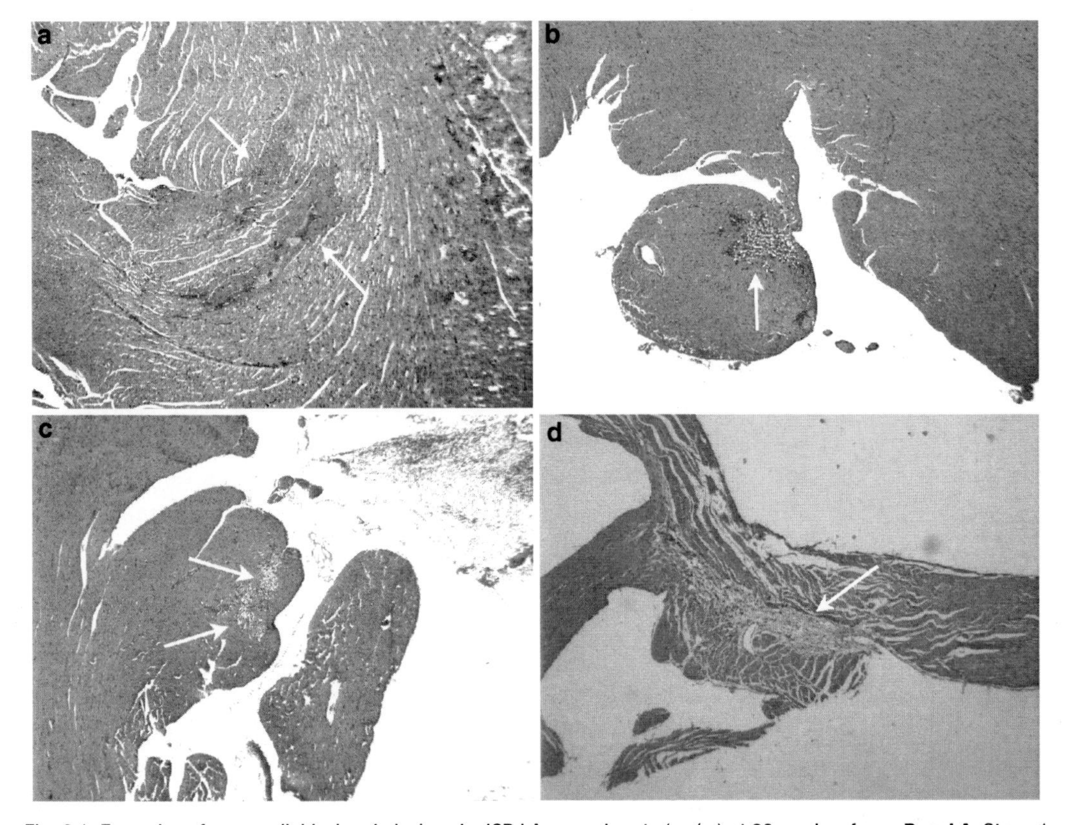

Fig. 2.1. Examples of myocardial ischemic lesions in JCR:LA-*cp* male rats (*cp/cp*) at 26 weeks of age. **Panel A**: Stage 1 lesion, area of necrosis without chronic inflammatory cell infiltration, in left ventricle. **Panel B**: Stage 2 lesion, area of active inflammatory cell activity and cell lysis, in lower trabecular muscle. **Panel C**: Stage 3 lesion, area of chronic inflammatory cell infiltration, without visible cell lysis, in trabecular muscle. **Panel D**: Stage 4, early scarred lesion with a small number of inflammatory cells or fibroblast, in upper perivalvular region of the heart. All images at × 20, H&E stained sections, with lesions indicated by *arrows.*

15. Lesions detected by SEM are identified and classified as areas of adherent fibrin; raised intimal lesions; areas of adherent macrophages; or areas of de-endothelialization (53). The aortic arch, after fixation and processing, retains its in vivo morphology in two semi-rigid segments composed of compound curves. It is not possible either to flatten the arch or to obtain planar images of the arch and areas of lesion that are suitable for quantification through image analysis. Thus, images of all lesions are recorded and each type of lesion is assigned a "severity score," based on area and character of the lesion. The scale ranges from 0 to 3, with 0 representing the absence of any lesions and 3 representing the most severe involvement, as seen in atherosclerotic control animals. The scores are summed for each animal.

Representative TEM images of atherosclerotic lesions in rats have been reported previously (54).

Acknowledgments

The advancement of Science builds on the contributions of those who went before us. The development of the laboratory rat can be traced to breeding and specialization of "fancy rats" in Kyoto, Japan dating to the seventeenth century (54). The modern experimental use of rats originated in The Wistar Institute in Philadelphia and from work of Henry Donaldson (55). Studies in our laboratory would not have been possible without the assistance of veterinarians David Secord and David Neil, senior technicians Dorothy Koeslag and Sandra Kelly, and many students and postdoctoral fellows. Essential financial support was provided by the Medical Research Council of Canada, the Heart and Stroke Foundation of Alberta and the Northwest Territories and a number of pharmaceutical firms.

References

1. Karsch, KR. (1992) Atherosclerosis – where are we heading? *Herz* **17**, 309–319.
2. Moore, S. (1981) Responses of the arterial wall to injury. *Diabetes* **30(S 2)**, 8–13.
3. Ross, R. (1999) Atherosclerosis: an inflammatory disease. *N Engl J Med* **340**, 115–126.
4. Yusuf, S, Reddy, S, Ôunpuu, S, et al. (2001) Global burden of cardiovascular diseases Part I: general considerations, the epidemiological transition, risk factors, and impact of urbanization. *Circulation* **104**, 2746–2753.
5. Proctor, SD, Russell, JC. (2006) Small animal models of cardiovascular disease: tools for study of roles of the metabolic syndrome, dyslipidemias and atherosclerosis. *J Cardiovasc Pathol* **15**, 318–330.
6. Davignon, J, Genest, J. (1998) Genetics of lipoprotein disorders. *Endocrin Metab Clinics of N Amer* **27**, 521–550.
7. Brown, MS, Goldstein, JL. (1996) Heart attacks: gone with the century? *Science* **272**, 629.
8. Ornskov, F. (1998) In Jacotot, B, Mathé, D, Fruchart, J-C, (eds.), *Atherosclerosis XI: Proceedings of the 11th International Symposium on Atherosclerosis.* Elsevier Science, Singapore, pp. 925–32.
9. Hegele, RA, Zinman, B, Hanley, AJ, et al. (2003) Genes, environment and Oji-Cree type 2 diabetes. *Clin Biochem* **36**, 163–170.
10. Razak, F, Anand, S, Vuksan, V, Davis, B, et al. (2005) Ethnic differences in the relationships between obesity and glucose-metabolic abnormalities: a cross-sectional population-based study. *Int J Obesity* **29**, 656–667.
11. Richardson, M, Kurowska, EM, Carroll, KK. (1994) Early lesion development in the aortas of rabbits fed low-fat, cholesterol-free, semipurified diet. *Atherosclerosis* **107**, 165–178.
12. Richardson, M, Fletch, A, Delaney, K, et al. (1997) Increased expression of vascular cell adhesion molecule-1 by the aortic endothelium of rabbits with *Pasteurella multocida* pneumonia. *Lab Anim Sci* **47**, 27–35.
13. Bolzán, AD, Bianchi, MS. (2002) Genotoxicity of streptozotoc. *Mutat Res* **512**, 121–134.
14. Yang, Y, Santamaria, P. (2006) Lessons on autoimmune diabetes from animal models. *Clin Sci* **110**, 627–639.
15. Zucker, LM, Zucker, TF. (1961) Fatty, a new mutation in the rat. *J Hered* **52**, 275–278.
16. Chua, SC Jr, White, DW, Wu-Peng, XS, et al. (1996) Phenotype of fatty due to Gln269Pro mutation in the leptin receptor (Lepr). *Diabetes* **45**, 1141–1143.
17. Amy, RM, Dolphin, PJ, Pederson, RA, et al. (1988) Atherogenesis in two strains of obese rats: the fatty Zucker and LA/N-corpulent. *Atherosclerosis* **69**, 199–209.
18. Peterson, RG. (2001) The Zucker Diabetic Fatty (ZDF) rat. In Sima, AAF, Shafrir, E,

(eds.), *Animal Models of Diabetes a Primer*. Harwood Academic Publishers, Amsterdam, pp. 109–128.

19. Schäfer, S, Steioff, K, Linz, W, et al. (2004) Chronic vasopeptidase inhibition normalizes diabetic endothelial dysfunction. *Euro J Pharmacol* **484**, 361–362
20. Oltman, CL, Richou, LL, Davidson, EP, et al. (2006) Progression of coronary and mesenteric vascular dysfunction in Zucker obese and Zucker Diabetic Fatty rats. *Am J Physiol Heart Circ Physiol* **291**, H1780–H1787.
21. Marsh, SA, Powell, PC, Agarwal, A, et al. (2007) Cardiovascular dysfunction in Zucker obese and Zucker diabetic fatty rats: role of hydronephrosis. *Am J Physiol Heart Circ Physiol.* **293**, H292–H298.
22. Koletsky, S. (1975) Pathologic findings and laboratory data in a new strain of obese hypertensive rats. *Am J Pathol* **80,** 129–142.
23. Wu-Peng, XS, Chua, SC Jr, Okada, N, et al. (1997) Phenotype of the obese Koletsky (*f*) rat due to Tyr763Stop mutation in the extracellular domain of the leptin receptor: evidence for deficient plasma-to-CSF transport of leptin in both the Zucker and Koletsky obese rat. *Diabetes* **46**, 513–518.
24. Russell, JC, Graham, SE, Dolphin, PJ. (1999) Glucose tolerance and insulin resistance in the JCR:LA-cp rat: effect of miglitol (Bay m1099). *Metabolism* **48**, 701–706.
25. Russell, JC, Bar-Tana, J, Shillabeer, G, et al. (1998) Development of insulin resistance in the JCR:LA-cp rat: role of triacylglycerols and effects of MEDICA 16. *Diabetes* **47**, 770–778.
26. Vance, JE, Russell, JC. (1990) Hypersecretion of VLDL, but not HDL, by hepatocytes from the JCR:LA-corpulent rat. *J Lipid Res* **31**, 1491–1501.
27. Russell, JC, Graham, SE, Richardson, M. (1998) Cardiovascular disease in the JCR:LA-cp rat. *Mol Cell Biochem* **188**, 113–126.
28. O'Brien, SF, Russell, JC, Davidge, ST. (1999) Vascular wall dysfunction in JCR:LA-cp rats: effects of age and insulin resistance. *Am J Physiol* **277**, C987–C993.
29. Proctor, SD, Kelly, SE, Russell, JC. (2005) A novel complex of arginine–silicate improves micro- and macrovascular function and inhibits glomerular sclerosis in insulin-resistant JCR:LA-cp rats, *Diabetologia* **48**, 1925–1932.
30. Richardson, M, Schmidt, AM, Graham, SE, et al. (1998) Vasculopathy and insulin resistance in the JCR:LA-cp rat. *Atherosclerosis* **138,** 135–146.
31. McCune, S, Park, S, Radin, MJ, et al. (1995) The SHHF/Mcc-facp: a genetic model of congestive heart failure. In Singal, PK, Beamish, RE, Dhalla, NS, (eds.), *Mechanisms of Heart Failure*. Kluwer Academic, Boston, MA, pp. 91–106.
32. Janssen, PML, Stull, LB, Leppo, MK, et al. (2002) Selective contractile dysfunction of left, not right, ventricular myocardium in the SHHF rat. *Am J Physiol Heart Circ Physiol* **284**, H772–H778.
33. Emter, CA, McCune, SA, Sparagna, GC, et al. (2005) Low-intensity exercise training delays onset of decompensated heart failure in spontaneously hypertensive heart failure rats. *Am J Physiol Heart Circ Physiol* **289**, H2030–H2038.
34. Portha, B. (2005) Programmed disorders of β-cell development and function as one cause for type 2 diabetes? The GK rat paradigm. *Diabetes Metab Res Rev* **21**, 495–504.
35. Matsumoto, T, Kakami, M, Noguchi, E, et al. (2007) Imbalance between endothelium-derived relaxing and contracting factors in mesenteric arteries from aged OLETF rats, a model of Type 2 diabetes. *Am J Physiol Heart Circ Physiol* **293**, H1480–H1490.
36. Kagota, S, Tanaka, N, Kubota, Y, et al. (2004) Characteristics of vasorelaxation responses in a rat model of metabolic syndrome. *Clin Exp Pharmacol Physiol* **31**, S54–S56.
37. Vine, D, Takechi, R, Russell, JC, et al. (2007) Impaired postprandial apolipoprotein-B48 metabolism in the obese, insulin-resistant JCR:LA-cp rat: increased atherogenicity for the metabolic syndrome. *Atherosclerosis* **190,** 282–290.
38. Mangat, R, Su, J, Scott, PG, Russell, et al. (2007) Chylomicron and apoB48 metabolism in the JCR:LA corpulent rat, a model for the metabolic syndrome. *Biochem Soc Trans* **35**, 477–481.
39. Poiley, SM. (1960) A systematic method of breeder rotation for non-inbred laboratory animal colonies. *Animal Care Panel* **10**, 159–161.
40. Russell, JC, Koeslag, DG, Amy, RM, et al. (1989) Plasma lipid secretion and clearance in the hyperlipidemic JCR:LA-corpulent rat. *Arteriosclerosis* **9**, 869–876.
41. Radomski, MW, Salas, E. (1995) Nitric oxide-biological mediator, modulator and factor of

injury: its role in the pathogenesis of athero-sclerosis. Atherosclerosis **118**, S69–S80.

42. De Lean, A, Munson, PJ, Rodbard, D. (1978) Simultaneous analysis of families of sigmoidal curves: application to bioassay, radioligand assay, and physiological dose-response curves. *Am J Physiol* **235**:E97–E102.
43. Leza, JC, Salas, E, Sawicki, G, et al. (1998) The effect of stress on homeostasis in JCR:LA-cp rats. Role of nitric oxide. *J Pharmacol Exp Ther* **28,** 1397–1403.
44. Russell, JC, Proctor, SD, Kelly, SE, Brindley, DN. (2008) Pair feeding-mediated changes in metabolism: stress response and pathophysilogy in insulin resistant, atherosclerosis-prone JCR:LA-cp rats. *Am J Physiol Endocrinol Metab,* **294**, E1078–1087.
45. Russell, JC, Ravel, D, Pégorier, J-P, et al. (2000) Beneficial insulin-sensitizing and vascular effects of S15261 in the insulin-resistant JCR:LA-cp rat. *J Pharmacol Exp Ther* **295,** 753–760.
46. Hanusch, C, Hoeger, S, Beck, GC. (2007) Anaesthesia of small rodents during magnetic resonance imaging. *Methods* **43,** 68–78.
47. Russell, JC, Amy, RM, Manickavel, V, et al. (1987) Insulin resistance and impaired glucose tolerance in the atherosclerosis prone LA/N-corpulent rat. *Arteriosclerosis* **7,** 620–626
48. Russell, JC, Graham, S, Hameed, M. (1994) Abnormal insulin and glucose metabolism in the JCR:LA-corpulent rat. *Metabolism* **43,** 538–543.
49. Russell, JC, Dolphin, PJ, Graham, SE, et al. (1998) Improvement of insulin sensitivity and cardiovascular outcomes in the JCR:LA-cp rat by d–fenfluramine. *Diabetologia* **41,** 380–389.
50. Russell, JC, Bar-Tana, J, Shillabeer, G, et al. (1998) Development of insulin resistance in the JCR:LA-cp rat: role of triacylglycerols and effects of MEDICA 16. *Diabetes* **47,** 770–778.
51. Russell, JC, Amy, RM, Graham, S, et al. (1993) Effect of castration on hyperlipidemic, insulin resistant JCR:LA-corpulent rats. *Atherosclerosis* **100,** 113–122.
52. Ferrario, F, Rastaldi MP. (2006) Histopathological atlas of renal diseases: diabetic nephropathy. *J Nephrol* **19,** 1–5.
53. Russell, JC, Amy, RM, Graham, SE, et al. (1995) Inhibition of atherosclerosis and myocardial lesions in the JCR:LA-cp rat by β,β'-tetramethylhexadecanedioic acid (MEDICA 16). *Arterioscler Thromb Vasc Biol* **15,** 918–923.
54. Mashimo, T, Voigt, B, Kuramoto, T, et al. (2005) Rat phenome project: the untapped potential of existing rat strains. *J Appl Physiol* **98,** 371–379.
55. Lindsay, JR. (1976) Historical Foundations. In Baker, HJ, Lindsay, JR, Weisbroth, SH (eds.), *The Laboratory Rat.* Academic Press, Orlando, pp. 1–36.

Chapter 3

Non-invasive Blood Pressure Measurement in Mice

Minjie Feng and Keith DiPetrillo

Abstract

Hypertension is a leading cause of heart attack, stroke, and kidney failure and represents a serious medical issue worldwide. The genetic basis of hypertension is well-established, but few causal genes have been identified thus far. Non-invasive blood pressure measurements are a critical component of high-throughput genetic studies to identify genes controlling blood pressure. Whereas this technique is fairly routine for blood pressure measurements in rats, non-invasive blood pressure measurement in mice has proven to be more challenging. This chapter describes an experimental protocol measuring blood pressure in mice using a CODA non-invasive blood pressure monitoring system. This method enables accurate blood pressure phenotyping in mice for linkage or mutagenesis studies, as well as for other experiments requiring high-throughput blood pressure measurement.

Key words: Tail-cuff, blood pressure, mice, high-throughput.

1. Introduction

High blood pressure (i.e., hypertension) is a substantial medical problem worldwide and is an important risk factor for heart disease, stroke, and kidney disease (1). Although many effective antihypertensive medications are available, most patients require multiple drugs to lower blood pressure (1), and many patients do not achieve adequate blood pressure lowering to the recommended levels. Thus, additional research into the causes of hypertension is imperative to identify new therapies to better treat hypertension.

Blood pressure is a highly heritable phenotype affected by multiple genes and environmental factors; identifying the genes causing hypertension could elucidate novel therapeutic targets. The genetic basis of hypertension has been investigated extensively in humans through genome-wide and candidate-gene association

K. DiPetrillo (ed.), *Cardiovascular Genomics*, Methods in Molecular Biology 573,
DOI 10.1007/978-1-60761-247-6_3, © Humana Press, a part of Springer Science+Business Media, LLC 2009

studies, as well as genome-wide linkage analyses. Despite the substantial effort made to identify genes underlying polygenic hypertension (*see* Cowley (2) for review), few causal genes have been identified to date.

One alternative approach to using humans to study the genetic basis of hypertension is to identify genes affecting blood pressure in model organisms, mainly rodents, and then test those genes for a role in human blood pressure control. For example, genetic studies in humans and rats successfully identified the genes encoding adducing (3) and 11β-hydroxylase (4) as important in blood pressure control. Because genetic analysis of blood pressure in rodents requires high-throughput blood pressure measurement methods, the American Heart Association Council on High Blood Pressure Research recommends non-invasive, tail-cuff blood pressure methods for mutagenesis or genetic linkage studies (5). Whereas tail-cuff blood pressure measurement in rats is well-established, tail-cuff blood pressure measurement in mice is more difficult and has hampered the use of mice for high-throughput genetic analysis of blood pressure. This chapter provides an experimental protocol for accurately measuring tail-cuff blood pressure in mice using the CODA non-invasive blood pressure system.

2. Materials

2.1. Mice

Mice can be purchased from several commercial vendors or bred internally. We do not recommend ear tagging for identifying mice to be used for tail-cuff blood pressure measurements because of the potential for the ear tag to catch on the restrainer while the animal is entering or exiting the restrainer. Ear notching, toe-clipping (mice < 12 days old), and tattooing are alternative methods for identifying mice intended for tail-cuff blood pressure measurements. All mice used for tail-cuff blood pressure measurements should have normal tails of adequate length.

2.2. CODA Non-invasive Blood Pressure System

The CODA non-invasive blood pressure system is available from Kent Scientific (Torrington, CT; www.kentscientific.com; **Fig. 3.1**) and can measure blood pressure in up to eight mice simultaneously. Several CODA systems can connect to a single computer, thus enabling measurement of up to 48 mice simultaneously from one computer. The CODA utilizes volume–pressure recording (VPR) technology to detect changes in tail volume that correspond to systolic and diastolic blood pressures. The VPR technology was validated by comparison to simultaneous radiotelemetry blood pressure measurements; the VPR method underestimates telemetry

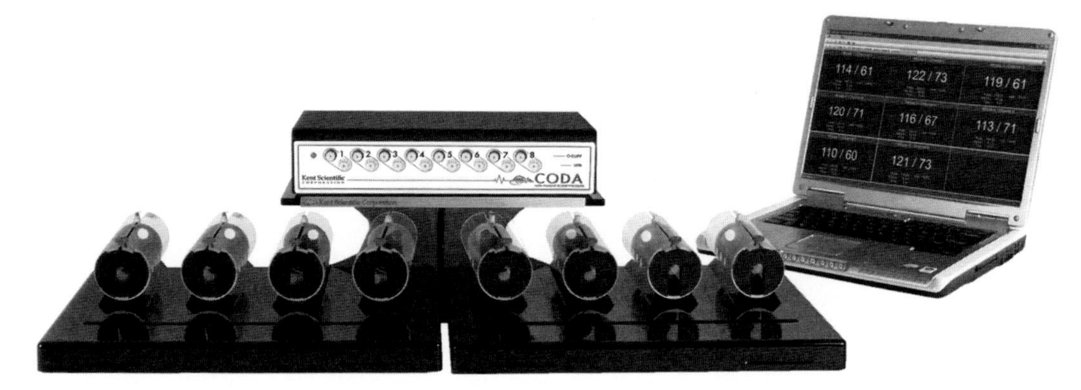

Fig. 3.1. CODA-8 non-invasive blood pressure system.

systolic blood pressure measurements by 0.25 mmHg and telemetry diastolic blood pressure measurements by 12.2 mmHg, on average (6; *see* **Note 1**).

2.3. Mouse Restrainer

Mice from most inbred strains will fit into the small or medium restrainers (**Fig. 3.2**), depending on age and body weight. We typically measure blood pressure in 8-week-old mice using small restrainers intended for mice under 25 g. The restrainer is designed to comfortably house the mouse while minimizing movement during the measurement session. The darkened nose cone limits the mouse's view and decreases the level of stress. Inside the restrainer, the mouse faces the nose cone with its nose protruding through a hole in the nose cone and its tail extends through the rear hatch of the restrainer. The nose cone should face the CODA box and the rear of the restrainer with the mouse's tail extended outward should face the user.

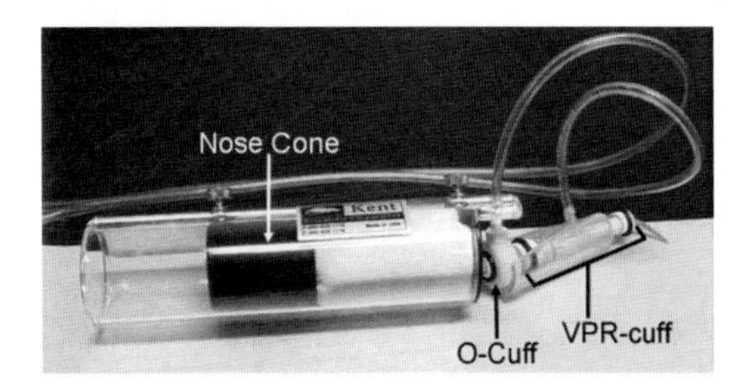

Fig. 3.2. Restrainer, occlusion cuff (O-cuff), and volume–pressure recording cuff (VPR-cuff) for the CODA non-invasive blood pressure monitoring system.

2.4. Mouse Cuff Sets

The CODA non-invasive blood pressure system relies on two tail-cuffs to measure blood pressure (*see* **Note 1**; **Fig. 3.2**), occlusion cuffs (O-cuff) and volume–pressure recording cuffs (VPR-cuff); the O-cuff is positioned near the base of the tail proximal to the

VPR-cuff (**Fig. 3.2**). The O-cuffs and VPR-cuffs should be connected to the CODA box prior to the experiment. Both the O-cuffs and VPR-cuffs require balloon bladders; for blood pressure measurements in mice, we use extra small O-cuff bladders and small VPR-cuff bladders.

2.5. Paper Towels

Scotch threefold paper towels are used for multiple purposes in measuring tail-cuff blood pressure in mice. Place paper towels underneath each restrainer to collect urine and feces during the measurement session. The paper towels are also very important for controlling the body temperature of the mice during the session (*see* **Note 2**). When the mice are too cold, use the paper towel to cover the mice to increase body temperature and blood flow to the tail. When the mice are too warm, cut the paper towel into 1.5 in.^{2}, wet with room temperature water, and place over the restrainer to cool the mice. Because sufficient blood flow to the tail without overheating is critical for tail-cuff blood pressure measurement, warming or cooling the mice with paper towels is a key component of the method.

3. Methods

3.1. Room Setup

It is important to create a warm, calm environment for the mice to obtain accurate blood pressure measurements. Ideally, the mice should be housed in a separate, nearby room to minimize transportation of the mice prior to the blood pressure measurement session. The proper room temperature is essential for accurate blood pressure measurements. If the temperature is too low, such as below 22°C (71.6°F), blood flow to the tail may be insufficient to detect changes in tail volume; a cold, steel table or a nearby air duct should be avoided. The room should also be insulated from extraneous noise by using a quiet location, placing signs to decrease traffic and noise outside the room, and by hanging noise reducing curtains (Unger Technologies Inc., Noblesville, IN, USA) at the entrances. In addition to keeping the room warm and quiet, the room should also be dimly lit with a desk lamp containing a 40 W bulb on each bench; the bulbs should face the ceiling so that the light is softened and dissipated throughout the room.

3.2. Machine Preparation

Turn on the heating pads and place the restrainers and paper towels on the platform at least 1 h prior to the starting blood pressure measurements to ensure that the equipment reaches the desired temperature before the measurement session begins.

Always turn on the computer before turning on the CODA box. Open the CODA Software and use the Device Manager (**Fig. 3.3**)

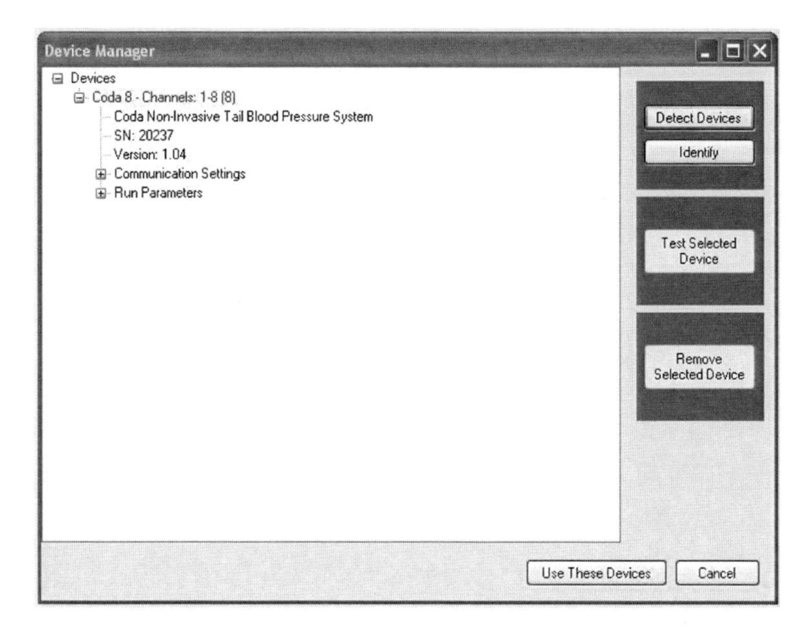

Fig. 3.3. Device Manager for CODA software. This screen opens when starting the CODA software and enables the user to test the CODA hardware, input the session parameters and animal identifications, and control the measurement session.

to check every channel required for your study (on the right-hand side, click "Test Selected Device" → "Select All" → "Test"). If cuff leaks are detected (**Fig. 3.4**), the corresponding bladders should be replaced as described in the user manual. Every channel in use should have an O-cuff and VPR-cuff attached, and all unused channels should be closed with the Port Cover Caps to reduce internal contamination of tubing and valves (*see* **Note 3**).

3.3. CODA Software Preparation

To start a new experiment within the CODA software, choose "File" → "New" → "Experiment" or use the "New Experiment" icon; to start a new session within an existing experiment, choose "File" → "New" → "Session" or use the "New Session" icon. These options will open the Experiment Wizard that allows you to set the number of cycles, input the animal identifications, and enter the measurement parameters for the session. The Basic Session Info screen in the Experiment Wizard (**Fig. 3.5**) will enable you to enter the session name, the number of acclimation and measurement cycles for the session, and the time between cycles. Our protocol consists of 10 acclimation cycles and 20 measurement cycles once daily for 5 days (maximum 100 blood pressure measurements), but this will vary for different experimental

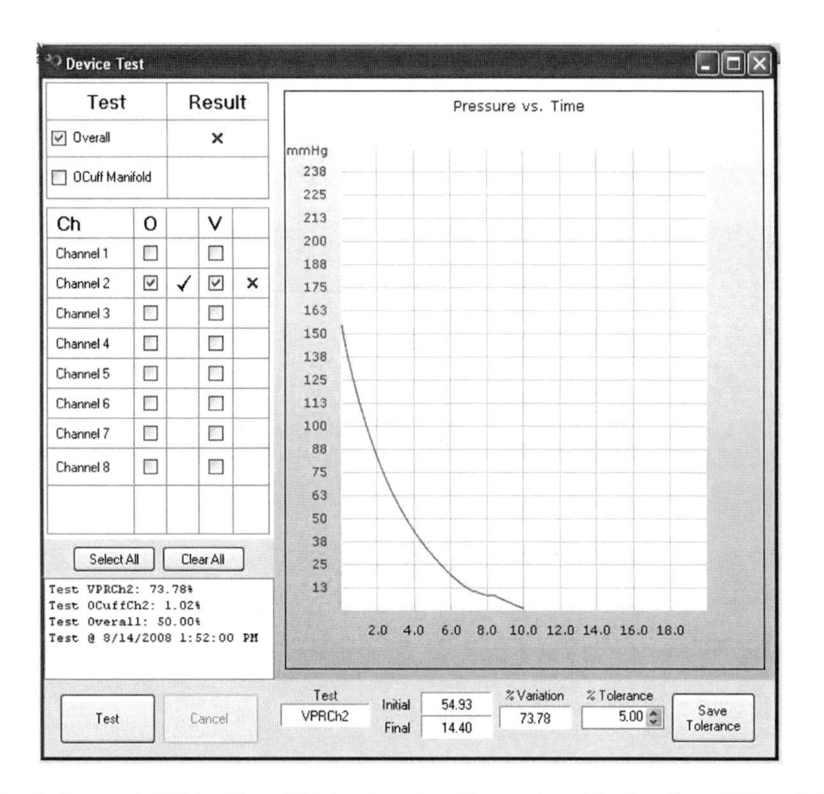

Fig. 3.4. Device Test screen in CODA software. This function allows the user to test the O-cuffs and VPR-cuffs for air leaks. Users can test all eight channels to confirm appropriate O-cuff and VPR-cuff function by clicking the "Select All" box, or test only a subset of channels by checking the appropriate boxes. In this example, the O-cuff on channel 2 passed the test, but the VPR-cuff on channel 2 failed the test and should be replaced.

designs. The Specimen Selection screen (**Fig. 3.6**) allows you to input the animal identifications and assign animals to each of the CODA channels during the session, and the Session Parameters (**Fig. 3.7**) screen enables you to set the deflation time, maximum occlusion pressure, and minimum tail volume.

3.4. Animal Handling

Accurate blood pressure measurement requires proper handling of the mice to ensure a relaxing environment. Blood flow in the tail will decrease and blood pressure will increase if the mouse is stressed, thus all mice should be gently guided, rather than be forced, into the restrainer. For a large mouse, adjust the nose cone to the far end of the restrainer to create a bigger space so that the mouse is more willing to enter (*see* **Note 4**). Some mice are reluctant to enter the restrainer, and it is sometimes helpful to fully remove the nose cone to create an open tube to entice the mouse to enter and then replace the nose cone once the mouse has partially entered the restrainer. Once the mouse has entered, gently hold the mouse in the restrainer and replace the rear hatch to contain the mouse within. The body of the mouse should not be compressed against the rear hatch of the restrainer; if this occurs,

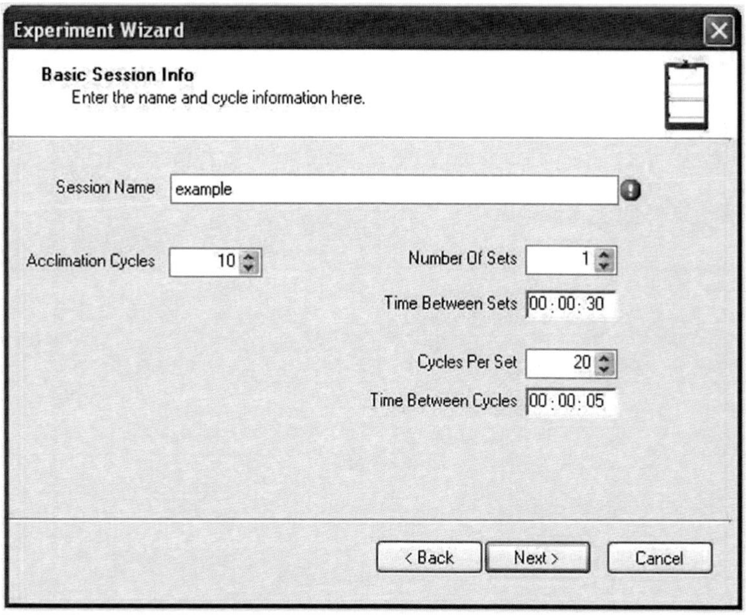

Fig. 3.5. Experiment Wizard, Basic Session Info screen, in CODA software. This function allows the user to input the session name, set the number of acclimation cycles, and enter the number of cycles (i.e., cycles per set) and time between cycles for the measurement session.

provide more space by either moving the nose cone forward or using a larger restrainer. Once the mouse enters the restrainer, the nose cone should be adjusted so that the mouse is comfortable but

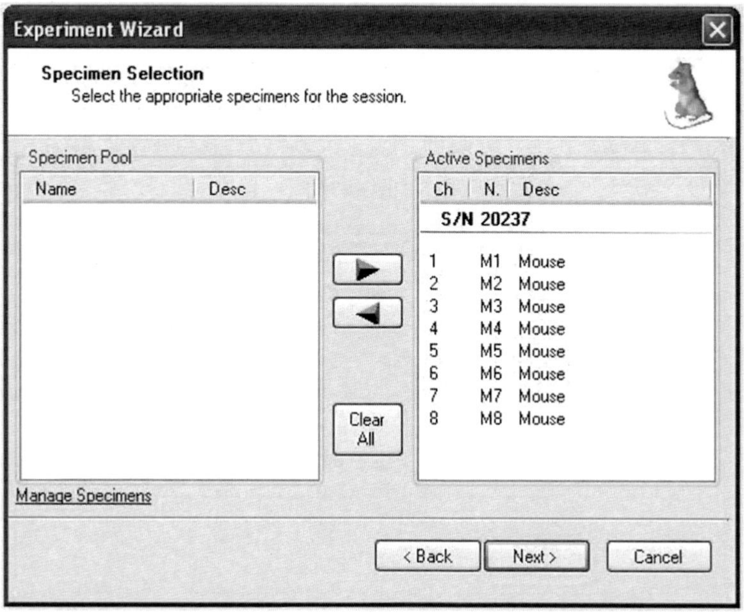

Fig 3.6. Experiment Wizard, Specimen Selection screen, in CODA software. This function allows the user to input the animal identification for the session.

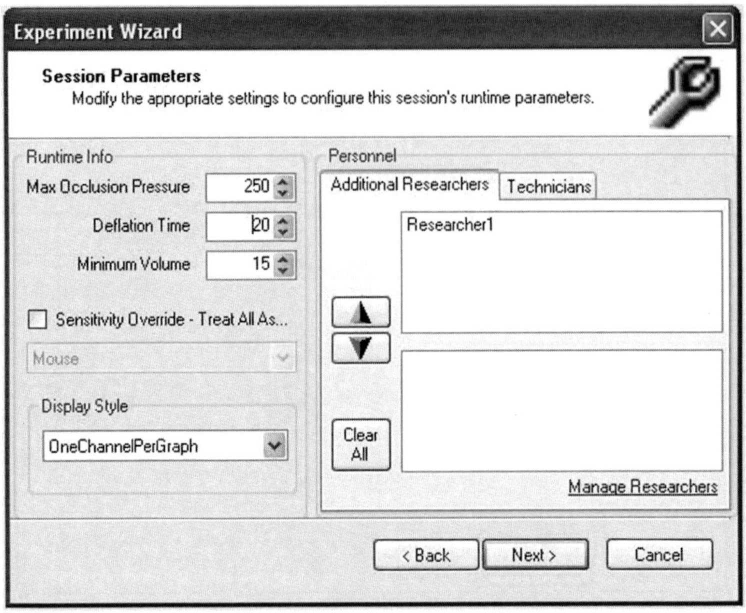

Fig. 3.7. Experiment Wizard, Session Parameters screen, in CODA software. This function allows the user to input the maximum O-cuff occlusion pressure, the O-cuff deflation time, and the minimum tail volume for a cycle to be accepted. The researcher's name can also be entered on this screen to track the personnel who performed individual measurement sessions.

not able to move excessively. However, if the restrainer is too large, the mouse will likely move during the measurement cycles, causing the cycles to be discarded.

3.5. Blood Pressure Measurement and Data Collection

Once all of the mice to be measured on the same CODA system are placed into their restrainers, allow the mice to acclimate to the restrainer for 5 min prior to initiating the blood pressure measurement protocol. This period allows the mice to relax and warm, which facilitates blood flow to the tail. Following the 5-min acclimation period, start the blood pressure measurement protocol. The mice are typically warmed by heating pads under the restrainer platform during the acclimation and measurement periods to ensure sufficient blood flow to the tail. However, mice from various strains will respond differently to heating, so it is very important to determine the appropriate amount and duration of heat for each strain (*see* **Note 5**). Monitor the animals closely throughout the blood pressure measurement protocol and remove them from the restrainers as soon as possible upon completing the measurement protocol.

3.6. Cleaning the CODA System After Use

After completing the measurement sessions for a day, the CODA system should be cleaned. The restrainers should be washed with soap in warm water and allowed to air dry overnight. Discard and replace the paper towel underneath each restrainer and wipe the platform and table clean with soap and warm water. Wipe the cuffs as needed.

3.7. Data Analysis

The CODA system measures systolic and diastolic blood pressures and calculates mean arterial pressures during each measurement cycle. After taking all of the blood pressure measurements for an experimental group (after day 5 in our typical protocol, for example), export the group data from the CODA system (*see* **Note 6**) and calculate the average and standard deviation of the replicate measurements for each blood pressure parameter for each mouse. Discard any individual reading of more than two standard deviations from the mean for an individual mouse and calculate a new average and standard deviation as the final data for the mouse.

4. Notes

1. During the measurement cycle, the VPR-cuff inflates first to push blood out of the tail. The O-cuff is inflated (maximum occlusion pressure set in Session Paramaters; **Fig. 3.7**) to block blood flow back into the tail and the VPR-cuff deflates to 30 mmHg pressure to remain snug around the tail to detect changes in tail volume. As the O-cuff deflates and the systolic blood pressure exceeds the O-cuff pressure, blood will flow into the tail and the VPR cuff will detect the increase in tail volume. The O-cuff pressure at the point when tail volume increases is the systolic blood pressure. As the O-cuff continues to deflate, the rate of tail volume change will increase until blood flow into and out of the tail equilibrates. The O-cuff pressure at the point when the rate of tail volume change no longer increases is the diastolic blood pressure. Because the increase in tail volume at the systolic point is more pronounced than the tail volume equalization at the diastolic point, it can be more readily identified to provide an accurate SBP measurement.
2. Monitor tail volume closely during the measurement session because tail volume is a good indicator of body temperature. Besides adjusting the environmental temperature by altering the heating pad settings or the ambient room temperature, there are important ways to regulate the temperature of each individual mouse. The mice tend to move excessively when too hot, so you may put wet paper towel squares on top of the restrainer to cool a mouse that is moving too much. When the

mice are too cold and there is insufficient blood flow to the tail, the signal line (the blue line) will be flat. In this case, cover the whole restrainer with a full paper towel to retain heat around the mouse. The mouse's body temperature is dynamic, so it is important to monitor the mice closely after covering the mice with a wet paper towel piece to cool them or a full paper towel to warm them; the best way to monitor their temperature response is by observing their tail volumes. As warm mice are cooled by a wet piece of paper towel, they will move less and their tail volume will decrease over consecutive, successful cycles. Once the tail volume approaches 40, you should remove the wet paper towel to prevent the mouse from becoming too cold. Conversely, as cool mice are warmed, the tail volume will increase over consecutive, successful cycles and the mice will start to move excessively if they get too hot. The heat threshold varies for mice from different strains, so observe both the tail volume and movement to determine when to remove the paper towel warming blanket.

3. When not in use for a few days (e.g., over the weekend), the cuffs should be partially sealed in a plastic bag, while remaining attached to the CODA system, to preserve the cuff bladders. If the machine is not in use for a week or longer, remove the cuffs from the CODA system, seal them in a plastic bag, and cover the channels with Port Cover Caps.

4. If the nose cone slides poorly within the groove of the restrainer, lubricate the restrainer with food grade silicon spray (Fisher Scientific; catalog # NC9186322) to facilitate nose cone movement.

5. Because mice from different strains tolerate heat differently, the level and duration of heating pad operation should be optimized for each strain. Start the CODA preparation at least 1 h prior to beginning measurement cycles with the heating pads turned on high to warm the restrainers and platforms for use during the measurement session. Mice from most strains will tolerate high heat during the 5-min relaxation period after entering the restrainers and during the 10 acclimation cycles. We typically reduce the heating pads to low once all mice on a single CODA system exhibit tail volumes greater than 30 or 40 and complete the rest of the measurement session with low heat. Mice from some strains may prefer high heat throughout the whole measurement session, whereas others may require the heating pads to be turned off at some point during the 20 measurement cycles. Kent Scientific now supplies heating pads with three heat settings to allow better temperature control. While not all mice on a single CODA will respond the same to heat, changing the

heating pad settings can accommodate most of the mice and the remaining mice can be comforted with either warming or cooling paper towels (*see* **Note 2**).

6. It is good practice to start a new database when initiating a large study. Multiple small studies or a single study spaced over several weeks or months can be recorded in the same database, but a large database can adversely affect the performance of the software (this problem has been corrected in software version 3). Back up the databases regularly onto an external hard drive.

References

1 Chobanian, AV, Bakris, GL, Black, HR, et al. (2003) Seventh Report of the Joint National Committee on Prevention, Detection, Evaluation, and Treatment of High Blood Pressure. Hypertension **42**(6), 1206–1252.

2 Cowley, AW. (2006) The genetic dissection of essential hypertension. Nat Rev Genet 7(11), 829–840.

3 Manunta, P, Bianchi, G. (2006) Pharmacogenomics and pharmacogenetics of hypertension: update and perspectives – the adducin paradigm. *J Am Soc Nephrol* **17**(4), S30–S35.

4 Pravenec, M, Kurtz, TW. (2007) Molecular genetics of experimental hypertension and the metabolic syndrome – from gene pathways to new therapies. Hypertension **49**(5), 941–952.

5 Kurtz, TW, Griffin, KA, Bidani, AK, et al. (2005) Recommendations for blood pressure measurement in humans and experimental animals – Part 2: blood pressure measurement in experimental animals – A statement for professionals from the subcommittee of professional and public education of the American heart association council on high blood pressure research. *Hypertension* **45**(2), 299–310.

6 Feng, MJ, Whitesall, S, Zhang, YY, et al. (2008) Validation of volume-pressure recording tail-cuff blood pressure measurements. Am J Hypertens **21**(12), 1288–1291.

Chapter 4

Direct Blood Pressure Monitoring in Laboratory Rodents via Implantable Radio Telemetry

Daniel A. Huetteman and Heather Bogie

Abstract

The ability to monitor and record precise blood pressure fluctuations in research animals is vital to research for human hypertension. Direct measurement of blood pressure via implantable radio telemetry devices is the preferred method for automatic collection of chronic, continuous blood pressure data. Two surgical techniques are described for instrumenting the two most commonly used laboratory rodent species with radiotelemetry devices. The basic rat procedure involves advancing a blood pressure catheter into the abdominal aorta and placing a radio transmitting device in the peritoneal cavity. The mouse technique involves advancing a thin, flexible catheter from the left carotid artery into the aortic arch and placing the telemetry device under the skin along the animal's flank. Both procedures yield a chronically instrumented model to provide accurate blood pressure data from an unrestrained animal in its home cage.

Key words: Blood pressure, telemetry, direct blood pressure, hypertension, experimental, implantable, chronic monitoring, rodent surgery, mouse, rat.

1. Introduction

It is difficult to overestimate the importance of hypertension research for the health and well-being of the human population. Any tools or techniques that advance that research are indispensable to scientists who study its mechanisms. Implantable radiotelemetry devices allow investigators to continuously monitor blood pressure and heart rate of laboratory animals without the stress artifacts associated with restraint or human interaction. The miniaturized, biocompatible radio telemetry devices are surgically implanted and the live physiologic data can be automatically collected by sophisticated, easy-to-use electronic data collection systems. Implantable radiotelemetry has improved significantly over

K. DiPetrillo (ed.), *Cardiovascular Genomics*, Methods in Molecular Biology 573,
DOI 10.1007/978-1-60761-247-6_4, © Humana Press, a part of Springer Science+Business Media, LLC 2009

the last 10–15 years and is now considered the state of the art for collecting a wide variety of physiologic parameters from freely moving animals.

Although this technology is available for all species commonly used in laboratory research, this article concentrates on the techniques used to implant blood pressure transmitters in mice and rats, the two animal species most commonly used in hypertension research. The rat technique involves placing a small cylindrical radio transmitter in the animal's peritoneal cavity and inserting a thin, flexible catheter into the abdominal aorta. Likewise, the transmitter model for the mouse operates in precisely the same manner, but the implantation technique differs in a few key points. The mouse transmitter is quite a bit smaller and is ideally placed under the skin along the animal's flank; the flexible catheter is subcutaneously routed to the neck and is advanced into the aorta via the left carotid artery. Both transmitters are capable of accurately monitoring the blood pressure of the animal remotely and can remain viable and undetected for the life of the animal.

2. Materials

2.1. Rat Surgery

2.1.1. Equipment

1. Binocular surgical magnification
2. Clipper for hair removal
3. Supplemental heating
4. PA-C40 transmitter
5. Isoflurane vaporizer

2.1.2. Instruments

1. Graefe Forceps – curved, serrated
2. Tissue Forceps – with teeth
3. Mayo Scissors – straight, 15 cm
4. Olsen-Hegar Needle Holder
5. Vessel Dilator – 45 degree angled tip, 13.5 cm long
6. Vessel Dilator – 10 degree angled tip, 11 cm long
7. Vessel Cannulation Forceps – 11 cm long
8. Wound Clip Applier and Wound Clips

2.1.3. Supplies

1. Disinfectant soap
2. 70% isopropyl alcohol
3. Sterile drape
4. Tape
5. Sterile gloves

6. Sterile gauze sponges – 4 × 4 in. (5 × 5 cm)
7. Sterile saline
8. Sterile cotton tip applicators
9. Surgical suture for vessel occlusion – non-absorbable, 4-0 or 5-0
10. Surgical suture for incision closure – non-absorbable, 4-0 or 5-0 with curved needle
11. Gel-loading micropipette tips
12. Syringe needle – 22 ga × 1 in.
13. Fiber Patch
14. Vetbond™ tissue adhesive
15. AM radio
16. Magnet
17. 1-cc Syringe

2.2. Mouse Surgery

2.2.1. Equipment

1. Binocular surgical magnification
2. Clipper for hair removal
3. Supplemental heating
4. PA-C10 transmitter
5. Isoflurane vaporizer

2.2.2. Instruments

1. Graefe Forceps – curved, serrated
2. Lexer-Baby Scissors – straight, 10 cm
3. Olsen-Hegar Needle Holder – extra delicate
4. Vessel Dilator – 45 degree angled tip, 13.5 cm long
5. Vessel Dilator – 10 degree angled tip, 11 cm long
6. Vessel Cannulation Forceps – 11 cm long
7. Graefe Forceps – curved, with teeth
8. Elastic Stay Hook – 3-mm sharp hook
9. Wound Clip Applier and Wound Clips

2.2.3. Supplies

1. Disinfectant soap
2. 70% isopropyl alcohol
3. Sterile drape
4. Tape
5. Sterile gloves
6. Sterile gauze sponges – 2 × 2 in. (5 × 5 cm)
7. Sterile saline
8. Sterile cotton tip applicators

9. Surgical suture for vessel occlusion – non-absorbable size 6-0
10. Surgical suture for incision closure – non-absorbable, 5-0 or 6-0 with curved needle
11. Gel-loading micropipette tips
12. Syringe needle – 25 ga × 5/8″
13. Vetbond™ tissue adhesive
14. AM radio
15. Magnet
16. 1-cc Syringe

3. Methods

Both procedures use Data Sciences International (DSI) implantable telemetry transmitters. The PA-C10 transmitter at 1.4 g is optimized for surgical implantation in mice and is the smallest and lightest pressure sensing device available. The PA-C40 is designed for rat-sized research animals and weighs 7.6 g. Both devices are designed for high fidelity blood pressure detection but can also be used to monitor biological pressure fluctuations almost anywhere in the body.

3.1. Rat: Abdominal Aorta Cannulation with Intraperitoneal Device Placement

The rat blood pressure telemetry device is positioned in the peritoneal cavity with the pressure sensing catheter advanced upstream into the abdominal aorta (*see* **Fig. 4.1**). The catheter insertion site is sealed with tissue adhesive and the transmitter body is sutured into the abdominal wall closure. A well-practiced surgeon can perform this procedure in 30–40 min. With effective post-surgical analgesia, a typical outbred rat will return to its pre-implantation body weight in 2–3 days and will return to normal circadian behavior patterns in 5–7 days.

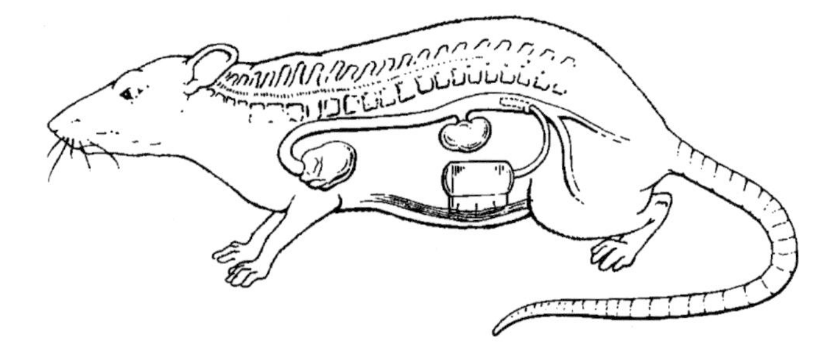

Fig. 4.1. Depiction of a laboratory rat instrumented with an implantable blood pressure telemetry device. The catheter is sealed into the abdominal aorta between the renal arteries and the iliac bifurcation and the transmitter is anchored to the abdominal wall closure.

3.1.1. Preparation

1. Administer surgical anesthesia for a 30–40 min procedure.
2. Shave and disinfect ventral abdomen.
3. Position the animal in dorsal recumbency.
4. Establish sterile surgical field and apply sterile draping material.
5. Open transmitter package, flood catheter trough with sterile saline to hydrate the catheter, 10 min is sufficient.

3.1.2. Surgical Site

1. Make a 3–4 cm midline abdominal incision through the skin and gently dissect the skin from the abdominal wall.
2. Incise the abdominal wall along the midline to provide good visualization of the aorta from the iliac bifurcation, cranial to the left renal vein.
3. Gently part the intestines using moistened cotton applicators; retract the intestines using moistened gauze sponges to allow good visualization of the descending aorta located along the dorsal body wall.
4. Carefully dissect the surrounding fat and connective tissue from the aorta using cotton applicators.
5. Clear excess tissue from the ventral surface of aorta to ensure good hemostasis following catheterization.

3.1.3. Preparation for Catheterization

1. Locate the left renal vein as it crosses over the anterior portion of the abdominal aorta (*see* **Fig. 4.2**). Using fine curved forceps, gently probe between the aorta and the vena cava just caudal to left renal vein; part the connective tissue between the aorta and the vena cava with the closed tips of the forceps.
2. Leaving the forceps in place between the aorta and vena cava, open the jaws of the forcep tips and grasp the end of a 4-0 suture tie and thread it between the aorta and the vena cava. This tie will be used to temporarily occlude blood flow at the time of vessel cannulation.
3. Place an additional occlusion suture between the vena cava and the aorta just cranial to the iliac bifurcation (*see* **Fig. 4.2**). This additional suture simplifies sealing of the catheter by improving the hemostasis of the vessel (*see* **Note 1**).

3.1.4. Aorta Catheterization

1. Prepare the device for implantation by removing it from the saline filled tray.
2. Using clean, gloved fingers, carefully remove the catheter's protective tip cover by alternating gentle traction and release. Do not apply excessive pressure to the catheter or grasp the catheter at the sensing region (*see* **Fig. 4.4**).

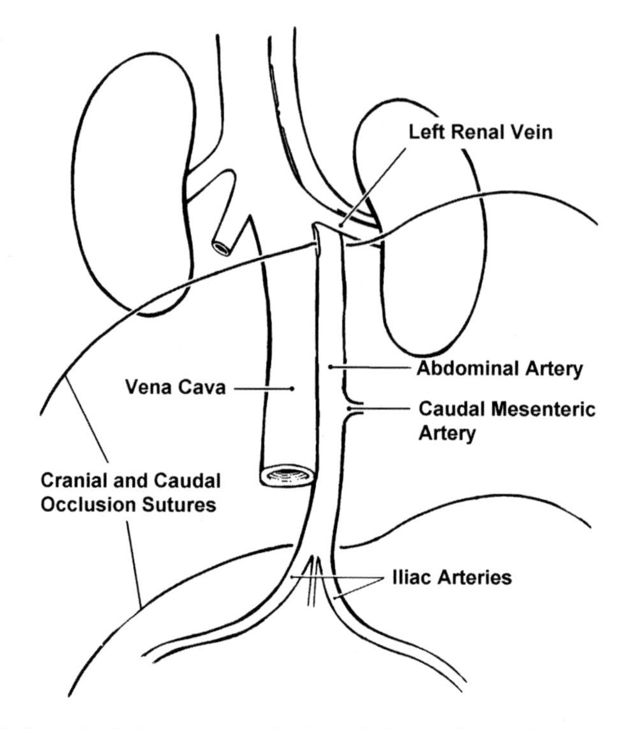

Fig. 4.2. Rat abdominal aorta preparation for catheter introduction. A temporary occlusion suture is placed just caudal to the left renal vein and an optional occlusion suture is positioned cranial to the iliac bifurcation.

3. Handle the exposed catheter tip with care to prevent contamination with fibers or glove powder.
4. Prepare a catheter-introducer by bending the beveled point of a 22-gauge needle. Grasp the bevel with a small needle holder and bend the pointed tip downward to a 90° angle such that the open part of the bevel is on the outside of the bend (*see* **Fig. 4.3**).
5. Restrict blood flow in the aorta by applying firm traction to the occlusion sutures. Place the transmitter in close proximity to the abdominal incision. Grasp the catheter proximal to the

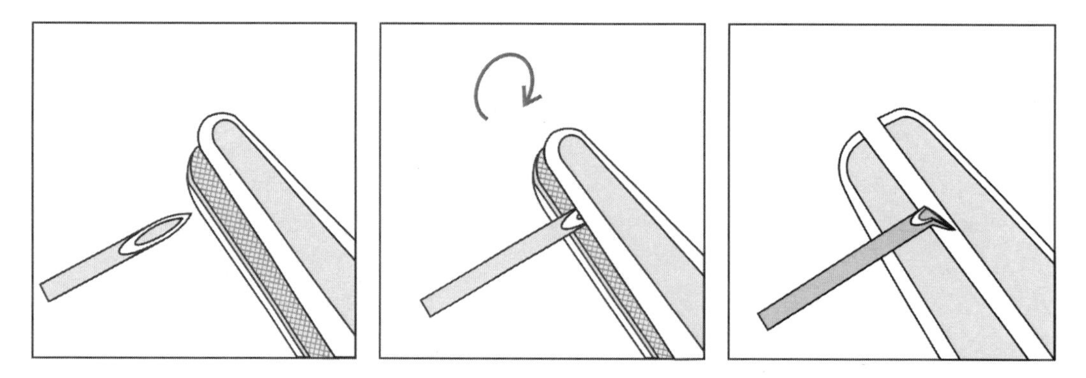

Fig. 4.3. A catheter introduction tool is prepared by bending the beveled tip of a hypodermic syringe needle.

overlap section (*see* **Fig. 4.4**) with vessel cannulation forceps. With the other hand use the pointed bevel of the bent needle to make a small puncture in the aorta just cranial to the iliac bifurcation (*see* **Note 2**), leaving the needle tip within the aorta lumen.

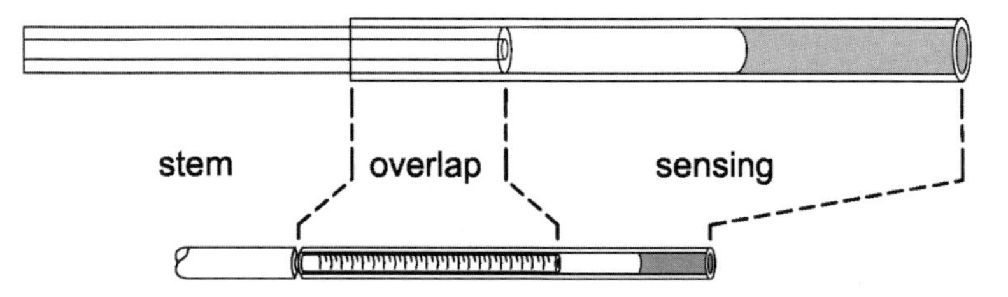

Fig. 4.4. Regions of the rodent blood pressure catheters. The rat-sized catheter (2.4 F) is pictured above and the mouse-sized catheter (1.2 F) is pictured below.

6. Using the bent needle as a catheter-introducer, guide the tip of the catheter beneath the needle and into the lumen of the vessel (*see* **Fig. 4.5**). Withdraw the needle and advance the catheter cranially until the distal tip of the catheter gently contacts the restriction at the anterior occlusion suture.

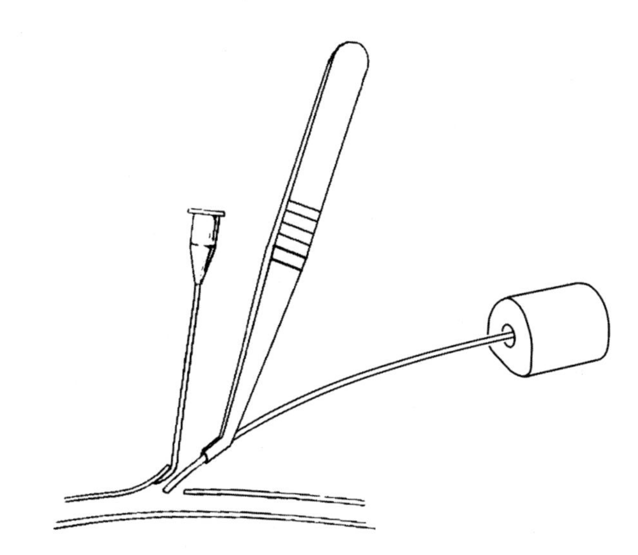

Fig. 4.5. Vessel cannulation technique. The catheter is held in a specialized pair of Vessel Cannulation Forceps. A bent-tipped syringe needle is used to incise the vessel wall and introduce the catheter into the artery.

7. Using cotton applicators clear the blood from immediate area. Thoroughly dry the puncture site and the surrounding tissue to ensure good bonding of the tissue adhesive (*see* **Note 3**).

8. Apply a tiny drop of Vetbond[1] tissue adhesive to the catheter and allow it to flow completely around the puncture site.
9. Allow 30 s for the adhesive to set. Slowly release the cranial occlusion suture and observe the catheterization site for blood leakage.
10. If bleeding occurs, re-occlude the vessel, clean the area with cotton applicators, and apply additional glue to the leakage point.
11. Repeat Steps 5 and 6 until hemostasis is maintained.
12. Prepare a small rectangle of fiber material approximately 5 × 7 mm. This is used to anchor the catheter until connective tissue has a chance to form (*see* **Fig. 4.6**).

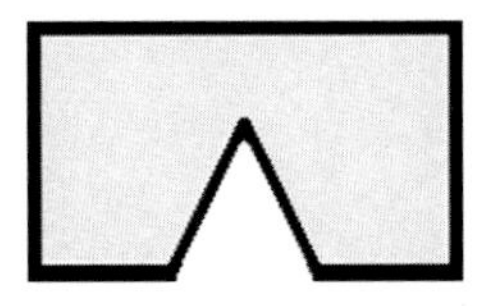

Fig. 4.6. Fiber patch for stabilization of the rat aorta catheter insertion site. The dimensions are approximately 5 × 7 mm.

13. Place the fiber patch over the catheterization site and apply additional adhesive to anchor the catheter to the surrounding tissues.
14. Without disturbing the catheter, remove all gauze sponges and retraction. Irrigate the abdominal cavity with warm saline and verify hemostasis of the catheter insertion site.

3.1.5. Device Placement

1. Gently restore the intestines to their original position.
2. Place the device body on top of the intestines, parallel to the long axis of the body with the catheter attachment directed caudally (*see* **Note 4**).
3. Close the abdominal wall incision with non-absorbable suture. Anchor the device by incorporating the longitudinal suture ridge on the device into the abdominal wall closure (*see* **Note 4**).

3.1.6. Surgical Recovery

1. Close the skin incision with staples or suture.
2. Discontinue surgical anesthesia.
3. Maintain supplemental warmth throughout the anesthetic recovery period. Monitor animal until fully recovered from anesthesia.

[1] Trademark of 3M Corporation.

4. Administer post-surgical analgesics as directed by your staff veterinarian.
5. Monitor the animal closely for the return of normal postures and behaviors.

3.2. Mouse: Carotid Artery Cannulation with Subcutaneous Device Placement

The mouse blood pressure device is ideally placed subcutaneously along the lateral flank between the forelimb and hind limb. The tip of the pressure sensing catheter is advanced from the left carotid artery and positioned in the free flowing blood of the aortic arch (*see* **Fig. 4.7**). A subcutaneous pocket is formed by blunt dissection from the neck incision down along the animal's flank. The

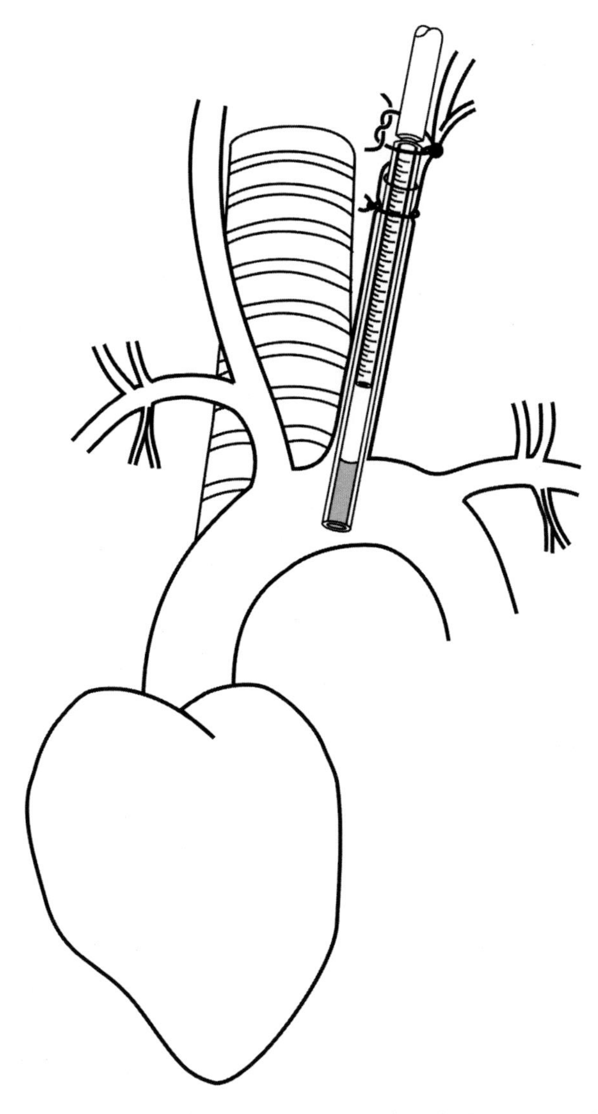

Fig. 4.7. Optimal placement of the blood pressure catheter in the mouse aortic arch. At least 2 mm of the sensing region of the catheter tip must be positioned in the free-flowing blood of the aorta.

choice of which side to place the device will be determined by the length of slack in the catheter, if the animal weighs less than 25.0 g place the transmitter on the right flank of the animal. If the animal weighs more than 25.0 g, place the transmitter on the left flank of the animal. A typical wild-type mouse will return to its pre-implantation body weight in 4–5 days and will return to normal circadian behavior patterns in 5–7 days after surgery.

3.2.1. Preparation

1. Administer surgical anesthesia for a 30–40 min procedure.
2. Loosely tape the mouse's forelimbs to the table.
3. Shave and disinfect ventral neck and upper chest.
4. Position the mouse in dorsal recumbency on the surgery table with the head closest to the surgeon. Provide supplemental warmth during surgery.
5. Establish a sterile surgical field and apply sterile draping material.
6. Open transmitter package and flood the catheter trough with sterile saline to hydrate the catheter, 10 min is sufficient.

3.2.2. Surgical Site

1. Using small surgical scissors, make a 1.5 cm midline incision through the skin overlying the trachea.
2. Carefully separate the mandibular glands using sterile cotton tip applicators.
3. Retract the left mandibular gland using an elastic stay hook (*see* **Note 5**).
4. Locate the carotid artery along the left side of the trachea using sterile cotton tip applicators. Using fine-tipped, curved forceps, carefully isolate the vessel from the surrounding tissue, making sure not to disturb the vagus nerve.

3.2.3. Preparation for Catheterization

1. Pass three pieces of 6-0 non-absorbable suture underneath the isolated artery section. The furthest cranial suture will be used to permanently ligate the carotid artery. The suture closest to the heart will be used to temporarily occlude blood flow and allow introduction of the catheter. The middle suture will be used to secure the catheter in the vessel (*see* **Fig. 4.8**).
2. Position the cranial ligation suture just proximal to the bifurcation of the interior and exterior carotid arteries. Tie a secure knot around the artery to permanently ligate the vessel (*see* **Note 6**).
3. Prepare a catheter-introducer by bending the beveled tip of a 25-gauge syringe needle. Hold the syringe needle with the beveled side facing up. Grasp just the beveled area of the needle with a needle holder and bend the tip downward to an angle of approximately 90° (*see* **Fig 4.3**; *see* **Note 7**).

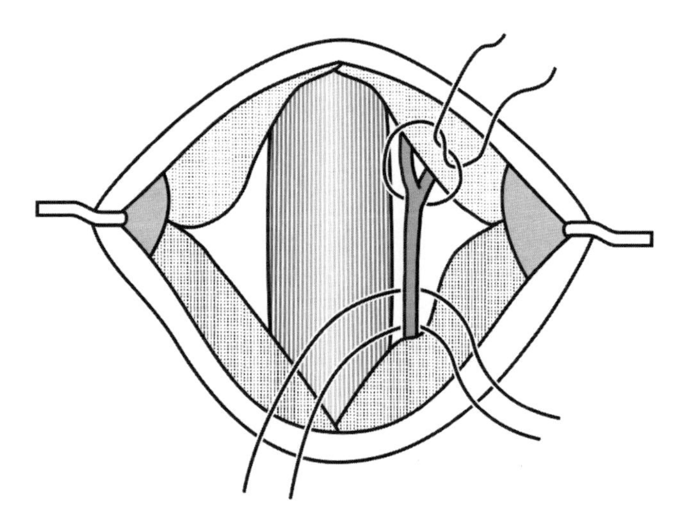

Fig. 4.8. Preparation of the mouse left carotid artery for catheter introduction. The cranial ligation suture is positioned at the bifurcation of the interior and exterior carotid arteries.

4. Remove the transmitter from the sterile package and transfer it to the sterile field (*see* **Note 8**).
5. Monitor the transmitter with an AM radio on and carefully remove the tip cover from the end of the catheter. Removal of the tip cover should be done by alternating gentle traction and release (*see* **Note 9**).

3.2.4. Carotid Catheterization

1. Gently apply tension to the occlusion sutures. This will elevate the artery and occlude blood flow.
2. Grasp the tip of the catheter just proximal to the sensing region (*see* **Fig. 4.4**) using a Vessel Cannulation Forceps (*see* **Note 8**).
3. Using the bent, 25-gauge needle as a catheter-introducer, puncture the carotid artery just proximal to the ligation suture and insert the catheter upstream toward the aorta (*see* **Fig 4.7**). Once the catheter is advanced into the vessel, withdraw the catheter-introducer.
4. Advance the catheter into the artery until it reaches the occlusion suture (*see* **Note 10**).
5. Release the tension on the occlusion suture and advance the catheter beyond the suture toward the aorta.
6. Continue to advance the catheter until at least 2 mm of the catheter sensing region (*see* **Fig. 4.4**) extends into the aortic arch (*see* **Fig 4.7**; *see* **Note 11**).
7. Tie the occlusion suture securely around the artery to seal the vessel wall around the catheter stem.
8. Release the tension on the ligation suture and tie the loose ends of the suture around the catheter stem to anchor it in place.

3.2.5. Device Placement

1. Insert small surgical scissors into the incision and form a subcutaneous pocket along the mouse's flank using blunt dissection (*see* **Fig. 4.9**; *see* **Notes 12 and 13**).

Fig. 4.9. Formation of a subcutaneous transmitter pocket using blunt dissection.

2. Once the pocket is formed, irrigate the pocket with warm, sterile saline and insert the transmitter. Orient the flat side of the PA-C10 transmitter against the body and the rounded side toward the skin (*see* **Fig. 4.10**; *see* **Note 14**).
3. Secure the catheter to the surrounding tissue using a few small drops of Vetbond tissue adhesive. This will ensure that the catheter will lie benignly under the skin (*see* **Note 15**).

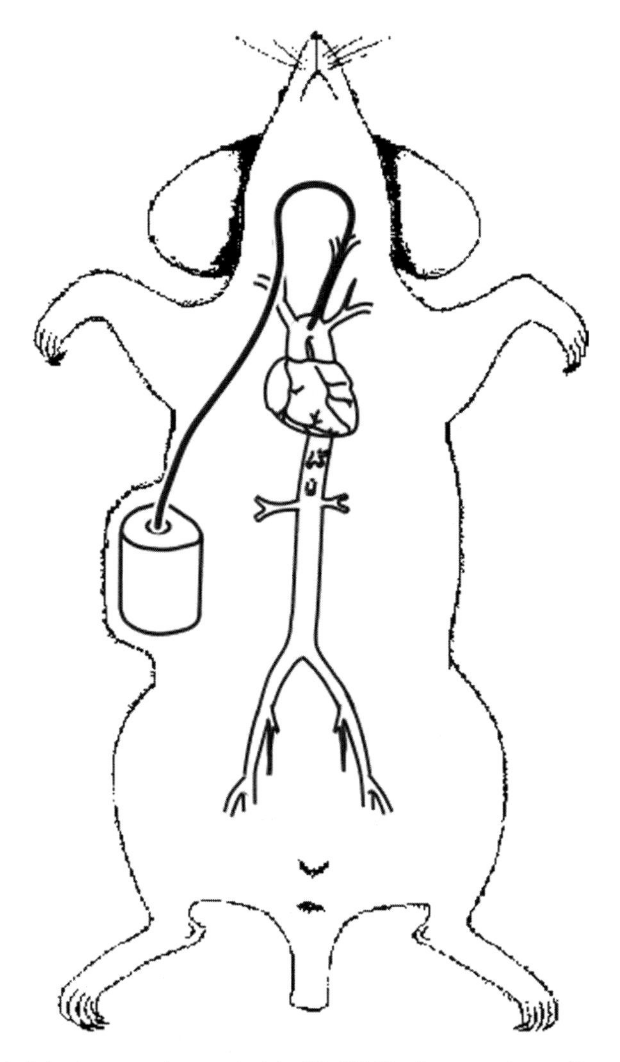

Fig. 4.10. Subcutaneous placement of the PA-C10 blood pressure transmitter.

4. Close the skin incision with 5-0 or 6-0 absorbable or non-absorbable suture. Once closed, seal the incision with Vet-bond tissue adhesive.

3.2.6. Surgical Recovery

1. Discontinue surgical anesthesia.
2. Maintain supplemental warmth throughout the anesthetic recovery. Monitor animal until fully recovered from anesthesia.
3. Administer post-surgical analgesia as directed by your staff veterinarian.
4. Monitor animal closely for the return of normal postures and behaviors.

4. Notes

1. When preparing for catheterization, the dual occlusion method is essential when using hypertensive animal models. If the caudal occlusion suture is not employed, residual back pressure from the extremities may make it difficult to properly seal the catheter in the vessel.
2. The bent needle technique allows the surgeon to make a puncture precisely matched to the diameter of the catheter. This provides for a much tighter seal around the catheter.
3. When catheterizing the aorta, it is important to stop all blood seepage from the vessel before applying the tissue adhesive.
4. The abdominal wall can be closed using a running suture pattern or a simple interrupted pattern. When anchoring the device, alternate one stitch through the suture hole and one stitch through the muscle alone. This anchors the device while providing secure closure of the abdominal wall.
5. Maintain tissue hydration with sterile saline throughout the entire procedure.
6. Make a loose knot in both the occlusion suture and the middle suture and position them as close to the clavicle as possible, this should isolate approximately 6 mm of the vessel.
7. The bent needle will be easier to manipulate if placed on the end of a 1-cc syringe.
8. Do not handle the transmitter by grasping the catheter. This may cause damage to the catheter or the pressure sensor.
9. Take care to prevent gel loss due to compression of the catheter or sudden release of the tip cover.
10. Position the middle suture around the artery and catheter. Secure the catheter by tightening the suture around the vessel. Releasing the catheter before it is secured may cause it to come out of the vessel.
11. Depending on the mouse strain and weight, the catheter may need to be advanced to different lengths to reach the proper position the aortic arch. Before performing any survival surgeries, determine the proper length to advance the catheter for each mouse model by measuring the distance from the carotid bifurcation to the aorta in several animals. This average distance plus 2–3 mm should be the proper length and can be verified by implanting a practice catheter (available from DSI) into a representative training animal. The animal can then be euthanized and the location of the catheter tip can be directly observed by gross dissection of the aortic arch.

12. The pocket should be formed such that the transmitter body will reside on the lateral flank of the animal, the more ventral the placement, the more likely the transmitter will interfere with normal behaviors.
13. If the pocket is not made large enough, the skin will stretch too tightly across the contours of the transmitter and pressure necrosis may result.
14. After inserting the transmitter, secure it using a small drop of Vetbond tissue adhesive. Draw the Vetbond up in a 1-cc syringe and insert the syringe into the pocket with the needle removed to dispense.
15. Use a micropipette tip to dispense the tissue adhesive.

Acknowledgments

We gratefully acknowledge the growing community of telemetry users around the world, their inventiveness and willingness to dialogue with their colleagues are responsible for the continuing effort to refine device implantation surgery for the benefit of our animal subjects.

5 Suggested Reading

5.1 Mouse Blood Pressure Articles

1. Barone, FC, Knudsen, DJ, Nelson, AH, et al. (1993) Mouse strain differences in susceptibility to cerebral ischemia are related to cerebral vascular anatomy. *J Cereb Blood Flow Metab* July; **13**(4), 683–692.
2. Brockway, BP, Mills, P, Kramer, K. (1998) Fully implanted radio-telemetry for monitoring laboratory animals. *Lab Anim* **27**, 40–45.
3. Butz, GM, Davisson, RL. (2001) Long-term telemetric measurement of cardiovascular parameters in awake mice: a physiological genomics tool. *Physiol Genomics* March 8; **5**(2), 89–97.
4. Carlson, SH, Wyss, M. (2000) Long-term telemetric recording of arterial pressure and heart rate in mice fed basal and high NaCL diets. *Hypertension* February; **35**(2), E1–E5.
5. Davisson, RL, Hoffmann, DS, Butz, GM, et al. (2002) Discovery of a spontaneous genetic mouse model of preeclampsia. *Hypertension* February; **39**(2 Pt 2), 337–342.
6. Fujii, M, Hara, H, Meng, W, et al. (1997) Strain-related differences in susceptibility to transient forebrain ischemia in SV-129 and C57black/6 mice. *Stroke* September; **28**(9), 1805–1810; discussion 1811.
7. Goecke, JC, Awad, H, Lawson, HC, et al. (2005) Evaluating postoperative analgesics in mice using telemetry. *Comp Med* February; **55**(1), 37–44.
8. Gross, V, Milia, AF, Plehm, R, et al. (2000) Long-term blood pressure telemetry in AT2 receptor-disrupted mice. *J Hypertens* July; 18(7), 955–961.
9. Knowles, JW, Esposito, G, Mao, L, et al. (2001) Pressure-independent enhancement of cardiac hypertrophy in natriuretic peptide receptor A-deficient mice. *J Clin Invest* April; **107**(8), 975–984.

10. Kramer, K, Voss, HP, Grimbergen, JA, et al. (2000) Telemetric monitoring of blood pressure in freely moving mice: a preliminary study. *Lab Anim* July; **34**(3), 272–280.
11. Mattson, DL. (1998) Long-term measurement of arterial blood pressure in conscious mice. *Am J Physiol* February; **274**(2 Pt 2), R564–R570.
12. Milia, AF, Gross, V, Plehm, R, et al. (2001) Normal blood pressure and renal function in mice lacking the bradykinin B(2) receptor. *Hypertension* June; **37**(6), 1473–1479.
13. Mills, PA, Huetteman, DA, Brockway, BP, et al. (2000) A new method for measurement of blood pressure, heart rate, and activity in the mouse by radiotelemetry. *J Appl Physiol* May; **88**(5), 1537–1544.
14. Sharp, J, Zammit, T, Azar, T, et al. (2003) Recovery of male rats from major abdominal surgery after treatment with various analgesics. *Contemp Top Lab Anim Sci* November; **42**(6), 22–27.
15. Swoap, SJ, Overton, MJ, Garber, G. (2004) Effect of ambient temperature on cardiovascular parameters in rats and mice: a comparative approach. *Am J Physiol Regul Integr Comp Physiol* August; **287**(2), R391–R396. Epub 2004 April 15.
16. Swoap, SJ, Weinshenker, D, Palmiter, RD, et al. (2004) Dbh (–/–) mice are hypotensive, have altered circadian rhythms, and have abnormal responses to dieting and stress. *Am J Physiol Regul Integr Comp Physiol* January; **286**(1), R108–R113. Epub 2003 September 11.
17. Tank, J, Jordan, J, Diedrich, A, et al. (2004) Clonidine improves spontaneous baroreflex sensitivity in conscious mice through parasympathetic activation. *Hypertension* May; **43**(5), 1042–1047. Epub 2004 March 29.
18. Uechi, M, Asai, K, Oska, M, et al. (1998) Depressed heart rate variability and arterial baroreflex in conscious transgenic mice with overexpression of cardiac Gsalpha. *Circ Res* March 9; **82**(4), 416–423.
19. Van Vliet, BN, Chafe, LL, Antie, V, et al. (2000) Direct and indirect methods used to study arterial blood pressure. *J Pharmacol Toxicol Methods* September–October; **44**(2), 361–373.
20. Whitesall, S E, Hoff, JB, Vollmer, AP, et al. (2004) Comparison of simultaneous measurement of mouse systolic arterial blood pressure by radiotelemetry and tail -cuff methods. *Am J Physiol Heart Circ Physiol* June; **286**(6), H2408–H2415. Epub 2004 February 12.
21. Zhu, Y, Bian, Z, Lu, P, et al. (2002) Abnormal vascular function and hypertension in mice deficient in estrogen receptor beta. *Science* January 18; **295**(5554), 505–508.

5.2 Rat Blood Pressure

Articles

22. Azar, T, Sharp, J, Lawson, D. (2005) Stress-like cardiovascular responses to common procedures in male versus female spontaneously hypertensive rats. *Contemp Top Lab Anim Sci* **44**(3), 25–30.
23. Bidani, AK, Griffin, KA, Picken, M, et al. (1993) Continuous telemetric blood pressure monitoring and glomerular injury in the rat remnant kidney model. *Am J Physiol* September; **265**(3 Pt 2), F391–F398.
24. Brockway, BP, Mills, PA, Azar, SH. (1991) A new method for continuous chronic measurement and recording of blood pressure, heart rate and activity in the rat via radio-telemetry. *Clin Exp Hypertens A* **13**(5), 885–895.
25. Buñag, RD, Butterfield, J. (1982) Tail-cuff blood pressure measurement without external preheating in awake rats. *Hypertension* November–December; **4**(6), 898–903.
26. Irvine, RJ, White, J, Chan, R. (1997) The influence of restraint on blood pressure in the rat. *J Pharmacol Toxicol Methods* November; **38**(3), 157–162.
27. Hassler, C R, Lutz, GA, Linebaugh, R, et al. (1979) Identification and evaluation of noninvasive blood pressure measuring techniques. *Toxicol Appl Pharmacol* February; **47**(2), 193–201.
28. Kramer, K, Kinter, LB. (2003) Evaluation and applications of radiotelemetry in small laboratory animals. *Physiol Genomics* May 13; **13**(3), 197–205. Review.
29. Kramer, K, Kinter, L, Brockway, BP, et al. (2001) The use of radiotelemetry in small laboratory animals: recent advances. *Contemp Top Lab Anim Sci* January; **40**(1), 8–16. Review.
30. Kramer, K, Remie, R. (2005) Measuring blood pressure in small laboratory animals. *Methods Mol Med* **108**, 51–62.

31. Lemmer, B, Mattes, A, Böhm, M, et al. (1993) Circadian blood pressure variation in transgenic hypertensive rats. *Hypertension* July; **22**(1), 97–101.
32. Liles, JH, Flecknell, PA. (1993) The effects of surgical stimulus on the rat and the influence of analgesic treatment. *Br Vet J* November–December; **149**(6), 515–525.
33. Nijsen, MJ, Ongenae, NG, Coulie, B, et al. (2003) Telemetric animal model to evaluate visceral pain in the freely moving rat. *Pain* September; **105**(1–2), 115–123.
34. Roughan, JV, Flecknell, PA. (2004) Behaviour-based assessment of the duration of laparotomy-induced abdominal pain and the analgesic effects of carprofen and buprenorphine in rats. *Behav Pharmacol* November; **15**(7), 461–472.
35. Schreuder MF, Fodor, M, van Wijk, JA, et al. (2006) Association of birth weight with cardiovascular parameters in adult rats during baseline and stressed conditions. *Pediatr Res* January; **59**(1), 126–130. Epub 2005 December 2.
36. Sharp, J, Zammit, T, Azar, T, et al. (2003) Recovery of male rats from major abdominal surgery after treatment with various analgesics. *Contemp Top Lab Anim Sci* November; **42**(6), 22–27.
37. St Stewart, LA, Martin WJ. (2003) Evaluation of postoperative analgesia in a rat model of incisional pain. *Contemp Top Lab Anim Sci* January; **42**(1), 28–34.
38. Swoap, SJ, Overton, MJ, Garber, G. (2004) Effect of ambient temperature on cardiovascular parameters in rats and mice: a comparative approach. *Am J Physiol Regul Integr Comp Physiol* August; 287(2), R391–R396. Epub 2004 April 15.
39. Van Vliet, BN, Chafe, LL, Antic, V, et al. (2000) Direct and indirect methods used to study arterial blood pressure. *J Pharmacol Toxicol Methods* September–October; **44**(2), 361–373.
40. Van Vliet, BN, Chafe, LL, Montani, JP. (2003) Characteristics of 24 h telemetered blood pressure in eNOS-knockout and C57Bl/6 J control mice. *J Physiol* May 15; **549**(Pt 1), 313–325. Epub 2003 March 28.
41. Waki, H, Katahira, K, Polson, JW, et al. (2006) Automation of analysis of cardiovascular autonomic function from chronic measurements of arterial pressure in conscious rats. *Exp Physiol* January; **91**(1), 201–213. Epub 2005 October 20.
42. Zhang, W, Wang, Z, Zhang, W. (2004) Characterizing blood pressure, heart rate, physical activity, circadian rhythm and their response to nos inhibitor and substrate in chf rats with telemetry. *Vascular Disease Prevention* July, **1**(2), 159–166.

Chapter 5

Measuring Kidney Function in Conscious Mice

David L. Mattson

Abstract

Mice lacking or over-expressing a gene of experimental interest have become important tools to understand the regulation of kidney function and water and electrolyte homeostasis. The use of mice in physiological studies is becoming more widespread, but there are still a number of technical limitations that preclude the full utilization of mouse models in renal research. The present chapter focuses upon a set of methods developed in our laboratory to quantify renal function in conscious mice. These measurements are based upon surgical instrumentation of mice with chronic indwelling arterial and venous catheters. This preparation permits direct measurement of arterial blood pressure, direct sampling of arterial and/or venous blood, intravenous or intra-arterial infusion of substances, and quantification of daily sodium balance. The advantage of these techniques is that all of these procedures can be performed in conscious mice freely moving in their home cages. As such, this in vivo preparation provides an assessment of physiological function in mice in their native state.

Key words: Mice, blood pressure, water and electrolyte excretion, sodium, kidney, phenotype, surgery, renin–angiotensin system.

1. Introduction

The experimental advantages provided by genetically manipulated mice have increased use of this species in many scientific studies. Evaluation of renal function is desired as a primary or secondary endpoint of many experiments. Assessing phenotypes related to renal function, however, is often handicapped by the relative fragility of mice as research animals compared to other species and the comparatively small volume of urine, tubular fluid, and plasma that can be obtained from mice. Despite the experimental advantages provided by genetic recombination in mice, the use of mice for renal function studies has been somewhat technique-limited. As a result, the desire to utilize genetically manipulated mice has led to

K. DiPetrillo (ed.), *Cardiovascular Genomics*, Methods in Molecular Biology 573,
DOI 10.1007/978-1-60761-247-6_5, © Humana Press, a part of Springer Science+Business Media, LLC 2009

the adaptation of techniques used in larger animals, the development and utilization of phenotyping methodologies based upon miniaturized electronics, and the increased use of non-invasive imaging techniques. Several comprehensive reviews of this subject have been previously published (1–6), and the reader is directed to these manuscripts for further information and alternative approaches.

The present chapter will focus upon a set of methods developed in our laboratory to quantify renal function in conscious mice (7–13). These methods are based upon surgical instrumentation of mice with chronic indwelling arterial and venous catheters. This preparation permits direct measurement of arterial blood pressure, direct sampling of arterial and/or venous blood, intravenous or intra-arterial infusion of substances, and quantification of daily sodium balance. The advantage of these techniques is that all of these procedures can be performed in conscious mice freely moving in their home cages. As such, this in vivo preparation provides an assessment of physiological function in mice in their native state.

2. Materials

1. Anesthetics – isoflurane (Midwest Veterinary Supply, Inc.; Burnsville, MN or Henry Schein; Denver, PA, USA) and/or sodium pentobarbital (Midwest Veterinary Supply, Inc. or Henry Schein)
2. Analgesics – buprenorphine hydrochloride (Midwest Veterinary Supply, Inc. or Henry Schein)
3. Antibiotics – cefazolin sodium (Midwest Veterinary Supply, Inc. or Henry Schein)
4. Isotonic saline (0.9% NaCl)
5. Tyrode's solution (145 mM NaCl, 6 mM KCl, 0.5 mM $MgCl_2$, 1 mM $CaCl_2$, 10 mM Glucose, 4.6 mm HEPES, pH 7.4)
6. Betadine (Midwest Veterinary Supply, Inc. or Henry Schein)
7. Heparin (Midwest Veterinary Supply, Inc. or Henry Schein)
8. Isoflurane anesthesia machine (Midwest Veterinary Supply, Inc. or Henry Schein)
9. Thermostatically controlled circulating water bath (Fisher or VWR) and surgical table (MCW Department of Physiology Machine Shop; water blanket can be substituted)

10. Stereomicroscope – 2–20 × final magnification; 10–20 cm working distance (Fisher or VWR)
11. Surgical lamp (Fiber-Optic Light Source; Fisher)
12. Catheters: Arterial and venous catheters consist of a 4–5 cm length of Micro-Renathane tubing (0.025 in. ID, 0.040 in. OD; Braintree Scientific) stretched over hot air to give a tip 300–500 μm in diameter. The Micro-Renathane tubing is coupled with a 23-gauge stainless steel pin (∼1 cm) to a 25-cm piece of polyethylene tubing (PE-50; 0.023 in. ID, 0.038 in. OD; Clay-Adams). A small piece of shrink tubing (3 M) is used to strengthen the junction between the two pieces of tubing. The catheter tip is illustrated in **Fig. 5.1**

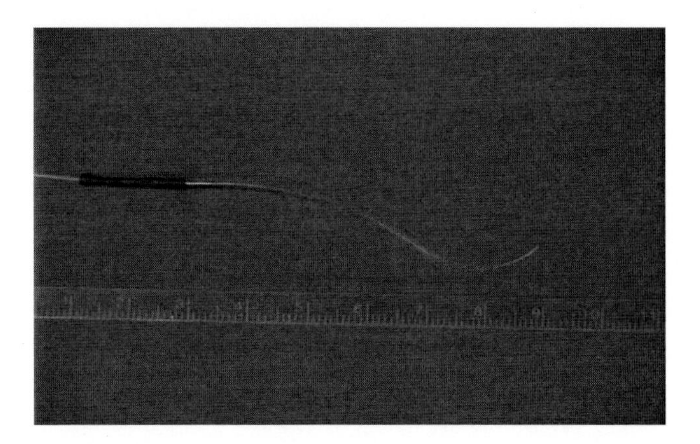

Fig. 5.1. Femoral arterial and venous catheters: a 4–5 cm length of Micro-Renathane tubing is stretched over hot air to give a tip 200–350 μm in diameter. The Micro-Renathane is coupled with a 23-ga stainless steel pin to a 25-cm piece of polyethylene tubing. A piece of shrink tubing holds the two pieces of tubing together over the pin.

13. Lightweight stainless steel spring (15 cm length, 0.3 cm diameter; Clay Adams) and 0.5 cm diameter stainless steel button (fabricated in the MCW Department of Physiology Machine Shop). The button and spring are illustrated in **Fig. 5.2**
14. Infusion swivels – single-channel infusion swivels (Instech Laboratories)
15. Syringe pumps (for delivery of 0.5–10 ml/day; Harvard Apparatus and multiple suppliers)
16. Pressure transducers (Cobe Laboratories, Lakewood, CO, USA)
17. General purpose amplifier (MCW Department of Physiology Machine and Electronics Shop)
18. Analog to Digital Convertor Board (Data Translation, Marlboro, MA, USA)

Fig. 5.2. Lightweight spring and stainless steel button: a 15-cm piece of lightweight stainless steel spring is attached to the back of the animal by sewing a 0.5-cm diameter stainless steel button into the strap muscles between the scapulae.

19. Personal computer
20. Syringes (various sizes; 0.5–10 ml)
21. Hypodermic needles (various sizes; 30–18 g)
22. Tygon tubing (20 mm ID × 60 mm OD)
23. Polyvinyl tubing (for infusions)
24. General surgical instruments (scissors, forceps, needle holders)
25. General surgical supplies (gauze, swabs, towels, silk, suture, etc.)
26. Steam autoclave
27. Stainless steel cages (4″ wide × 6″ deep × 6″ high) with interchangeable bedding pans and urine collection funnels (Suburban Surgical, Chicago, IL, USA)
28. Liquid AIN-76A animal chow (AIN-76A; Dyets, Bethlehem, PA, USA)
29. Liquid chow feeders (Dyets, Bethlehem, PA, USA)

3. Methods

Multiple approaches can be considered to examine renal function in mice. A number of non-invasive or minimally invasive measurements can be obtained from mice to quickly obtain an index of renal function. These measures generally involve obtaining a urine or blood sample (*see* **Note 1**). A large number of invasive procedures to examine renal functional parameters have also been effectively performed in mice. Non-survival procedures to quantify renal excretory function (clearance and micropuncture studies) and to assess renal hemodynamics (in vitro and in vivo measurements of blood flow regulation) are routinely performed (*see* **Note 2**). In addition, invasive procedures as a component of long-term studies have also been useful, particularly experiments in which mice were chronically instrumented for long-term blood pressure measurements, intravenous or renal infusion, or sequential blood sampling. This chapter will provide specific details for the measurement of parameters related to renal function in conscious, chronically catheterized mice.

3.1. Animal Care and Surgical Preparation

1. Mice are preanesthetized with isoflurane (1.5–2.5% v/v) and then administered pentobarbital sodium (50 mg/kg ip) to induce anesthesia (*see* **Note 3**). Mice are maintained in a deep plane of anesthesia as evidenced by a lack of palpebral and withdrawal reflexes. Supplemental anesthesia with pentobarbital (5–25 mg/kg ip) is administered as needed. The mice are shaved in the areas to be surgically prepared, the skin is swabbed with betadine, and the animals are placed on a thermostatically controlled warming table.
2. All surgery is performed using aseptic technique.
3. The animals are placed on their backs. The hind legs are gently extended and secured in place with tape; this is necessary to provide access to the femoral artery and vein.
4. A cutaneous incision (~2 cm) is made in the femoral triangle. Adherent fat and connective tissue are gently dissected to isolate the femoral artery and vein from each other and from the femoral nerve (*see* **Note 4**).
5. Starting with the femoral vein, the two vessels are cannulated. Three 6–0 silk ligatures with loose knots are individually placed around the femoral vein or artery. The distal ligature is permanently tightened to occlude blood flow through the vessel from the distal portions of the limb (*see* **Note 5**).
6. The proximal portion of the vessel is temporarily occluded (*see* **Note 6**), and a small incision (approximately 1/3 through the vessel at a 45° angle) is placed in the vessel, between the proximal and distal ligatures, with Vannas scissors.
7. The tapered end of the catheter (**Fig. 5.1**) is inserted into the vessel. This is accomplished by gently lifting the incised portion of the vessel open with the tip of a Dumont #7 forceps. The tapered end of the catheter, which is also held with a forceps, is then inserted into the vessel. Once the tip of the catheter has been inserted into the vessel, the middle ligature is tightened to temporarily secure the catheter in place. The proximal ligature is then loosened and the catheter is advanced into the vessel.

8. The tip of the venous catheter is typically advanced into the initial portion of the inferior vena cava; the arterial catheter is advanced to the bifurcation of the distal aorta (*see* **Note 7**). The catheters are filled with isotonic saline containing 100 U/ml heparin and sealed with a 23-ga pin; the catheters are only opened during periods of infusions, blood sampling, or pressure recording. All ligatures are tied around the vessel and the cannula in order to secure it in place.
9. The catheters are subcutaneously tunneled with a trocar (a 10 cm length of 16-ga stainless steel tubing with a 45° bevel on one end) from the region of the femoral triangle to the back of the neck where they are exteriorized in a lightweight spring that is attached to a stainless steel button. The button contains four holes (**Fig. 5.2**) that are used to suture it into the strap muscles at the back of the neck.
10. The incisions are closed with absorbable suture. The mice are administered antibiotics (cefazolin, 25 mg/kg, im; twice per day for 2 days) to prevent infection and analgesics (buprenorphine, 0.05 mg/kg, sc; twice per day for 2 days) to alleviate postsurgical discomfort. The mice are continuously monitored and kept on a heated warming table until they are fully recumbent.
11. The mice are placed in stainless steel cages with food and water available ad libitum. The catheters, exteriorized in the lightweight spring, are accessed at the top of the cage. The spring is attached to a lightweight swivel to permit the animal to move freely while infusions or blood pressure recordings are performed (**Fig. 5.3**).

3.2. Intravenous Infusion in Conscious Mice (see Note 8)

1. Surgically prepared mice are permitted to recover from surgery for 3–5 days prior to experiments.
2. A sterile syringe with a blunted 23-ga needle is filled with sterile saline containing vehicle (or drug) and is loaded into a syringe pump. An appropriate length of tygon tubing (30 mm ID) is strung from the syringe to the swivel, the dead space is flushed with the infusate, and the intravenous catheter is attached to the swivel.
3. The desired infusion rate is set and the intravenous infusion is continued for the duration of the experiment.

3.3. Measurement of Arterial Blood Pressure in Conscious Mice (see Note 9)

1. Mice are surgically prepared with indwelling catheters as described in **Section 3.1** and are permitted to recover from surgery for 3–5 days prior to experiments.
2. The arterial catheter is connected with tygon tubing (30 mm ID) to a solid state pressure transducer and a general-purpose amplifier. Amplified data is fed into an analog-to-digital convertor and to a computer. The computerized data-acquisition

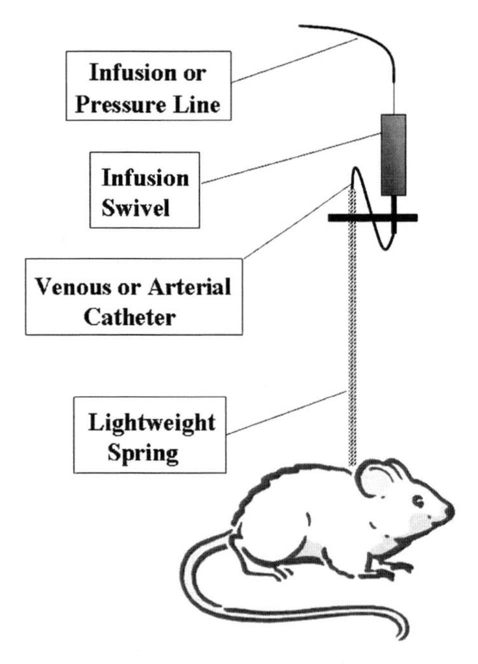

Fig. 5.3. Schematic depicting a conscious, instrumented mouse. Chronic indwelling catheters are implanted in arteries and veins, tunneled subcutaneously, and exteriorized in a lightweight spring at the back of the neck. The spring is suspended from the top of the cage and attached to a lightweight swivel device that permits continuous infusion or arterial blood pressure measurements.

software reduces pulsatile blood pressure signals collected at 500 Hz to periodic (1 min) averages of systolic, diastolic, and mean arterial blood pressure and heart rate (HR).

3. Arterial pressure is typically measured for a 3-h period at the same time each day. In general, the first 30 min of the recording are discarded since the MAP and HR values tend to be artificially elevated due to disturbances in the animal room associated with the technician's presence. The final 2 h of the recording period will be retained.
4. Following the recording period, the arterial catheter is flushed with saline containing 100 U/ml heparin and sealed.

3.4. Quantification of Sodium Balance in Conscious Mice (see Note 10)

1. Mice are surgically prepared with indwelling catheters as described in **Section 3.1** and are permitted to recover from surgery for 3–5 days prior to experiments.
2. An intravenous infusion is begun as described in **Section 3.2**. Mice are fed sodium-free liquid chow and administered tap water ad libitum. All of the sodium intake is provided by intravenous infusion. As described in **Note 5**, mice are infused with 1–6 ml/day of isotonic saline to provide a daily sodium intake equivalent to that observed on a low, normal, or high NaCl diet.

3. Daily urine output is collected in a stainless steel funnel assembly on the bottom of the cage. The total volume collected and the time of collection are tabulated. The funnel assembly is carefully washed with a fixed volume (50 ml), which is collected. The sodium and potassium concentrations of the urine and wash are measured with a flame photometer, the daily sodium output is calculated, and the value is compared to the sodium intake from the infusion. This method provides an accurate measurement of sodium intake, sodium output, and sodium balance (intake minus output).

3.5. Quantification of Circulating Hormones (see Note 11)

1. Mice are surgically prepared with indwelling catheters as described in **Section 3.1** and are permitted to recover from surgery for 3–5 days prior to experiments.
2. As described in **Note 6**, systemic hemodynamics are very sensitive to changes in blood volume. To minimize changes in blood volume, donor blood is simultaneously infused intravenously as arterial blood is drawn. Littermate (donor) mice are deeply anesthetized with sodium pentobarbital (50 mg/kg ip) and whole blood is drawn from the apex of the heart with a 23-ga needle into an equal volume of Tyrode's solution containing 100 U/ml sodium heparin. The suspension is centrifuged at 2,000 × *g* for 2 min, the supernatant is discarded, and an excess volume of Tyrode's solution is added to wash the cells. The cells are gently inverted to bring them into suspension and spun again at 2,000 × *g*. The supernatant is again discarded and this process is repeated three more times. The washed red blood cells are then resuspended in artificial plasma (151 mM NaCl, 2 mM K_2HPO_4, pH 7.4, equilibrated with 95% O_2/5% CO_2) with a final hematocrit of approximately 40%. The donor blood is kept on ice until it is used, whereupon it is warmed to body temperature and used in the push–pull experiment.
3. Arterial blood is drawn by uncapping the arterial line and permitting the blood to drip into a microcentrifuge tube. Simultaneously, an equal volume of donor blood is infused intravenously to maintain arterial blood volume and pressure. Following the blood collection and infusion of donor blood, the arterial and venous lines are flushed with saline containing 100 U/ml heparin and sealed.
4. The collected blood samples are spun at 2,000 × *g*, and the plasma is collected for further analysis.

4. Notes

1. The present chapter provides a description of a select number of invasive methods utilized to assess renal function in mice. It is important for the reader to understand that there are a number of non-invasive or minimally invasive measures that can be obtained from mice to quickly obtain an index of renal function. These measures generally involve nothing more invasive than obtaining a urine or blood sample or administering an anesthetic agent. Approaches of this type are often used as screening tools and provide the rationale for the more comprehensive studies described in this chapter.

   One of the most basic indices of renal function that can be obtained from humans or animals is a blood sample for the quantification of plasma creatinine or blood urea nitrogen (BUN). Since the body's excretion of creatinine and BUN is primarily dependent on glomerular filtration, an elevation of either of these parameters indicates reduced renal filtration capacity and may indicate underlying renal disease. The strength of this technique is the ability to obtain a reasonably accurate indicator of kidney function from a fairly small amount of blood. The method does have some limitations; interanimal and interassay variablity can limit the sensitivity of this method to detect significant changes in kidney function, and fairly substantial changes in glomerular filtration rate are often necessary before significant changes in circulating levels of creatinine or BUN are reproducibly detected. Furthermore, such assays require blood sampling. Though a blood sample can be obtained relatively simply by retro-orbital, tail vein, or saphenous vein sampling, some manipulation of the animal is required.

   A second non-invasive indicator of renal function is the assessment of albumin or total protein excretion in the urine. The development of albuminuria/proteinuria is a widely used index of kidney damage; the healthy kidney will normally excrete only minimal amounts of protein. This assessment can be performed on samples of urine collected over a period of up to 24 h or more or can be obtained from "spot sampling" of urine. It is generally more accurate to obtain a timed collection of urine in order to accurately quantify the protein excretion rate. A spot sample of urine could be obtained, however, and the ratio of protein (or albumin) to creatinine in the sample can be utilized as an indicator of disease.

A third non-invasive methodology is the use of imaging techniques to assess indices of renal function. Functional magnetic resonance imaging (MRI) techniques have been utilized to obtain indices of renal function in humans and animals (14). Several of these methodologies have been applied to mouse models in the study of renal function. Blood oxygen level-dependent (BOLD) MRI has been used to assess changes in renal cortical and medullary tissue oxygen levels in mice (15). As such, BOLD MRI can provide a non-invasive index of tubular function and blood flow in the kidney. Contrast-enhanced magnetic resonance imaging with gadolinium (16) or a dendrimer-based contrast agent (17) was utilized to visualize kidney development and renal damage in mice. The degree of renal tubular damage assessed by MRI was correlated with changes in serum creatinine and BUN, confirming the usefulness of these MRI techniques.

Though the MRI methodologies are all "non-invasive" in that the abdomen is not opened and the animals can survive the procedure to be studied at a later date, it should be noted that the MRI images are obtained in anesthetized mice (14–17). This is necessary to minimize movement artifacts that interfere with the MRI signal. Moreover, vascular access is often required to administer the contrast agent and/or any other drugs utilized during these procedures. As such, these types of procedures require a degree of proficiency in animal handling and surgical instrumentation in addition to the unique expertise and equipment required to perform MRI studies.

2. A variety of invasive approaches that have been utilized in rats, rabbits, dogs, and other animals have also been used to quantify whole kidney, renal vascular, and epithelial function in anesthetized mice. Classical clearance experiments, in which timed urine collections are obtained and indices of renal function derived from the excretion and clearance of water, electrolytes, and other markers of renal function, have been successfully performed to assess renal function in mice (1–3, 18–20). In such experiments, the animals are anesthetized, vascular access is achieved for blood sampling and infusion, and timed urine collections are made. In larger animals, the ureters are individually cannulated. In mice, the bladder itself is typically cannulated to facilitate the collection. The renal clearance rate of inulin, *para*-amino hippuric acid, or other substances can also be measured to quantify glomerular filtration rate, renal blood flow, or other indices of renal function. In addition to standard clearance techniques, miniaturized flow probes for measurement of renal blood flow (21, 22) and laser-Doppler flowmetry to assess regional renal blood flow changes (23) have been used to assess renal hemodynamics in anesthetized mice.

Studies to assess tubular and vascular function at the level of individual tubules and vessels have also been performed in mice. Micropuncture and micropressure measurements as well as specialized preparations to assess vascular function have been successfully completed (2, 18, 21, 24–27). A common issue in performing these studies is the small size of mice and the relatively small fluid volumes available for collection. This problem is especially apparent in micropuncture studies where distal nephron fluid flow rates are very low, making accurate collection of fluid from these segments especially difficult. Despite these problems, the genetically manipulated mouse has proven to be a valuable model for testing questions of interest.

3. An important consideration with any surgical procedure is the influence of anesthesia on the animals and the potential effects of anesthesia on the experimental results. It is often assumed that a mouse will behave and respond to anesthesia in a manner similar to that observed with rats or other species. As observed with other species, mice (and different strains of mice) demonstrate a unique sensitivity to different anesthetic agents. In general, mice are very sensitive to anesthetics and exhibit cardiovascular and respiratory depression that can adversely alter experimental results (2–4, 28, 29). Rao and Verkman have provided a useful summary of the doses and side-effects of many common anesthetics when used in mice (4).

   The choice of anesthetics is dependent on the user and the particular application. For recovery surgery, we have found that isoflurane anesthesia (2–3% v/v) is quite reliable and elicits a fast-acting, stable anesthesia that is rapidly reversible. The reversibility of anesthesia is critical in survival surgery protocols, particularly when blood vessels in the leg are catheterized. We have observed that animals that are able to rapidly obtain sternal recumbency following a catheterization surgery recover much faster and the success of the procedure is greatly increased. Since the use of inhalation anesthetics requires a nose cone, isoflurane is not ideal for surgical procedures involving manipulation of the head and neck. In such cases, we have found that sodium pentobarbital (25–50 mg/kg ip) is a suitable anesthetic, though the time for recovery is notably longer compared to isoflurane.

   Even with a stable plane of anesthesia, the administration of different anesthetic agents may lead to a dramatic reduction in arterial pressure. This is somewhat mitigated by the administration of supplemental oxygen (2, 23), but, in general, the anesthetic regimen must be modified and carefully monitored for the application. In addition, the use of a heat source such as a thermostatically controlled surgical table is essential for any procedure in anesthetized mice in order to maintain a constant body temperature. The small body size of mice leads to a rapid loss of core body temperature during anesthesia with accompanying cardiovascular and respiratory depression.

4. It is important to minimize trauma to the femoral nerve during the dissection of the artery and vein and the cannulation procedure. The success of this procedure is largely dependent on the mouse regaining use of the catheterized limb as soon as possible following recovery from anesthesia. We believe that the use of the instrumented limb upon recovery promotes the development of the collateral circulation. Collateral blood flow is essential for perfusion of the hindlimb since the femoral artery and vein are permanently ligated in this procedure.

5. The mouse femoral vein is highly distensible with a low luminal pressure and is therefore easily collapsed if twisted or compressed. Since it is much more difficult to catheterize a collapsed vessel, care must be taken to avoid unnecessary stress during the dissection and when placing the ligatures on this vessel. It is highly recommended that the vein be cannulated prior to the artery since the artery has a much higher luminal pressure and does not collapse as easily as the vein. In case the artery becomes constricted, a single drop of 2% lidocaine hydrochloride will usually dilate the vessel sufficiently for cannulation. The excess lidocaine solution in the surgical field should be quickly removed by absorption with a cotton swab to minimize adverse effects on the femoral nerve.

6. The proximal portion of the vessel is occluded in a user-dependent fashion. The use of a small vascular clamp ("Bull-dog Clamp") miniaturized for mouse work is preferred by some investigators. This is a useful method, though all but the smallest of these clamps are still rather large. Other investigators place tension upon the proximal ligature (by gently pulling the ligature and anchoring it with tape or a hemostat) to occlude the vessel. Care must be taken not to overstretch and tear the vessel when using this method. A third method is to temporarily occlude the vessel by tying an overhand knot in the ligature; the knot is then untied when the cannula has been successfully inserted into the vessel. This method is useful but requires care when loosening the knot. Regardless of the preferred method, occlusion of the proximal portion of the vessel is important to minimize unnecessary bleeding when the vessel is incised. This is absolutely essential for the arterial cannulation since the blood is under high pressure in the arterial system, and the animal will rapidly lose a large amount of blood if the proximal portion of the vessel is not securely occluded.

7. Though the blood vessels can be visualized with the naked eye, most investigators prefer to work with a stereomicroscope (2–20 × final magnification with an appropriate working distance) or other magnifying device to help visualize the vessels and the surgical field.
8. The ability to administer substances in an acute or chronic setting is often necessary in experimental preparations. For acute protocols using anesthetized mice, intravenous or intra-arterial infusion is a relatively simple matter and can be performed as part of the routine protocol. A number of strategies have been used for long-term administration of substances or to administer agents to conscious mice in the short-term. For substances that are orally absorbed, the addition of the desired agent to the food or drinking water is a simple, non-invasive solution for long-term administration. In studies in which oral administration strategies are employed, it is important to monitor the intake of food or water before and after the addition of the drug in order to quantify intake of the drug and to ensure that the animals continue to eat and drink. It may also be necessary to adjust the concentration of the drug in the food or water to maintain a constant administration rate if eating habits change.

   Substances can also be administered through implanted osmotic minipump devices designed to deliver substances subcutaneously, intraperitoneally, or intravenously for extended periods of time (30). This method requires a substance that will remain stable in solution. Furthermore, the investigator should take steps to confirm that the drugs to be delivered are administered at a stable rate.

   The most invasive method for continuous delivery of drug or other substances is intravenous infusion. This requires surgical implantation of a catheter and a means to constantly deliver the infused substance as described in **Fig. 5.1**. We have successfully used this approach for the long-term administration of a number of different agents for up to 35 days (7–9, 12, 13). Though this method provides the most stable rate of administration and permits blood sampling, this is the most technically demanding of the methods listed and may not be the best choice for all applications.
9. The chronic catheterization of the femoral artery in this technique permits the direct quantification of arterial blood pressure. We have utilized the direct measurement of blood pressure in anesthetized and conscious mice with implanted catheters and also performed studies in which arterial pressure was measured directly with implanted telemetry transmitters (*see* **Chapter 4**). The telemetry system is generally regarded as

the state-of-the-art approach for blood pressure measurements in mice. This method requires a fairly straightforward surgical implantation and can continuously monitor arterial pressure in conscious animals. The major drawback to this approach is the expense of the proprietary hardware and software necessary to make this measurement. In addition, the implanted telemetry devices have a limited battery life and must be periodically refurbished by the manufacturer.

A second method used to measure arterial pressure in mice is the implanted catheter technique. We have used this approach to record arterial blood pressure for periods up to 35 days following surgery (10). In addition, we have used this approach 24 h per day for recording of arterial pressure (11). As described above, implanted catheters in conscious mice have the advantage of permitting blood sampling and infusions. In general, this system also requires specialized experimental equipment (appropriate caging and recording systems) and involves a more invasive surgical approach. This method is not as widely used as telemetry applications; moreover, the maintenance of the mice following this surgery requires a great deal of time and effort. Direct measurements of arterial blood pressure have also been obtained from anesthetized mice. The adverse effects of anesthesia generally preclude this method from widespread acceptance as a primary method of blood pressure measurement, though the results may be sufficient in some circumstances.

Finally, tail-cuff plethysmography is an indirect measurement of arterial blood pressure that has been utilized by a number of investigators [(31, 32); *see* **Chapter 3**]. This method is a convenient tool for screening since it is noninvasive and can be performed repeatedly without the need for a direct intervention. The physical restraint of the animals that is necessary to carry out these measurements and the limited time in which measurements can be taken are the primary criticisms of this technique.

10. Experiments examining kidney function frequently require not only the quantification of daily excretion rates, but also the determination if the mouse is in a negative or positive 24 h sodium and/or water balance during an experimental manipulation. This particular task has proven particularly challenging because of difficulty in completely accounting for the intake and output of fluids and electrolytes. When one considers the difficulties associated with the accurate measurement of the intake and output of solute and water in small animals (accounting for evaporative losses, spillage of food, dripping water bottles, and fecal losses of water and electrolytes), it is difficult to accurately quantify intake and output.

The use of specially designed metabolic cages increases the efficiency of fluid and electrolyte balance measurements, though the above-mentioned errors cannot be fully eliminated.

One alternative we have used to assess sodium balance is to prepare mice with chronic indwelling venous catheters, feed them a sodium-free chow, infuse the daily sodium intake intravenously in isotonic saline, and collect the excreted urine (12). This preparation permits the precise determination of sodium intake, since all sodium is delivered intravenously. This method provides accurate collection of sodium intake since the steady-state collection of excreted sodium with this preparation averaged 100±3% of intake in 20 mice over a 10-fold range of sodium intake (the intake range of mice eating chow containing 0.4–4.0% NaCl). An example of this technique is illustrated in **Fig. 5.4** from an experiment in which we assessed mean arterial blood pressure, daily sodium intake and output, and calculated daily sodium balance in conscious mice as sodium intake was increased from approximately 200 μEq/day (the sodium intake when mice consume a 0.4% NaCl diet) to ~500 μEq/day for 3 days, and finally to ~1000 μEq/day (the sodium intake when mice consume a 4.0% NaCl diet). These data illustrate the ability of this technique to resolve the retention of sodium on the first day of the transition to a new sodium intake with the attainment of neutral sodium balance thereafter. This example also demonstrates that these transitions in sodium intake and excretion rate occurred in the absence of changes in systemic arterial blood pressure.

11. In studies examining the regulation of renal/cardiovascular function, it is often necessary or desirable to quantify the circulating levels of different hormones that may influence kidney function. Since the quantification of plasma hormones usually requires a volume of blood greater than the minimal volume needed to assess creatinine or BUN, this is a prime example of the difficulties presented by the small size (and small blood volume) of mice. As has been previously reviewed, the terminal collection of trunk blood is one option to assess circulating hormones (3). The major drawback to this approach is that the stress placed upon the animal during handling prior to anesthesia, the direct and indirect effects of anesthetics, and the surgical manipulation required to access trunk blood can all significantly alter the circulating levels of different hormones. In an attempt to avoid these confounding problems, we developed a method to assess the plasma levels of renin, angiotensin II, and aldosterone in freely moving, conscious mice as described above (7–9). This arrangement

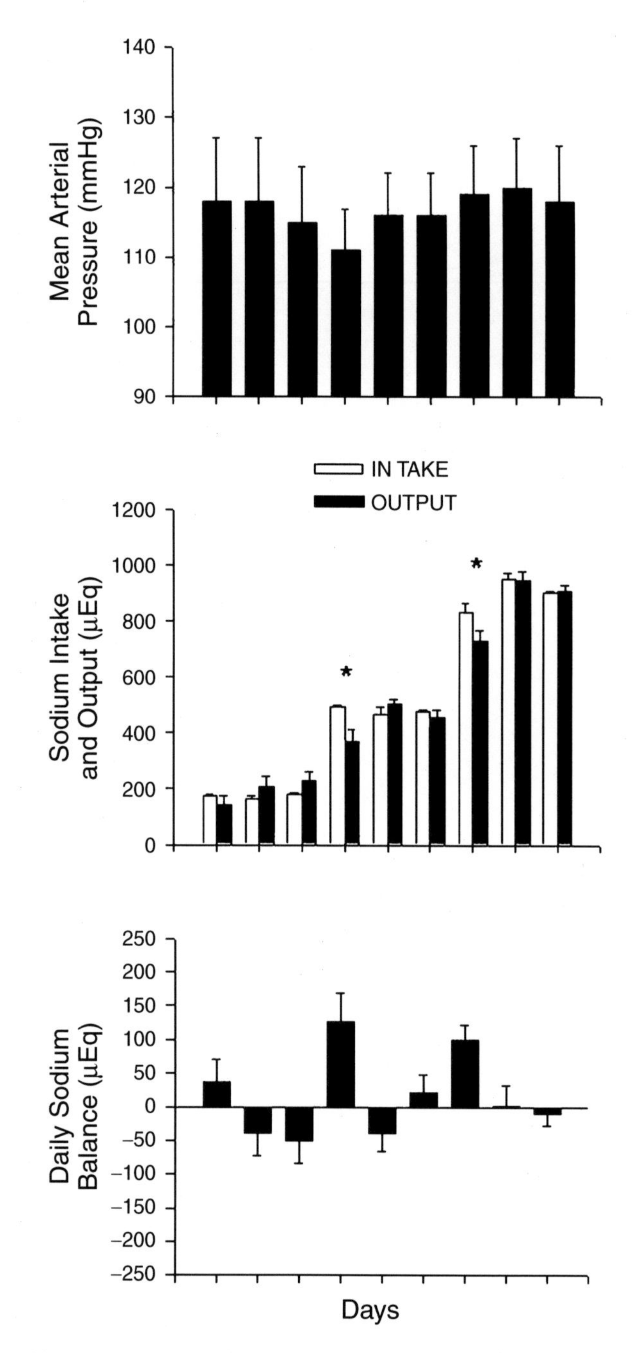

Fig. 5.4. Mean arterial blood pressure, sodium intake and output, and daily sodium balance in chronically instrumented, conscious mice as sodium intake was increased from approximately 150 to 900 μ Eq/day. *Indicates output significantly different from the intake for that day ($P<0.05$). Redrawn from reference (12).

permits the mouse to move freely about his home cage in an undisturbed state while an investigator can gain access to arterial or venous blood (10–12).

In order to assess the circulating components of the renin–angiotensin–aldosterone axis from the same animal under different physiological conditions, it was necessary to obtain as much as 500 μl of blood. Since even large mice (30 g) have a small blood volume (~2 ml), the withdrawal of 500 μl of blood has a significant impact on arterial blood pressure. In preliminary experiments, we observed that blood pressure significantly decreased after the withdrawal of ~50–100 μl of blood; withdrawing larger volumes would clearly have a greater effect on cardiovascular dynamics and stimulate a cascade of neural and endocrine reflexes. To maintain total blood volume and minimize disturbances in arterial pressure during the blood withdrawal protocols, the arterial blood sampling in this protocol was performed with the simultaneous intravenous infusion of donor blood. The donor blood was drawn from littermate donors, the donor red blood cells (RBCs) were washed with Tyrode's solution, the RBCs were resuspended in artificial plasma, and the donor blood was infused intravenously during the arterial blood withdrawal. This technique was utilized in order to control the possible changes in systemic hemodynamics that would lead to alterations in the levels of different components of the renin–angiotensin system.

An example of the utility of this sampling protocol is illustrated in **Fig. 5.5** in which the plasma angiotensin II and aldosterone levels were quantified in conscious mice maintained on a low or high dietary NaCl intake or during the continuous intravenous infusion of angiotensin II or aldosterone. This figure illustrates the usefulness of this methodology as a means to collect plasma for quantification of circulating hormones. The technique was capable of detecting the physiological suppression of angiotensin II and aldosterone when salt intake was increased from low to high levels as well as the significant increase in angiotensin II or aldosterone that occurred during the exogenous infusion of these agents. For the quantification of circulating factors from conscious mice, this method is a useful approach.

One drawback with the simultaneous intravenous infusion of donor blood during arterial blood sampling is the likelihood of diluting the plasma and thereby lowering the absolute level of the hormones measured. Though this is clearly possible, especially with the larger volumes of blood drawn in order to quantify plasma aldosterone and angiotensin II, the experiments were performed in a paired fashion. Any error

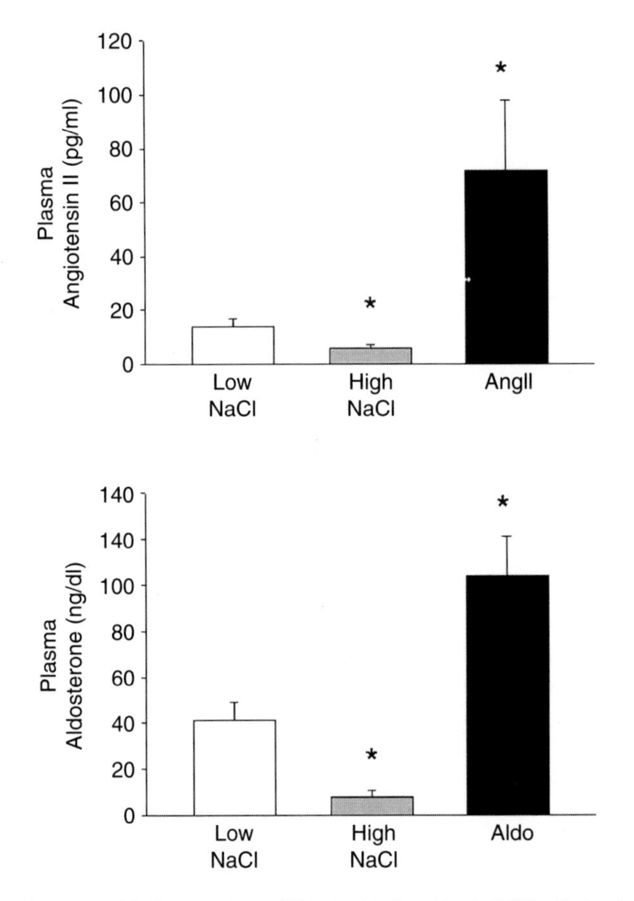

Fig. 5.5. *Upper panel*: influence of low (200 μEq/day) and high (1000 μEq/day) sodium intake and intravenous angiotensin II (AngII, 40 ng/kg/min) infusion on plasma AngII concentration in conscious mice. *Lower panel*: influence of low and high sodium intake and intravenous infusion of exogenous aldosterone (Aldo, 50 ng/kg/min) on plasma Aldo in conscious mice. $*P < 0.05$ vs. low NaCl. Redrawn from reference (8) with permission from the American Physiological Society.

introduced by infusion of donor blood should therefore be consistent between the different conditions in which the plasma samples were drawn. Another major drawback with this method is the labor-intensive nature of the measurement as well as the fairly specialized caging systems necessary to house the animals for this preparation. Nonetheless, this approach does provide a method to assess changes in circulating factors in the blood of conscious, unrestrained mice under different physiological conditions.

References

1. Gross, V, Luft, FC. (2002) Adapting renal and cardiovascular physiology to the genetically hypertensive mouse. *Semin Nephrol* **22**, 172–179.
2. Lorenz, JN. (2002) A practical guide to evaluating cardiovascular, renal and pulmonary function in mice. *Am J Physiol* **282**, R1565–R1582.

3. Meneton, P, Ichikawa, I, Inagami, T, et al. (2000) Renal physiology of the mouse. *Am J Physiol* **476**, F339–F351.
4. Rao, S, Verkman, AS. (2000) Analysis of organ physiology in transgenic mice. *Am J Physiol* **279**, 1–18.
5. Vallon, V. (2003) In vivo studies of the genetically modified mouse kidney. Nephron Physiol **94**, 1–5.
6. Vallon, V, Traynor, T, Barajas, L, et al. (2001) Feedback control of glomerular vascular tone in neuronal nitric oxide synthase knockout mice. *J Am Soc Nephrol* **12**, 1599–1606.
7. Cholewa, BC, Mattson, DL. (2005) Influence of elevated renin substrate on angiotensin II and arterial blood pressure in conscious mice. *Experimental Physiol* **90**, 607–612.
8. Cholewa, BC, Mattson, DL. (2001) Role of the renin-angiotensin system during alterations in sodium intake in conscious mice. *Am J Physiol* **281**, R987–R993.
9. Cholewa, BC, Meister, CJ, Mattson, DL. (2005) Importance of the renin-angiotensin system in the regulation of arterial blood pressure in conscious mice and rats. *Acta Physiol Scand* **183**, 309–320.
10. Mattson, DL. (1998) Long-term measurement of arterial blood pressure in conscious mice. *Am J Physiol* **274**, R564–R570.
11. Mattson, DL. (2001) Comparison of arterial blood pressure in conscious mice. *Am J Hypertens* **14**, 405–8.
12. Mattson, DL, Krauski, KR. (1998) Chronic sodium balance and blood pressure response to captopril in conscious mice. *Hypertension* **32**, 923–928.
13. Mattson, DL, Meister, CJ. (2006) Sodium sensitivity of arterial blood pressure in L-NAME hypertensive but not eNOS knockout mice. *Am J Hypertens* **19**, 327–9.
14. Prasad, PV. (2006) Functional MRI of the kidney: tools for translational studies of pathophysiology or renal disease. *Am J Physiol* **290**, F958–F974.
15. Li, LP, Ji, L, Lindsay, S, et al. (2007) Evaluation of intrarenal oxygenation in mice by BOLD MRI on a 3.0T human whole-body scanner. *J Magn Reson Imaging* **25**, 635–638.
16. Chapon, C, Franconi, F, Roux, J, et al. (2005) Prenatal evaluation of kidney function in mice using dynamic contrast-enhanced magnetic resonance imaging. *Anat Embryol* **209**, 263–267.
17. Kobayashi, H, Kawamoto, S, Jo, S-K, et al. (2002) Renal tubular damage detected by dynamic micro-MRI with a dendrimer-based magnetic resonance contrast agent. *Kidney Int* **61**, 1980–1985.
18. Lorenz, JN, Baird, NR, Judd, LM, et al. (2002) Impaired renal NaCl absorption in mice lacking the ROMK potassium channel, a model for Type II Bartter's Syndrome. *J Biol Chem* **277**, 37871–3780.
19. Obst, M, Gross, V, Janke, J, et al. (2003) Pressure natriuresis in AT_2 receptor–deficient mice with L-NAME hypertension. *J Am Soc Nephrol* **14**, 303–310.
20. Zhao, D, Navar, LG. (2008) Acute angiotensin II infusions elicit pressure natriuresis in mice and reduce distal fractional sodium reabsorption. *Hypertension* **52**, 137–142.
21. Hashimoto, S, Huang, Y, Briggs, J, et al. (2006) Reduced autoregulatory effectiveness in adenosine 1 receptor-deficient mice. *Am J Physiol* **290**, F888–F891.
22. Just, A, Arendshorst, WJ. (2007) A novel mechanism of renal blood flow autoregulation and autoregulatory role of A1 adenosine receptors in mice. *Am J Physiol* **293**, F1489–F1500.
23. Mattson, DL, Meister, CJ. (2005) Renal cortical and medullary blood flow responses to L-NAME and angiotensin II in wild-type, nNOS null mutant, and eNOS null mutant mice. *Am J Physiol* **289**, R991–R997.
24. Harrison-Bernard, L, Cook, AK, Oliverio, MI, et al. (2003) Renal segmental microvascular responses to ANG II in AT1A receptor null mice. *Am J Physiol* **284**, F538–F545.
25. Schnermann, J. (1999) Micropuncture analysis of tubuloglomerular feedback regulation in transgenic mice. *J Am Soc Nephrol* **10**, 2614–2619.
26. Sonnenberg, H, Honrath, U, Chong, CK, et al. (1994) Proximal tubular function in transgenic mice overexpressing atrial natriuretic factor. *Can J Physiol Pharmacol* **72**, 1168–1170.
27. Wulff, P, Vallon, V, Huang, DY, et al. (2002) Impaired renal Na^+ retention in the sgk1-knockout mouse. *J Clin Invest* **110**, 1263–1268.
28. Janssen, BJA, De Celle, T, Debets, JJM, et al. (2004) Effects of anesthetics on systemic hemodynamics in mice. *Am J Physiol* **287**, H1618–H1624.
29. Rieg, T, Richter, K, Osswald, H, et al. (2004) Kidney function in mice: thiobutabarbital versus a-chloralose anesthesia. *Naunyn-Schmiedeberg's Arch Pharmacol* **370**, 320–323.

30. Theeuwes, F, Yum, SI. (1976) Principles of the design and operation of generic osmotic pumps for the delivery of semisolid or liquid drug formulations. *Ann Biomed Eng* **4**, 343–353.

31. Krege, JH, Hodgin, JB, Hagaman, JR, et al. (1995) A noninvasive computerized tail-cuff system for measuring blood pressure in mice. *Hypertension* **25,** 1111–1115.

32. Kurtz, TW, Griffin, KA, Bidani, AK, et al. (2005) Recommendations for blood pressure measurement in humans and experimental animals Part 2: blood pressure measurement in experimental animals. *Hypertension* **45**, 299–310.

Chapter 6

Ischemic Stroke in Mice and Rats

Aysan Durukan and Turgut Tatlisumak

Abstract

Ischemic stroke occurs most often in the territory of the middle cerebral artery (MCA) in humans. Since its description in rats more than two decades ago, the minimally invasive intraluminal suture occlusion of MCA is an increasingly used model of stroke in both rats and mice due to its ease of inducing ischemia and achieving reperfusion under well-controlled conditions. This method can be used under the guidance of laser-Doppler flowmetry to ascertain the magnitude of occlusion or reperfusion and to decrease the rate of subarachnoid hemorrhage. Ninety minutes of transient ischemia in the territory of MCA results in substantial and reproducible ischemic lesions in both the striatum and the cortex, with characteristics of lesion core and penumbra. Thus, this model is applicable to neuroprotective drug studies, including ischemic brain lesion evaluation (either in vivo with magnetic resonance imaging or post-mortem with brain tissue staining) and neurological status (motor deficits simply assessed by a six-point neurological score scale) as outcome parameters.

Key words: Brain, rat, mouse, stroke, focal ischemia, magnetic resonance imaging.

1. Introduction

We pose the questions, but the answers come from experimental animals. Since we became capable of modifying their genes, mice now are used to answer even more specific questions. Stroke is a devastating, complex, and yet hardly deciphered disease that continues to inspire researchers to explore its underlying pathophysiological mechanisms. In clinical stroke research, very large patient group sizes are required to avoid confounding effects of the diversity (etiology, duration, localization, and severity of ischemia, co-existing systemic diseases, patient gender, and age). Consequently, obtaining answers from humans is difficult, expensive, time consuming, and even unethical in certain conditions. In

K. DiPetrillo (ed.), *Cardiovascular Genomics*, Methods in Molecular Biology 573,
DOI 10.1007/978-1-60761-247-6_6, © Humana Press, a part of Springer Science+Business Media, LLC 2009

contrast, when stroke is induced in laboratory animals, many of these variables are eliminated and experiments can easily be replicated with low cost and less time. In this chapter, we will describe how to model transient focal cerebral ischemia in rats and mice and how to evaluate the outcome. The model is based on the occlusion of the middle cerebral artery (MCA) by a monofilament suture via intraluminal approach. It is minimally invasive, including a simple neck surgery that is quickly performed, and delicately simulates malignant MCA infarction in humans. Furthermore, application of this method to both rats and mice enables researchers to address ischemic stroke using transgenic technology. Laser-Doppler flowmetry (LDF) provides information on the cortical cerebral blood flow (CBF) to determine whether occlusion and reperfusion succeeded. Once the stroke is induced, the next step is to evaluate its consequences. Assessment of ischemic lesion size is the basic but inevitable outcome measure. To this end, a traditional method is post-mortem staining of brain slices, but it is increasingly replaced with in vivo lesion detection by magnetic resonance imaging (MRI). The second outcome parameter is the neurological status of the animal, which is simply assessed by testing motor capability.

MCA occlusion in this model leads to large and reproducible infarcts with predictable sizes and small variability, involving both cortical and white matter structures. Further, substantial brain edema enables researchers to study this aspect. Substantial decreases in infarct size and edema volume, improvement in neurological scores, and decrease in mortality, all serve as positive outcome measures when intervention efficacy is tested.

2. Materials

2.1. Animals

Normotensive adult male Wistar rats weighing 300 to 350 g and NMRI adult male mice weighing 28 to 34 g are used (*see* **Note 1**). The animals are maintained on a 12-h light/12-h dark cycle and have free access to food and water (ad libitum) before and after the surgical procedures. Approval of Animal Research Committee of the study protocol is required prior to study initiation.

2.2. Tools for Surgical Procedures

1. Surgical microscope (Wild/Leica M650)
2. One scalpel size 3 with no. 10 blade
3. One pair of small scissors for blunt dissection
4. Two pairs of fine-tipped dissecting forceps for microsurgery, curved, 100 mm

5. One pair of DeBakey forceps
6. One pair of microsurgical scissors
7. One needle holder
8. Cotton-tipped applicators (to remove small quantities of blood from the surgical field)
9. 3-0 Surgilon, 2.0 metric, black braided nylon, pre-cut sutures (Sherwood Davis & Geck, USA)
10. 5-0 Surgilon, 1.0 metric, black braided nylon, pre-cut sutures (Sherwood Davis & Geck, USA)
11. 3-0 Visorb, 30 in. sutures (CP Medical, USA)

    Major instruments required for surgery can be seen in **Fig. 6.1**. They are kept in a tray suitable for repeated autoclaving (en bloc).

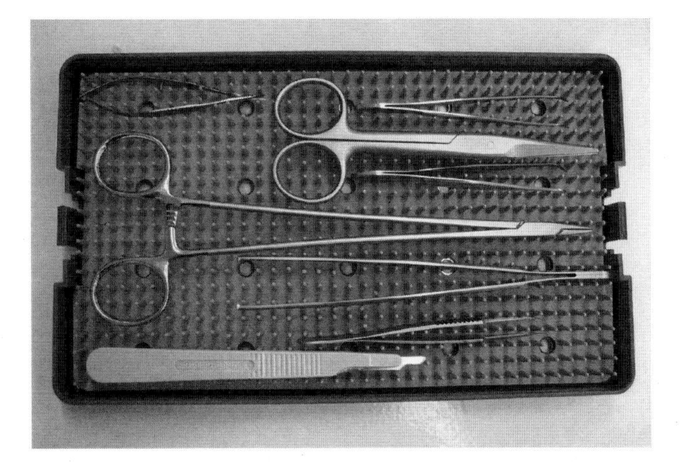

Fig. 6.1. Surgical tools in the sterilization tray.

2.3. Preparing Vessel Occluders and Cannulas

1. 4-0 Ethilon II, 1.5 metric, blau polyamid 6 monofil (Ethicon, USA)
2. 5-0 Ethilon II, 1 metric, blau polyamid 6 monofil (Ethicon)
3. Low-viscosity silicone (Provil L, Bayer Dental, Leverkusen, Germany)
4. PE-50 catheter (0.58 mm of inner diameter, 0.965 mm of outer diameter; BD Intramedic Polyethylene Tubing, Clay Adams, MD, USA)
5. PE-10 catheter (0.28 mm of inner diameter, 0.61 mm of outer diameter; BD Intramedic Polyethylene Tubing, Clay Adams, MD, USA)
6. Bruker–Turk counting chamber (0.0025 mm^2, Superior, Marienfeld, Germany)

2.4. Tools for Measurement of Physiological Parameters

1. Rectal temperature probe (Therm Alert TH-8, Physitemp, Clifton, NJ, USA)
2. Blood pressure meter (Olli Blood Pressure Meter 533, Kone Oy, Espoo, Finland)
3. Blood gas analyzer (AVL OPTI, Roche Oy, Helsinki, Finland)
4. Glucometer (Accutrend Sensor, Roche Diagnostics, Espoo, Finland)

2.5. Assessing Stroke

1. LDF: CBF is measured on-line by the BF/F/0.5 bare-fiber flexible probe of the Oxy-Flow device (Oxford Optronix, Oxford, UK).
2. MRI: Magnetic resonance imaging is performed with a 4.7-Tesla scanner (PharmaScan, Bruker BioSpin, Germany) using a 90-mm shielded gradient that is capable of producing maximum gradient amplitude of 300 mT/m with 80-μs rise time. A linear birdcage RF coil with an inner diameter of 38 mm is used for rats and 23 mm for mice.
3. Neurological examination: A six-point scale evaluating motor deficits (*see* **Note 2**) is applied.

2.6. Tissue Processing

1. Sodium barbiturate: 60 mg/mL (Mebunat, Orion, Finland)
2. Ice-cold saline: 200 mL for rats, 20 mL for mice
3. Serum infusion set ending with a blunted 20-ga needle for rats, 22-ga needle for mice
4. One pair of small scissors for blunt dissection
5. One needle holder
6. One pair of rongeurs
7. Spatula
8. Acrylic brain-cutting matrix for coronal, 2-mm thick slices in rats and 1-mm thick slices in mice (Zivic Instruments, Pittsburgh, PA, USA)
9. Tissue-Tek (Sakura Finetek, Inc., Tokyo, Japan)
10. 10% phosphate-buffered formaldehyde
11. 2% triphenyltetrazolium chloride (TTC) in saline
12. Laboratory scale
13. Digital camera (Sony, Tokyo, Japan)
14. Computer image analysis software, Image J 1.36b (National Institutes of Health, USA)

3. Methods

3.1. Preparation of Animals

General anesthesia is induced by intraperitoneal injection of ketamine hydrochloride (75 mg/kg for both rats and mice) and by subcutaneous injection of medetomidine hydrochloride (0.5 mg/kg for rats and 1 mg/kg for mice; *see* **Note 3**). The room temperature is kept at 25 to 27°C. The animal is placed and kept on its back during surgery and continuously warmed by a heating blanket. When necessary (i.e., when rectal temperature falls below 36.0°C) animals are additionally heated with a thermo-regulated heating lamp positioned 30 to 35 cm above the body (*see* **Note 4**).

3.2. Surgical Procedures

3.2.1. Cannulation of Femoral Artery and Vein

1. The hair on the left medial thigh (femoral region) is shaved, and then the skin is disinfected with alcohol 70%.
2. A 1.5 to 2-cm incision, guided by palpation of femoral artery pulsation, is made over the groin region to reach the femoral vessels.
3. Femoral vessels are exposed by blunt dissections and separated from each other and from the femoral nerve (**Fig. 6.2**).

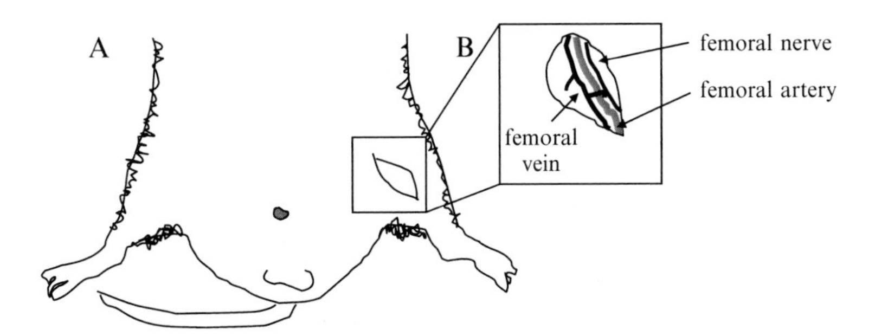

Fig. 6.2. Schematic drawing of the preparation for femoral artery and vein cannulation in the rat. (**A**) Site of incision on the left groin. (**B**) Magnified view of the locations of the femoral vein, artery, and nerve in the dissected area.

4. For the femoral vein catheterization, a loose ligature is placed proximally and stretched, while the vein is permanently occluded by a ligature placed distally.
5. A small incision is made in the vein using microsurgical scissors.
6. The proximal ligature is released and a blunt-tipped PE-50 catheter is inserted 2 to 3 cm (for mice femoral vessel cannulation, a PE-10 catheter is suitable).
7. The catheter is secured by tightening the proximal ligature.
8. Same procedures are repeated for the cannulation of the femoral artery except that the catheter has a sharp tip (*see* **Note 5**).

3.2.2. Middle Cerebral Artery Occlusion (MCAO)

The model was originally described by Koizumi et al. (1) and later modified (2). The intraluminal occluder for rats is prepared from 4-0 monofilament nylon suture.

1. The suture is cut to 5 cm length.
2. The tip is rounded to 0.330 to 0.360 mm of diameter by heating near the tip of a hot (approximately 300°C) soldering iron under a microscope.
3. One to two cm from the tip is coated with low-viscosity silicone (*see* **Note 6**).
4. The tip is measured under microscope on a Bruker–Turk counting chamber (0.0025 mm^2), and sutures with diameters of 0.380 to 0.400 mm are chosen as occluders.
5. For mouse, occluders are prepared from 5–0 monofilament nylon sutures with its tip rounded (by the method described above) to a final 0.150 to 0.200 mm diameter (**Fig. 6.3**).

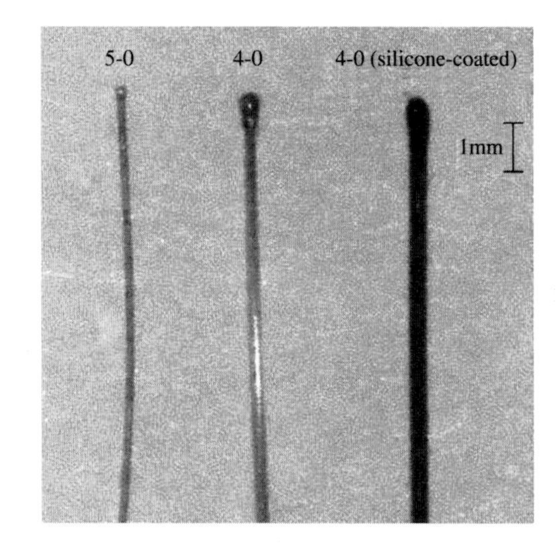

Fig. 6.3. Sutures used for occluding the middle cerebral artery.

6. The right common carotid artery (CCA) and the right external carotid artery are exposed through a ventral midline neck incision and further blunt dissection between the sternohyoid, sternomastoid, and omohyoid muscles (*see* **Note 6**).
7. The proximal portion of the CCA and the origin of the external cerebral artery are ligated with a 3-0 thread.
8. The internal carotid artery (ICA) is loosely ligated with a 3-0 thread, but no branches of the ICA are ligated.
9. The CCA is loosely ligated with a 6-0 silk thread 3 mm below the bifurcation to mark the arteriotomy site and prevent bleeding. The arteriotomy is made just proximal to the 6-0 silk thread.

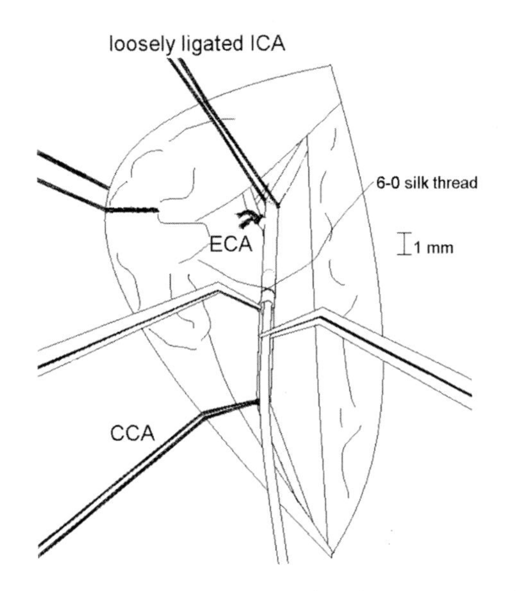

Fig. 6.4. Schematic drawing of suture insertion into the common carotid artery. The external carotid artery (ECA) and the common carotid artery (CCA) are ligated. The internal carotid artery (ICA) is loosely ligated and stretched. A silicone-coated, 4–0 nylon monofilament is inserted into the CCA from the arteriotomy site 3 mm below the carotid bifurcation. Note that 6–0 silk thread prevents bleeding from the arteriotomy.

10. The occluder is carefully advanced intracranially (**Fig. 6.4**), approximately 17 to 19 mm from the bifurcation in rats and 9 to 11 mm in mice.
11. At this point, mild resistance indicates that the tip of the occluder is lodged into the proximal anterior cerebral artery and occluding the origins of the ipsilateral MCA and the posterior communicating artery (**Fig. 6.5**).

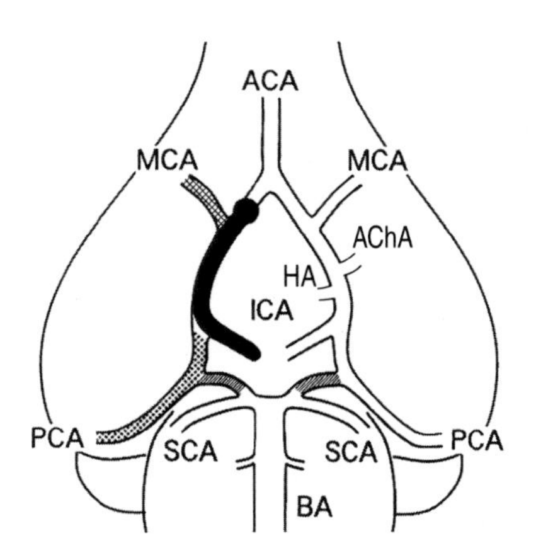

Fig. 6.5. Suture occluder in the circle of Willis. Note that the body of the suture, if large, would occlude the hypothalamic and anterior choroidal branches of the internal carotid artery. ACA, anterior cerebral artery; AChA, anterior choroidal artery; BA, basilar artery; HA, hypothalamic artery; ICA, internal carotid artery; MCA, middle cerebral artery; PCA, posterior cerebral artery; SCA, superior cerebellar artery.

12. The loosely ligated 3-0 and 6-0 threads are then ligated firmly and the animal is turned to prone position.

3.3. Measurement of Physiological Parameters

The following physiological measurements are obtained before induction of ischemia, throughout MCAO, and 15 to 30 min after reperfusion:

1. Blood is sampled from the femoral artery catheter for analysis of blood gases (pH, $PaCO_2$, and PaO_2) and glucose (*see* **Notes 7 and 8**); 0.1 to 0.2 mL of blood is usually sufficient. Baseline, during ischemia, and after reperfusion measurements are appropriate.
2. Rectal temperature is continuously monitored with a rectal probe inserted 4 cm deep from the anal ring (*see* **Note 4**).
3. Mean arterial blood pressure is continuously monitored via the femoral artery catheter (*see* **Note 9**) and recorded repeatedly (e.g., every 15 or 30 min).

3.4. Assessment of Stroke

3.4.1. Laser-Doppler Flowmetry

1. The scalp is incised in the midline and the skull is exposed.
2. In rats, the skull area over the ipsilateral MCA region (between 1.0 and 2.5 mm posterior and 6.0 mm lateral from the bregma) is thinned by a dental drill. In mice, no drilling is necessary and probe location is chosen over the territory supplied by the proximal part of the MCA representing the core region (2 mm caudal to bregma and 3 to 4 mm lateral to the midline).
3. The probe is attached with a cyanoacrylate tissue adhesive to the skull at a place that represents baseline CBF and fixed to that place.
4. The CBF signal is then obtained from the same place throughout the entire experiment via a data collection software (Perisoft Version 1.3; Perimed, Inc.) (*see* **Note 10**).

3.4.2. Magnetic Resonance Imaging

1. Diffusion-weighted image (DWI) scans are acquired using a spin-echo echo-planar imaging sequence with three different *b* values (repetition time/echo time = 4,000/80 ms, matrix size = 128 × 128; *b* values: *b* 0 = 0.4, *b* 1 = 1,280, and *b*2 = 2,342 s/mm^2, diffusion is measured in the read-out gradient direction).
2. T1 measurements are obtained with an inversion recovery snapshot – fast low-angled shot (repetition time/echo time = 2.2/1.4 ms, 12 inversion delay from 140 to 3,230 ms, flip angle = 5°, matrix size = 128 × 128). This sequence with inversion delay of 1,826 ms provides T1-weighted image.
3. Fluid-attenuated inversion recovery images are acquired with rapid acquisition–relaxation enhancement sequence (repetition time/echo time = 10,000/38.6 ms, inversion time = 1,800 ms, matrix size = 256 × 128, echo train length = 16, number of averages = 1).

4. The field-of-view of all sequences in rat imaging is 40 × 40 mm and in mouse imaging 20 × 20 mm.
5. Slice thickness is chosen as 2 mm in rats and 1 mm in mice.
6. Rectal temperature during imaging is maintained at 37°C by using an MRI-compatible heating pad (Gaymar Industries, Orchard Park, NY, USA; *see* **Note 11**).

3.4.3. Neurological Examination

Sensorimotor evaluation of both rats and mice is performed before the sacrifice of the animals, most often 24, 48, or 72 h after ischemia–reperfusion with a six-point scoring scale (adopted from Ref. (3)), which includes 0, normal; 1, contralateral paw paresis; 2, 1 plus decreased resistance to lateral push; 3, 2 plus circling behavior; 4, no spontaneous walking with depression of consciousness; and 5, death (*see* **Note 2**).

3.4.4. Tissue Processing

Transcardiac perfusion requires a simple surgical procedure following euthanasia that is performed in 2 to 3 min immediately after respiration has ceased, but while heart is still beating (*see* **Note 12**).

1. Animals are euthanized with an overdose of sodium barbiturate (60 mg/kg, intraperitoneally).
2. First, the abdominal cavity is opened by a midline incision (from the umbilicum to the xiphoid process) and the diaphragm is dissected.
3. While the xiphoid process is held with needle holders, a V-shaped incision is made by cutting both sides of the ribcage toward the axilla region.
4. The xiphoid process is pulled rostrally to expose the chest cavity.
5. The connective tissue around the heart is broken so that the heart can be held between the fingers with the apex exposed. The heart will still be beating at this point.
6. A perfusion needle (with a fairly large bore, but a blunt tip) is inserted into the left ventricle through the apex and advanced toward the aorta; at this point, the tip of the needle can be seen in the aortic arch.
7. The right atrium is cut to allow venous blood to drain out while ice-cold saline (200 mL for rats or 20 mL for mice) is infused into the aorta at approximately 100 mmHg of pressure.
8. The fluid draining out of the heart should become clear in a few minutes before the perfusion ends.
9. After decapitation with either sharp scissors or a guillotine, the brain is removed.
10. While taking care to avoid damaging the brain, gently insert rangeurs into the foramen magnum to first grip then break the skull by pulling up and laterally.

11. The remaining skull is removed piece by piece until the entire dorsal surface of the brain is visible.
12. From the bottom, the cranial nerves are severed with a spatula and brain is lifted on the spatula and removed through the skull opening prepared with the rangeurs.
13. After washing the brain with saline, six to eight coronal slices (2-mm thick in rats, 1-mm thick in mice) are cut with the brain-cutting matrix.
14. To keep fresh frozen tissue samples, the slices are embedded in Tissue-Tek (Sakura Fientek, Inc., Tokyo, Japan) before being frozen in liquid nitrogen. Otherwise, the tissue slices are fixed by immersion in 10% formaldehyde (*see* **Note 13**).
15. TTC of 2% concentration is used for staining to further evaluate ischemic lesion size (*see* **Note 14**). Slices are bathed in the TTC solution at 37°C for 15 min.
16. The TTC-stained brain slices are mounted on scale (mm) paper and photographed with a digital camera for further planimetric lesion size analyses.

3.4.5. Calculating Volumes of Brain Infarction and Swelling

Ischemic lesion size can be calculated either in vivo (MRI-based) or post-mortem (TTC staining-based; *see* **Note 14**). For in vivo lesion size determination, DWI (until 24 h) or T2-WI (starting at 48 h) sequences are used to calculate the lesion area in each slice (**Fig. 6.6**). Post-mortem lesion size determination uses TTC, a yellowish powder and that is colorless when dissolved in saline at approximately 37°C. It is reduced by enzymes in functioning

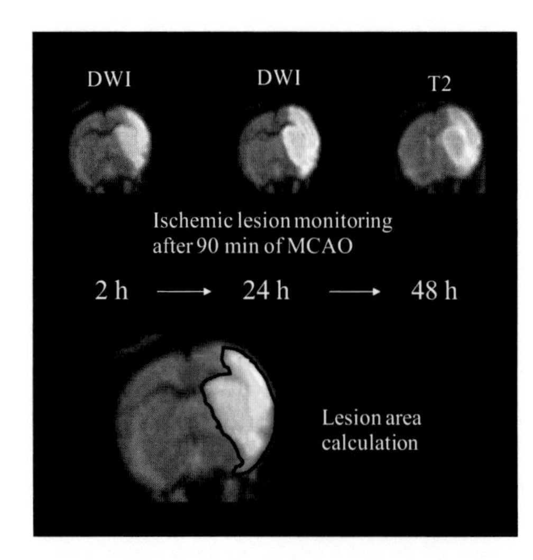

Fig. 6.6. Ischemic lesion monitoring by magnetic resonance imaging; 90 min of middle cerebral artery occlusion induces a large ischemic lesion. The size of the lesion is fairly similar at 2, 24, and 48 h post-reperfusion. For area calculation, the lesion is manually outlined. DWI, diffusion-weighted imaging; MCAO, middle cerebral artery occlusion.

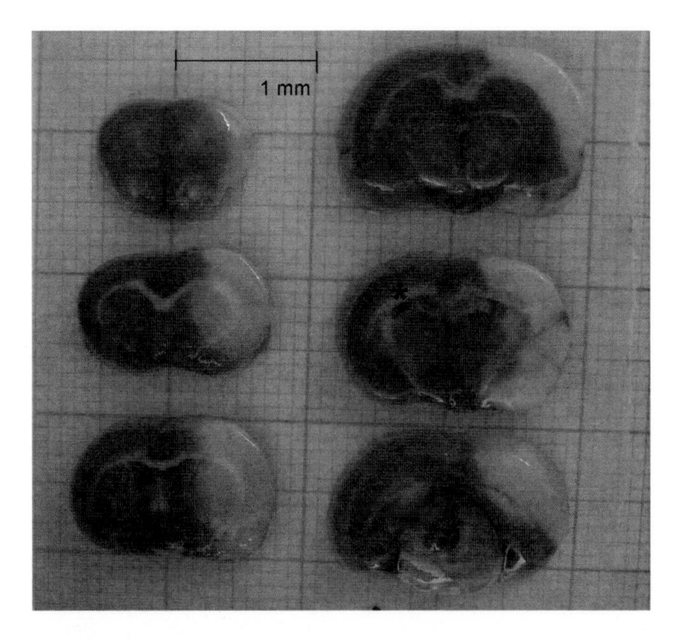

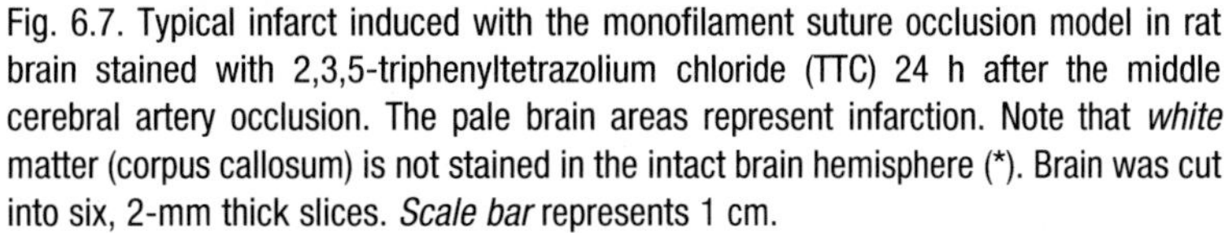
Fig. 6.7. Typical infarct induced with the monofilament suture occlusion model in rat brain stained with 2,3,5-triphenyltetrazolium chloride (TTC) 24 h after the middle cerebral artery occlusion. The pale brain areas represent infarction. Note that *white* matter (corpus callosum) is not stained in the intact brain hemisphere (*). Brain was cut into six, 2-mm thick slices. *Scale bar* represents 1 cm.

mitochondria (especially by succinate dehydrogenase) to stain the intact brain regions dark (formazan) red. Infarcted regions with damaged, non-functioning mitochondria do not stain and remain white (**Fig. 6.7**), which allows easy demarcation of ischemic lesions from the intact brain areas.

1. Ischemic lesions on MR images are manually outlined by using ParaVision software (Bruker BioSpin, Ettlingen, Germany).
2. Uncorrected lesion volume is the sum of all lesion areas multiplied by slice thickness.
3. Since enlargement of infarcted tissue by edema results in overestimation of infarct volume, a corrected infarct volume is calculated to compensate for the effect of brain edema. The corrected infarct area in a slice is calculated by subtracting the area of normal tissue in the ipsilateral hemisphere from the total area of the contralateral hemisphere.
4. Multiplying the corrected infarct area by the slice thickness and summing the areas from all the slices give the corrected infarct volume (3).
5. The percentage of the hemisphere affected by ischemia can be calculated by the following equation: {[total lesion volume − (ipsilateral hemisphere volume − contralateral hemisphere volume)] / contralateral hemisphere volume} × 100 (4).

6. Percentage of brain swelling is derived from volumetric growth of the ischemic hemisphere compared to the intact hemisphere and is calculated by the following equation: [(right hemisphere volume/left hemisphere volume) − 1] × 100 (5).

4. Notes

1. Among rat strains, Wistar and Sprague-Dawley rats are the most commonly used in stroke studies due to reproducible and desirable size of infarct volumes achieved by the intraluminal suture MCAO model described here. However, certain rat strains (i.e., spontaneously hypertensive rats and Fisher-344) may be more sensitive to MCAO and exhibit more extensive infarct volumes (6–8). Surprisingly, within the same strain (e.g., Sprague-Dawley rats), vendor differences may greatly affect ischemia susceptibility resulting in up to a three-fold increase in ischemic lesion volume (9). Strain differences can also affect the amount of benefit from reperfusion or neuroprotective drugs. Sprague-Dawley rats respond better to reperfusion than Wistar rats (10), and some neuroprotective drugs (e.g., calcium antagonist isradipine) reduce infarct size to a greater extent in Wistar rats than in Sprague-Dawley rats (8), while others (e.g., non-competitive *N*-methyl-D-aspartate antagonist dizocilpine, MK-801) act in the opposite way (7, 8).

   Gerbils are not recommended for preclinical stroke studies because many pharmacological agents show protection in gerbils, but not in other species (11).

   Female rats sustain smaller infarcts after MCAO than male rats, but this gender influence on infarct size is lost when female animals are ovariectomized at an early age (12). Moreover, recent findings suggest a sex-related dichotomy in the apoptotic response to ischemia; it appears that ischemia activates mainly the caspase-dependent pathway of apoptotic cell death in females, but primarily the caspase-independent pathway in males (13–15).

   To conclude, the choice of gender and strain of the animal used for focal ischemia studies may greatly affect the results, and the model used in each laboratory must be validated.

2. Simple measures of motor function are available in rodent models and such evaluations can be completed within a few minutes (16–18). Tests to examine the effects of focal stroke on more refined sensorimotor function include limb placing, beam walking, grid walking, rotarod, the sticky label test, and

the staircase test (19). The SIRPA procedure (20), which lasts only several minutes, can be used as a composite behavioral end-point. Behavioral assessment is very sensitive for detecting efficacy of a neuroprotective drug (21, 22).

3. Systemic anesthesia with the combination of ketamine and medetomidine is easy to apply by injection, is well tolerated by the animals, is inexpensive, and does not require artificial ventilation. This method provides 60 to 90 min of deep anesthesia that is usually enough for surgical procedures. Additionally, catalepsy-like immobilization lasts 2 to 3 h, which enables long-lasting MRI. When better control of the duration and depth of the anesthesia is required, the choice should be inhalation anesthetics (23), such as halothane plus nitrous oxide (via a face-mask in spontaneously breathing rats or endotracheal tube under mechanical ventilation) or isoflurane (via face-mask). It is still debatable whether commonly used anesthetics possess a degree of neuroprotection (24). Whichever method or anesthetic is used, giving exactly the same anesthetics to all animals in all groups will avoid intergroup bias.
4. Even small differences in brain temperature during ischemia affect the extent of neuronal injury (25), and mild to moderate brain hypothermia confers marked neuroprotection. Temperature can be monitored in various ways: (1) direct brain temperature measurement (with fine thermocouples inserted into the brain parenchyma by a microsurgical technique); (2) indirect brain temperature measurement (by a probe placed in the epidural space or under the temporal muscle); and (3) body temperature monitoring by rectal temperature recording. Rectal, subtemporal muscle, and brain temperatures correlate well in rats (26). Because measurement of body temperature by a rectal probe is the easiest of these methods, it has been widely accepted and used. To accurately measure rectal temperature, the probe should be inserted to a depth of several centimeters, whereas lesser insertion may underestimate the temperature. Spontaneous hyperthermia is a complication of long-lasting MCAO and appears to be associated with hypothalamic injury (27). When reperfusion follows ≥ 2 h after MCAO, presence of a fanning device to counteract hyperthermia is required in the laboratory. Hypothermia can be a side effect of anesthesia (low metabolic rate), especially if the animal has a severe neurological sequelae and is unable to move. Heating blankets and lamps are used to treat hypothermia.
5. Femoral artery cannulation is more difficult than femoral vein cannulation and requires more attention and slower application. The difficulty arises from the narrower lumen and higher resistance of the artery. Arterial cannula should have a slightly

sharpened tip, whereas a blunt tip is inserted easily into the femoral vein. Both arterial and venous catheters must be filled with heparinized saline (20–24 U/mL) before insertion and flushed occasionally with small amounts of same solution (200–300 μL) to maintain catheter patency after insertion.

Limb ischemia may follow arterial catheterization due to insufficient collateral circulation. Repair of arteriotomy is technically difficult and time-consuming; thus it is not worthwhile when the experimental duration is short. Limb ischemia can easily be prevented by inserting the catheter more distally to the profunda artery. Performing the femoral vein occlusion before the femoral artery cannulation may enrich the collateral circulation. Additionally, avoiding manipulation of the arterial wall and removing the blood extravasated during the surgery can prevent vasospasm, a potential reason for limb ischemia in subacute and later periods (28).

In some experiments, femoral vein cannulation is not required and systemic injections can easily be made via the tail vein. Tail vein injection is the intravenous injection method best tolerated by small animals. The technique is considered difficult by beginners, but is easy to apply once it is learned. The success lies in the degree of tail vein dilation by heating and/or proximal compression. It is much harder to perform rapid tail vein injection in rats than in mice due to the thick tail skin and the higher injection pressure required to overcome vascular resistance. In mice, retroorbital (retrobulbar plexus) injection is an alternative method that is easy to apply, but it is not recommended for multiple injections because severe tissue damage is unavoidable.

6. Intraluminal suture occlusion of MCA involves a simple surgery which lasts 10 to 15 min when performed by an experienced surgeon. This simplicity reduces potential intersurgeon variability, with a good reproducibility even among investigators with varying degrees of experience (29).

   When isolating the CCA, care should be taken not to touch or damage the vagal nerve, which runs next to the carotid artery. Intraluminal suture occlusion by Koizumi's method requires CCA ligation (1), therefore the depth of ischemia is increased compared to Longa's method, in which the occluder is introduced into the external carotid artery while the CCA is left intact (16). The suture must be advanced gently into the ICA. Insertion length in rats is approximately 17–19 mm and in mice 9 to 11 mm from the carotid bifurcation. Rat MCA occluders can be both 3–0 and 4–0 nylon sutures, either coated or uncoated. For MCA occlusions in mice, 5–0 or 6–0 nylon sutures with a blunted tip are most commonly used (30–32).

Resistance felt before the full length of the occluder has been inserted can indicate a narrow canal ostium on the skull base or narrowing of the ICA at the branching point of vessels other than MCA (i.e., hypothalamic artery, anterior choroidal artery) (33, 34). In either case, no infarct will be seen in the MCA territory. Advancing the suture against the resistance inevitably results in vessel rupture and subarachnoid hemorrhage, and is, indeed, a suitable model for subarachnoid hemorrhage (35).

In some cases, although suture is advanced to an appropriate length expected to occlude MCA, no resistance is felt. This indicates that suture lies in the lumen of the ACA and further insertion may rupture the ACA, with subsequent subarachnoid hemorrhage. Therefore, resistance should not be the only parameter determining the location of the suture tip, and the inserted suture length must be carefully examined during the procedure. Guidance by laser-Doppler flow measurements increases the rate of successful MCA occlusion and decreases the rate of vessel rupture (36).

Another important factor affecting the outcome of MCAO is the suture itself. Silicone-coated sutures induce larger and more consistent infarcts compared to uncoated sutures (37, 38), with the penumbral region preserved for a longer time (39). Filament diameter correlates well with the size of the infarct (40), and there are continuous efforts to optimize suture design and the methodology of the intraluminal suture MCAO technique (36, 41, 42).

7. Because glycemia differences may affect the susceptibility of the brain to ischemia, monitoring of blood sugar levels during experiments is a crucial factor (43). Some researchers suggest pre-experimental fasting.
8. These parameters should be closely monitored, particularly in artificially ventilated animals because depth of anesthesia and artificial respiratory rate represent additional sources of variability (44). Most arterial gas measurement devices deliver sodium, potassium, calcium, and glucose values from the same blood sample.
9. Fall of mean arterial pressure below 70 to 80 mmHg in rats may exacerbate the ischemic lesion and must be corrected with intravenous saline infusion.
10. Guidance by laser-Doppler flowmetry ensures appropriate MCAO (>75% drop of baseline value) and reperfusion (recovery to >40–50% of baseline value 5–10 min after reperfusion). A severe and sustained drop in flow after MCAO may result from subarachnoid hemorrhage (45). If no continuous monitoring is available, a single measurement immediately

after MCAO may detect decreased flow due to reflex vasoconstriction instead of a real occlusion, hence, the measurement must be repeated in several minutes.

11. Induction of brain ischemia in an MRI scanner is called "in-bore" occlusion. The in-bore version of the suture occlusion method is easy to perform, is almost as successful as MCAO induced outside of the MRI scanner, and reperfusion is available in-bore (46). Rectal temperature monitoring coupled with a feed-back regulated, MRI-compatible heating method (a blanket including circulating warm water) is necessary to sustain normal body temperature during imaging. Unless the animal has a severe neurological deficit that compromises respiratory capacity, MRI is safe and allows up to 2 to 3 h of continuous data collection with a single dose of anesthesia.

    Diffusion-weighted sequencing is able to detect and fully depict ischemic lesions in the hyperacute phase of ischemia. Lesions depicted with MRI correlate highly with the location and volume of pathologically confirmed cerebral infarctions (47). Edema leads to a space-occupying effect and a strong horizontal displacement of midline structures on the MR images. Gerriets et al. (48) showed a good correlation between MRI- and TTC staining-based hemispheric lesion volumes with edema correction.

12. Cardiac perfusion should be performed under approximately 100 mmHg pressure. Higher perfusion pressures can burst nasal capillaries and fluid will leak from the nose. The needle should be placed through the apex of the heart into the proximal aorta. Heart beating at the beginning of the perfusion helps cold saline to circulate through the vasculature. As perfusion progresses, the animal's tissue and eye color will fade. Inappropriate perfusion will leave blood within the vasculature and interfere with tissue evaluations. When clear fluid starts flowing out of the punctured right atrium, cardiac perfusion can be terminated.

13. Fixation prevents tissue degradation and preserves both structure and tissue antigens by forming cross-linking bonds between tissue components. For light and fluorescence microscopy, brain slices can also be immersion fixed in 2 to 4% paraformaldehyde at 4°C for 24 h. Perfusion fixation (fixative is infused instead of saline during transcardiac perfusion) is superior to immersion fixation in most instances (49, 50) and preferable for morphological and immunocytochemistry studies (51).

14. TTC staining intensity depends on the TTC concentration used (52), but even 0.05% TTC solution delineates ischemic lesions as well as 1% TTC (53). TTC reliably detects infarcted tissue 24 h after MCAO, and data support broader use of

TTC staining from 6 to 72 h after MCAO (54–57). The reliability of TTC staining within 6 h has not been extensively studied, but one study found that TTC staining up to 4 h after ischemic insult underestimates the lesion size (58), likely because not enough mitochondria are damaged to create sharp contrast between normal and injured tissue. After 72 h, pathophysiologic inflammatory responses (e.g., leukocytes with intact mitochondria migrate into the infarcted region) may obscure the TTC staining demarcation line at the periphery of the infarction. When necessary, transcardiac TTC perfusion can be an alternative to histological examination up to 4 h after permanent ischemia (59). Transcardiac TTC perfusion is occasionally used in rodent stroke studies (60–62) and was shown to be equally reliable as immersion TTC in primates (55). When combined with fixation (i.e., paraformaldehyde infusion) (59), this method may save considerable time.

A disadvantage of TTC staining arises from the fact that white matter contains a low density of mitochondria, resulting in false-positive, pale staining of normal white matter (including the corpus callosum) that is difficult to distinguish from infarcted tissue (**Fig. 6.7**). Observations from our laboratory suggest that mild ischemia (45 min or shorter durations of MCAO) results in less pale (pinkish) infarctions 24 h after MCAO that are slightly more difficult to outline manually. The explanation might be that mild ischemia induces mainly apoptosis-mediated cell death and the evolution of infarction is slower and yet incomplete at 24 h after ischemic insult (63).

Hematoxylin–eosin staining is more time-consuming and expensive than TTC, but is widely used as the gold-standard histological method to verify infarction in both early and delayed periods of focal ischemia (64).

Acknowledgments

This work was supported in part by the Helsinki University Central Hospital, the Finnish Academy of Sciences, and the European Union (grant no: FP7 202213).

References

1. Koizumi, J, Yoshida, Y, Nakazawa, T, et al. (1986) Experimental studies of ischemic brain edema: 1. A new experimental model of cerebral embolism in rats in which recirculation can be introduced in the ischemic area. *Jpn J Stroke* **8**, 1–8.
2. Takano, K, Latour, LL, Formato, JE, et al. (1996) The role of spreading depression in

focal ischemia evaluated by diffusion mapping. *Ann Neurol* **39**, 308–318.

3. Tatlisumak, T, Takano, K, Carano, RA, et al. (1998) Delayed treatment with an adenosine kinase inhibitor, GP683, attenuates infarct size in rats with temporary middle cerebral artery occlusion. *Stroke* **29**, 1952–1958.
4. Tatlisumak, T, Carano, RA, Takano, K, et al. (1998) A novel endothelin antagonist, A-127722, attenuates ischemic lesion size in rats with temporary middle cerebral artery occlusion: a diffusion and perfusion MRI study. *Stroke* **29**, 850–857.
5. Strbian, D, Karjalainen-Lindsberg, ML, Tatlisumak, T et al. (2006) Cerebral mast cells regulate early ischemic brain swelling and neutrophil accumulation. *J Cereb Blood Flow Metab* **26**, 605–612.
6. Duverger, D, MacKenzie, ET. (1988) The quantification of cerebral infarction following focal ischemia in the rat: influence of strain, arterial pressure, blood glucose concentration, and age. *J Cereb Blood Flow Metab* **8**, 449–461.
7. Oliff, HS, Marek, P, Miyazaki, B, et al. (1996) The neuroprotective efficacy of MK-801 in focal cerebral ischemia varies with rat strain and vendor. *Brain Res* **731**, 208–212.
8. Sauter, A, Rudin, M. (1995) Strain-dependent drug effects in rat middle cerebral artery occlusion model of stroke. *J Pharmacol Exp Ther* **274**, 1008–1013.
9. Oliff, HS, Weber, E, Eilon, G, et al. (1995) The role of strain/vendor differences on the outcome of focal ischemia induced by intraluminal middle cerebral artery occlusion in the rat. *Brain Res* **675**, 20–26.
10. Walberer, M, Stolz, E, Muller, C, et al. (2006) Experimental stroke: ischaemic lesion volume and oedema formation differ among rat strains (a comparison between Wistar and Sprague-Dawley rats using MRI). *Lab Anim* **40**, 1–8.
11. Stroke Therapy Academic Industry Roundtable. (1999) Recommendations for standards regarding preclinical neuroprotective and restorative drug development. *Stroke* **30**, 2752–2758.
12. Alkayed, NJ, Harukuni, I, Kimes, AS, et al. (1998) Gender-linked brain injury in experimental stroke. *Stroke* **29**, 159–165.
13. McCullough, LD, Zeng, Z, Blizzard, KK, et al. (2005) Ischemic nitric oxide and poly (ADP-ribose) polymerase-1 in cerebral ischemia: male toxicity, female protection. *J Cereb Blood Flow Metab* **25**, 502–512.
14. Du, L, Bayir, H, Lai, Y, et al. (2004) Innate gender-based proclivity in response to cytotoxicity and programmed cell death pathway. *J Biol Chem* **279**, 38563–38570.
15. Renolleau, S, Fau, S, Goyenvalle, C, et al. (2007) Specific caspase inhibitor Q-VD-OPh prevents neonatal stroke in P7 rat: a role for gender. *J Neurochem* **100**, 1062–1071.
16. Longa, EZ, Weinstein, PR, Carlson, S, et al. (1989) Reversible middle cerebral artery occlusion without craniectomy in rats. *Stroke* **20**, 84–91.
17. Menzies, SA, Hoff, JT, Betz, AL. (1992) Middle cerebral artery occlusion in rats: a neurological and pathological evaluation of a reproducible model. *Neurosurgery* **31**, 100–106.
18. Bederson, JB, Pitts, LH, Tsuji, M, et al. (1986) Rat middle cerebral artery occlusion: evaluation of the model and development of a neurologic examination. *Stroke* **17**, 472–476.
19. Hunter, AJ, Hatcher, J, Virley, D, et al. (2000) Functional assessments in mice and rats after focal stroke. *Neuropharmacology* **39**, 806–816.
20. Rogers, DC, Fisher, EM, Brown, SD, et al. (1997) Behavioral and functional analysis of mouse phenotype: SHIRPA, a proposed protocol for comprehensive phenotype assessment. *Mamm Genome* **8**, 711–713.
21. Yamaguchi, T, Suzuki, M, Yamamoto, M. (1995) YM796, a novel muscarinic agonist, improves the impairment of learning behavior in a rat model of chronic focal cerebral ischemia. *Brain Res* **669**, 107–114.
22. Kawamata, T, Alexis, NE, Dietrich, WD. (1996) Intracisternal basic fibroblast growth factor (bFGF) enhances behavioral recovery following focal cerebral infarction in the rat. *J Cereb Blood Flow Metab* **16**, 542–547.
23. Zausinger, S, Baethmann, A, Schmid-Elsaesser, R. (2002) Anesthetic methods in rats determine outcome after experimental focal cerebral ischemia: mechanical ventilation is required to obtain controlled experimental conditions. *Brain Res Brain Res Protoc* **9**, 112–121.
24. Kirsch, JR, Traystman, RJ, Hurn, PD. (1996) Anesthetics and cerebroprotection: experimental aspects. *Int Anesthesiol Clin* **34**, 73–93.
25. Busto, R, Dietrich, WD, Globus, MY. (1987) Small differences in intraischemic brain temperature critically determine the

extent of ischemic neuronal injury. *J Cereb Blood Flow Metab* 7, 729–738.

26. Hasegawa, Y, Latour, LL, Sotak, CH. (1994) Temperature dependent change of apparent diffusion coefficient of water in normal and ischemic brain of rats. *J Cereb Blood Flow Metab* **14**, 383–390.
27. Li, F, Omae, T, Fisher, M. (1999) Spontaneous hyperthermia and its mechanism in the intraluminal suture middle cerebral artery occlusion model of rats. *Stroke* **30**, 2464–2470.
28. Choi, JM, Shin, YW, Hong, KW. (2001) Rebamipide prevents periarterial blood-induced vasospasm in the rat femoral artery model. *Pharmacol Res* **43**, 489–496.
29. Takano, K, Tatlisumak, T, Bergmann, AG. (1997) Reproducibility and reliability of middle cerebral artery occlusion using a silicone-coated suture (Koizumi) in rats. *J Neurol Sci* **153**, 8–11.
30. Clark, WM, Lessov, NS, Dixon, MP. (1997) Monofilament intraluminal middle cerebral artery occlusion in the mouse. *Neurol Res* **19**, 641–648.
31. Ardehali, MR, Rondouin, G. (2003) Microsurgical intraluminal middle cerebral artery occlusion model in rodents. *Acta Neurol Scand* **107**, 267–275.
32. Li, J, Henman, MC, Doyle, KM. (2004) The pre-ischaemic neuroprotective effect of a novel polyamine antagonist, N1-dansyl-spermine in a permanent focal cerebral ischaemia model in mice. *Brain Res* **1029**, 84–92.
33. He, Z, Yamawaki, T, Yang, S. (1999) Experimental model of small deep infarcts involving the hypothalamus in rats: changes in body temperature and postural reflex. *Stroke* **30**, 2743–2751.
34. He, Z, Yang, SH, Naritomi, H. (2000) Definition of the anterior choroidal artery territory in rats using intraluminal occluding technique. *J Neurol Sci* **182**, 16–28.
35. Strbian, D, Durukan, A, Tatlisumak, T. (2008) Rodent models of hemorrhagic stroke. *Curr Pharm Des* **14**, 352–358.
36. Schmid-Elsaesser, R, Zausinger, S, Hungerhuber, E. (1998) A critical reevaluation of the intraluminal thread model of focal cerebral ischemia: evidence of inadvertent premature reperfusion and subarachnoid hemorrhage in rats by laser-Doppler flowmetry. *Stroke* **29**, 2162–2170.
37. Laing, RJ, Jakubowski, J, Laing, RW. (1993) Middle cerebral artery occlusion without craniectomy in rats. Which method works best? *Stroke* **24**, 294–297.
38. Shimamura, N, Matchett, G, Tsubokawa, T. (2006) Comparison of silicon-coated nylon suture to plain nylon suture in the rat middle cerebral artery occlusion model. *J Neurosci Methods* **156**, 161–165.
39. Bouley, J, Fisher, M, Henninger, N. (2007) Comparison between coated vs. uncoated suture middle cerebral artery occlusion in the rat as assessed by perfusion/diffusion weighted imaging. *Neurosci Lett* **412**, 185–190.
40. Abraham, H, Somogyvari-Vigh, A, Maderdrut, JL. (2002) Filament size influences temperature changes and brain damage following middle cerebral artery occlusion in rats. *Exp Brain Res* **142**, 131–138.
41. Gerriets, T, Stolz, E, Walberer, M. (2004) Complications and pitfalls in rat stroke models for middle cerebral artery occlusion: a comparison between the suture and the macrosphere model using magnetic resonance angiography. *Stroke* **35**, 2372–2377.
42. Ma, J, Zhao, L, Nowak, TS, Jr. (2006) Selective, reversible occlusion of the middle cerebral artery in rats by an intraluminal approach Optimized filament design and methodology. *J Neurosci Methods* **156**, 76–83.
43. Ginsberg, MD, Busto, R. (1989) Rodent models of cerebral ischemia. *Stroke* **20**, 1627–1642.
44. Macrae, I. (1992) New models of focal cerebral ischaemia. *Br J Clin Pharmacol* **34**, 302–308.
45. Woitzik, J, Schilling, L. (2002) Control of completeness and immediate detection of bleeding by a single laser-Doppler flow probe during intravascular middle cerebral artery occlusion in rats. *J Neurosci Methods* **122**, 75–78.
46. Li, F, Han, S, Tatlisumak, T. (1998) A new method to improve in-bore middle cerebral artery occlusion in rats: demonstration with diffusion- and perfusion-weighted imaging. *Stroke* **29**, 1715–1719.
47. Mack, WJ, Komotar, RJ, Mocco, J, et al. (2003) Serial magnetic resonance imaging in experimental primate stroke: validation of MRI for pre-clinical cerebroprotective trials. *Neurol Res* **25**, 846–852.
48. Gerriets, T, Stolz, E, Walberer, M, et al. (2004) Noninvasive quantification of brain edema and the space-occupying effect in rat stroke models using magnetic resonance imaging. *Stroke* **35**, 566–571.

49. Palay, SL, Mc, G-RS, Gordon, S, Jr. (1962) Fixation of neural tissues for electron microscopy by perfusion with solutions of osmium tetroxide. *J Cell Biol* **12**, 385–410.
50. Adickes, ED, Folkerth, RD, Sims, KL. (1997) Use of perfusion fixation for improved neuropathologic examination. *Arch Pathol Lab Med* **121**, 1199–1206.
51. Liesi, P. (2006) Methods for analyzing brain tissue, In Handbook of Experimental Neurology. Tatlisumak, T, and Fisher, M, (eds.), Cambridge University Press, Cambridge, pp. 173–180.
52. Mathews, KS, McLaughlin, DP, Ziabari, LH, et al. (2000) Rapid quantification of ischaemic injury and cerebroprotection in brain slices using densitometric assessment of 2,3,5-triphenyltetrazolium chloride staining. *J Neurosci Methods* **102**, 43–51.
53. Joshi, CN, Jain, SK, Murthy, PS. (2004) An optimized triphenyltetrazolium chloride method for identification of cerebral infarcts. *Brain Res Protoc* **13**, 11–17.
54. Bederson, JB, Pitts, LH, Germano, SM, et al. (1986) Evaluation of 2,3,5-triphenyltetrazolium chloride as a stain for detection and quantification of experimental cerebral infarction in rats. *Stroke* **17**, 1304–1308.
55. Dettmers, C, Hartmann, A, Rommel, T, et al. (1994) Immersion and perfusion staining with 2,3,5-triphenyltetrazolium chloride (TTC) compared to mitochondrial enzymes 6 hours after MCA-occlusion in primates. *Neurol Res* **16**, 205–208.
56. Li, F, Irie, K, Anwer, MS, et al. (1997) Delayed triphenyltetrazolium chloride staining remains useful for evaluating cerebral infarct volume in a rat stroke model. *J Cereb Blood Flow Metab* **17**, 1132–1135.
57. Okuno, S, Nakase, H, Sakaki, T. (2001) Comparative study of 2, 3, 5-triphenyltetrazolium chloride (TTC) and hematoxylin-eosin staining for quantification of early brain ischemic injury in cats. *Neurol Res* **23**, 657–661.
58. Benedek, A, Moricz, K, Juranyi, Z. (2006) Use of TTC staining for the evaluation of tissue injury in the early phases of reperfusion after focal cerebral ischemia in rats. *Brain Res* **1116**, 159–165.
59. Park, CK, Mendelow, AD, Graham, DI. (1988) Correlation of triphenyltetrazolium chloride perfusion staining with conventional neurohistology in the detection of early brain ischaemia. *Neuropathol Appl Neurobiol* **14**, 289–298.
60. Chan, YC, Wang, MF, Chen, YC, et al. (2003) Long-term administration of Polygonum multiflorum Thunb. reduces cerebral ischemia-induced infarct volume in gerbils. *Am J Chin Med* **31**, 71–77.
61. Yang, DY, Wang, MF, Chen, IL, et al. (2001) Systemic administration of a water-soluble hexasulfonated C(60) (FC(4)S) reduces cerebral ischemia-induced infarct volume in gerbils. *Neurosci Lett* **311**, 121–124.
62. Izumi, Y, Roussel, S, Pinard, E, et al. (1991) Reduction of infarct volume by magnesium after middle cerebral artery occlusion in rats. *J Cereb Blood Flow Metab* **11**, 1025–1030.
63. Lee, SH, Kim, M, Kim, YJ, et al. (2002) Ischemic intensity influences the distribution of delayed infarction and apoptotic cell death following transient focal cerebral ischemia in rats. *Brain Res* **956**, 14–23.
64. Garcia, JH, Liu, KF, Ho, KL. (1995) Neuronal necrosis after middle cerebral artery occlusion in Wistar rats progresses at different time intervals in the caudoputamen and the cortex. *Stroke* **26**, 636–642.

Chapter 7

Mouse Surgical Models in Cardiovascular Research

Oleg Tarnavski

Abstract

Mouse models that mimic human diseases are important tools for investigating underlying mechanisms in many disease states. Although the demand for these models is high, there are few schools or courses available for surgeons to obtain the necessary skills. Researchers are usually exposed to brief descriptions of the procedures in scientific journals, which they then attempt to reproduce by trial and error. This often leads to a number of mistakes and unnecessary loss of animals. This chapter provides comprehensive details of three major surgical procedures currently employed in cardiovascular research: aortic constriction (of both ascending and transverse portions), pulmonary artery banding, and myocardial infarction (including ischemia–reperfusion). It guides the reader through the entire procedure, from the preparation of the animal for surgery until its full recovery, and includes a list of all necessary tools and devices. Due consideration has been given to the pitfalls and possible complications in the course of surgery. Adhering to our recommendations should improve reproducibility of the models and bring the number of the animal subjects to the minimum.

Key words: Cardiac hypertrophy, heart failure, surgical procedure.

1. Introduction

The preferred rodent species for conducting cardiovascular research is shifting from the rat to the mouse. The main advantage of using the rat is its bigger size ($\sim 10\ \times$ bigger than a mouse). However, the appearance of more sophisticated microdissecting microscopes and other imaging devices, high-caliber microsurgical instruments, miniature catheters for hemodynamic measurements, etc., has made microsurgery in the mouse as feasible as in the rat (1). Mouse models have gained popularity for a number of reasons, including small size, rapid gestation age (21 days), relatively low maintenance costs, convenience in housing and handling, and less compound requirements for pharmacological studies.

K. DiPetrillo (ed.), *Cardiovascular Genomics*, Methods in Molecular Biology 573,
DOI 10.1007/978-1-60761-247-6_7, © Humana Press, a part of Springer Science+Business Media, LLC 2009

Moreover, the mouse genome has been extensively characterized and gene-targeted (i.e., knockout) and transgenic overexpression experiments are more commonly performed using mice rather than rats (1). It is now clear that similar pathways regulate the development of the heart and vasculature in mice and humans (2).

Although the demand for animal models in cardiovascular research is extremely high, there are few schools or courses that specifically provide training in surgical procedures entailing mice. Also, scientific publications rarely share details of surgical procedures. Investigators are left to their own devices in this regard and often need to work out the surgical details by trial and error. This inevitably leads to a number of mistakes and unnecessary loss of animals.

This chapter provides comprehensive details of the techniques required to perform cardiac surgery in the mouse, guiding the reader through the whole procedure, from the initial handling of the animal for anesthesia to its full recovery after surgery. Specifically, a detailed description of the intubation of the mouse and three major surgical procedures used in cardiovascular research have been provided: aortic constriction (left ventricular pressure-overload model), pulmonary artery constriction (right ventricular pressure-overload model), and myocardial infarction (including ischemia–reperfusion model). Emphasis has been placed on technical procedures and not on scientific results per se with the techniques. A number of helpful hints are presented to facilitate the researcher's success in surgery. These are complimented by the inclusion of pictures and detailed descriptions of all recommended equipment and devices needed for surgery.

Aortic banding (constriction) is an established method to induce left ventricular hypertrophy in mice. The condition develops from the elevated blood pressure in the left ventricular chamber and mimics similar syndromes in humans which occur in aortic stenosis or as a consequence of chronic systemic hypertension. A band can be placed on either the ascending, transverse, or descending portion of the aorta depending on the study design. In this chapter, only the ascending (AAC) and transverse (TAC) aortic constrictions will be described as they represent the most popular methods for effecting left ventricular hypertrophy. Both methods have been described (with some modifications) in a number of papers (3–7), but few technical details are provided. The main difference between these two models is determined by the location of the occlusion and hence the regional anatomy involved. In the AAC model, a ligature is placed around the aorta close to its origin. This leads to rather high pressure build-up in the left ventricular chamber, with subsequent functional and morphological changes in the heart over the course of several hours to days. In the TAC model, a ligature is placed around the aortic arch between the innominate and left carotid arteries. In this case, the innominate artery acts as a shunt which relieves the pressure in the left ventricle. The resulting hypertrophy develops more gradually compared to the AAC model. The choice of method should be determined by the study objectives.

The pulmonary artery banding model mimics pulmonary coarctation (pulmonary stenosis) in humans, which is usually a congenital syndrome. It may be present as an isolated form, but is commonly associated with other cardiovascular abnormalities, such as tetralogy of Fallot. It also may be seen as a sequelae of congenital rubella, Williams', Noonan's, or Alagille's syndrome (8).

In cardiovascular research, pulmonary artery banding has been used as a model of right ventricle hypertrophy and failure (9). In small animals, the pulmonary banding model has not been as widely used in research as the aortic banding model, presumably reflecting the lower prevalence and therefore investigative interest in right ventricular hypertrophy relative to left ventricular hypertrophy and failure (10), as well as the complexities involved in the surgical procedure itself.

Myocardial infarction (MI) remains one of the leading causes of fatalities worldwide. In animal models this condition has been mimicked by ligation of the left coronary artery. Reperfusion of the ischemic heart has become an important therapeutic intervention for MI. However, the process of reperfusion, although beneficial in terms of myocardial salvage, also results in cell death and scar formation due to a process known as reperfusion injury (2). The underlying mechanisms involved in injury following ischemia–reperfusion are yet not well understood.

The murine model of myocardial infarction and ischemia–reperfusion has been widely described in the literature (11–16). However, techniques for correctly visualizing the left anterior descending (LAD) coronary artery and properly ligating it can be rather challenging. In this chapter, technical details which should aid in the success and reproducibility of this surgical model are provided.

This chapter is intended to be a useful reference for the uninitiated researcher to avoid pitfalls, reproducibly perform surgeries, and reduce animal usage. Experts may still gain some new insight into surgical techniques that they may not have considered previously.

2. Materials

2.1. Mice

Mice of any strain can be used for the surgical procedures described in this text. The strains most often used include FVB/N and C57BL/6. Surgical procedures may be performed on transgenic, knock-out, and wild-type mice. Adult, male mice at 11–12 weeks of age are preferred since the developmental growth of the heart is complete at this time. The weight of mice could range between 25 and 30 g. Female mice can also be used, but are not preferred (*see* **Note 1**).

2.2. Surgical Tools

For all of the operations described below, six basic surgical tools are necessary.

1. Curved forceps (Roboz, cat #RS-5228) used for intubation (the angle is slightly increased to 45%).
2. Slightly curved forceps (Roboz, cat #RS-5136).
3. Foerster curved forceps (Roboz, cat #RS-5101), curvature is increased to 90%.
4. Scissors with large rings (Roboz, cat #RS-5850).
5. Chest retractor (Roboz, cat #RS-6510), the two proximal pairs of teeth (closest to the junction) should be removed with the other two pairs left in place.
6. Needle holder (Roboz, cat #RS-6410).

Occasionally some other tools are needed, these are described further in the text.

2.3. Sutures and Needles

1. 6-0 Nylon suture (Ethicon, cat #1859G) for the chest cavity and skin closure.
2. 6-0 Vicryl (absorbable suture, Ethicon, cat #J499G) for suturing the muscles.
3. 7-0 Silk (Ethicon, cat #K809H) for manipulations on the vessels.
4. BV-1 tapered needle with 7-0 silk suture (Ethicon, cat #K809H) is used for ligating the coronary artery.

2.4. Special Devices

1. Device to hold the needle for banding procedures: two metal rods (Harvard Apparatus, cat #AH 50-4415 and #AH 50-4407) connected by a two-ball joint connector (Harvard Apparatus, cat #AH 50-4431) (**Fig. 7.1**).
2. L-shaped blunted needle (modified 1-ml insulin syringe with 28G needle) to create a passageway underneath the aortic arch and pulmonary trunk.
3. "Wire and snare" device to place the suture underneath the aortic arch and pulmonary trunk (*see* **Note 12**).

2.5. Respiratory Support, Heating Systems

1. Mouse volume-controlled ventilator 687 series (Harvard Apparatus).
2. Heating systems (homeothermal blankets from Stoelting, cat #50300) needed to maintain the animal body temperature in the physiological range for the ischemia–reperfusion procedure.
3. 60-W lamp or a regular pharmacy heating pad at a low setting to maintain the body temperature at recovery.

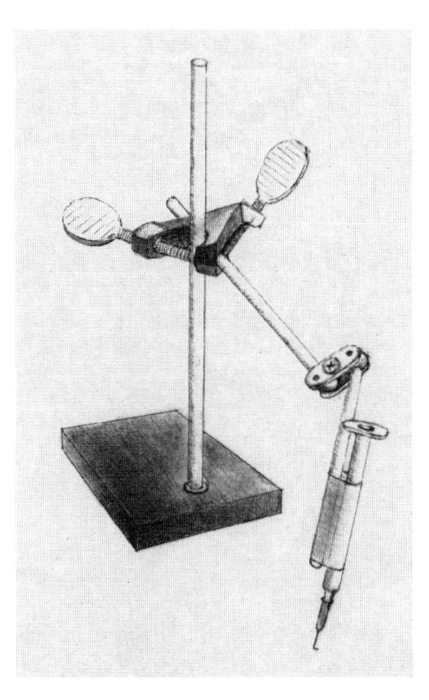

Fig. 7.1. Construct to hold needle for the pressure-overload model. Figure has been previously published (22).

2.6. Optical Device and Illumination

1. Binocular lens (SCS Limited, cat #H10L1.75 ×) with a 1.75 magnification.
2. Power light with flexible horns (MVI, cat #MVI-DGN).

2.7. Hair Removal

1. Animal hair clipper (Moser by Wahl, mini ARCO) to shave the chest.

3. Methods

3.1. Anesthesia and Analgesia, Intubation and Ventilation, Presurgical Preparation

1. Anesthesia with pentobarbital (Nembutal) 70 mg/kg in a solution of 15 mg/ml; presurgical analgesia is not recommended (*see* **Note 2**).
2. Shave the chest with the hair clipper.
3. Position the animal on the surgical stage for the subsequent intubation, thread 3-0 silk behind the front incisors, pull taut, and fix with tape (*see* **Note 3**, **Fig. 7.2A**).
4. Intubate the animal with 20-ga catheter, attached to the extender (*see* **Note 4**, **Fig. 7.2B**).
5. Connect the intubation tube to the ventilator (687 series, Harvard Apparatus), start ventilation (*see* **Note 5**).

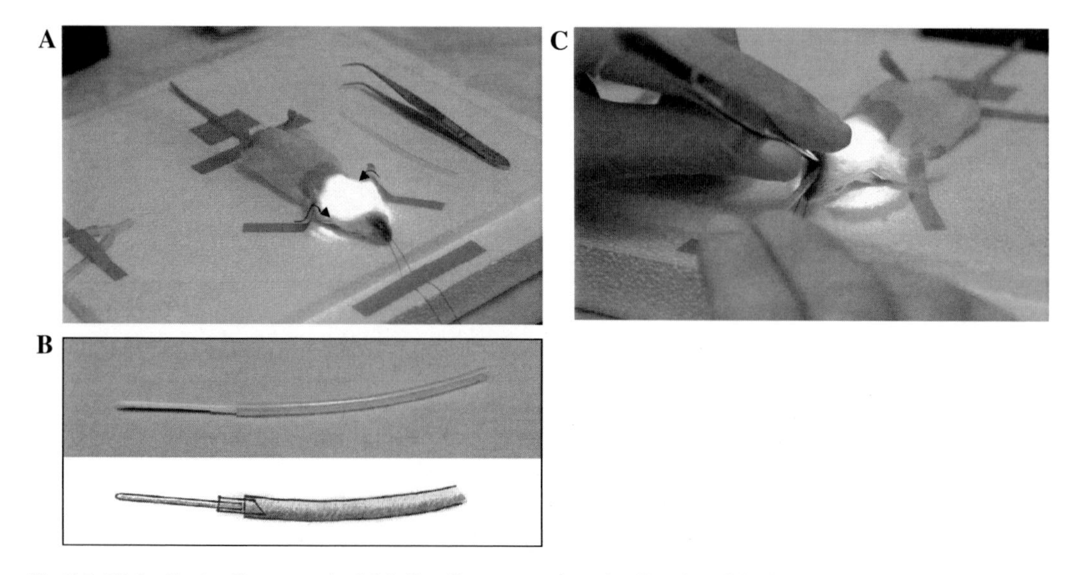

Fig. 7.2. (**A**) Positioning the mouse for intubation (the *arrows* show the direction of the flaps of tape). (**B**) Intubation tube, made from a 20-ga catheter, with the connector (PE-190 tubing) and the extender (PE-240 tubing). (**C**) Position of the operator's hands for the mouse intubation (the head of the mouse is at eye level of the operator). Figure has been previously published (22).

6. Perform the procedure at the room temperature (except for the ischemia–reperfusion) (*see* **Note 6**).
7. Perform the surgical field preparation (*see* **Note 7**).

3.2. Aortic Banding (Pressure-Overload Model)

The surgical approach is similar in both AAC and TAC models.

1. To supplement the animal with local anesthesia inject 0.1 ml of 0.2% lidocaine subcutaneously at the surgical site (*see* below).
2. Make a transverse 5-mm incision of the skin with scissors 1–2 mm higher than the level of the "armpit" (with paw extended at 90°) 2 mm away from the left sternal border (*see* **Note 8, Fig. 7.3A**).
3. Separate two layers of thoracic muscle (*see* **Note 8**).
4. Separate intercostal muscles in the 2d intercostal space (*see* **Note 8**).
5. Insert the chest retractor to facilitate the view (*see* **Note 9**).
6. Pull the thymus and surrounding fat behind the left arm of the retractor. Gently pull pericardial sac and attach it to both arms of the retractor (*see* **Note 9**).
7. The great vessels and upper part of the left atrial appendage should then be observable. From this point the further steps differ in AAC and TAC models and will be described separately.

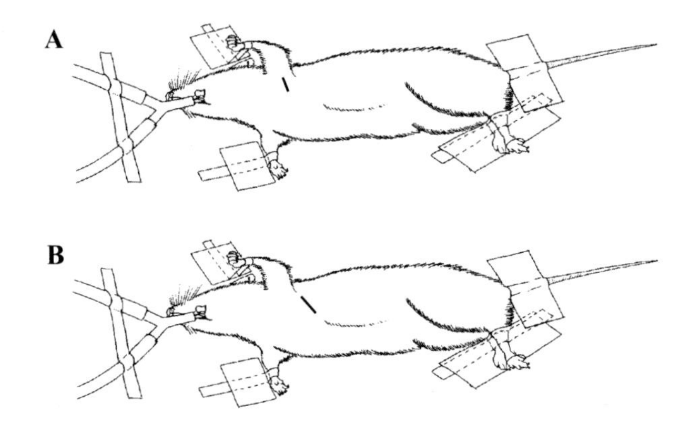

Fig. 7.3. Fixation of the mouse for surgery and location of the incision for (**A**) pressure-overload model and pulmonary artery banding (second intercostal space), and (**B**) myocardial infarction model (fourth intercostal space). Figure has been previously published (22).

3.2.1. Ascending Aortic Constriction (AAC)

1. With Foerster curved forceps, bluntly dissect the ascending portion of the aorta on its lateral side from the pulmonary trunk (*see* **Note 10**, **Fig. 7.4A**).
2. Position Foerster curved forceps from the medial side under the ascending aorta, catch the 7-0 silk suture on the opposite side and move it underneath the aorta (*see* **Note 10**).
3. Tie a loose double knot to create a loop 7–10 mm in diameter (*see* **Note 10**).
4. Position a needle of proper size into the loop (*see* **Note 10**).

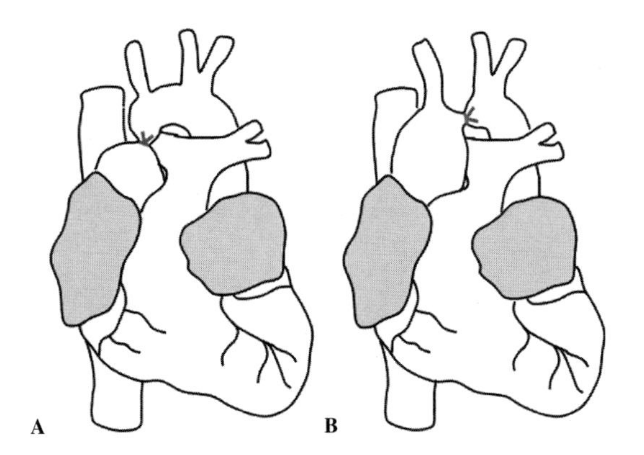

Fig. 7.4. Position of the band in (**A**) Ascending aortic constriction, and (**B**) Transverse aortic constriction. In the latter, the band is placed on the aortic arch, the innominate artery works as a shunt, and the resulting pressure overload is less stressful with more gradual hypertrophy.

5. Tie the loop around the aorta and needle and secure with the second knot; immediately remove the needle to provide a lumen with a stenotic aorta. Make another knot to secure the tie (*see* **Note 10**).
6. Remove the chest retractor and re-inflate the lungs (*see* **Note 11**).
7. Close the chest wound layer-by-layer (*see* **Note 11**).
8. For performing a sham surgery *see* **Note 12**.

3.2.2. Transverse Aortic Constriction (TAC)

1. Bluntly separate the intimate link between the thymus and pericardial sac with slightly curved forceps to demonstrate the aortic arch, then further bluntly dissect the aortic arch from the surrounding tissues and vessels (*see* **Note 13**, **Fig. 7.4B**).
2. With blunted needle, create a passageway underneath the aortic arch (*see* **Note 13**).
3. With special "wire and snare" device, deliver a 7-0 silk suture underneath the aortic arch between the innominate and left carotid arteries (*see* **Note 13**).
4. Tie a loose double knot to create a loop 7–10 mm in diameter (*see* **Note 13**).
5. Position a needle of proper size into the loop (*see* **Note 13**).
6. Tie the loop around the aorta and needle, and secure with the second knot; immediately remove the needle to provide a lumen with a stenotic aorta. Make another knot to secure the tie (*see* **Note 10**).
7. Remove the chest retractor and re-inflate the lungs (*see* **Note 11**).
8. Close the chest wound layer-by-layer (*see* **Note 11**).

3.3. Pulmonary Artery Banding (Right Ventricular Pressure-Overload Model)

Pulmonary artery banding is similar in principle to the aortic banding. However, while performing this surgery, be advised of certain challenges (*see* **Note 14**). Some unique tools and devices are employed to aid in the success of this surgery (**Fig. 7.5**).

Access to the great vessels is obtained via the second intercostal space in the same manner as described for aortic banding.

1. To supplement the animal with local anesthesia inject 0.1 ml of 0.2% lidocaine subcutaneously at the surgical site (*see* below).
2. Make a transverse 5-mm incision of the skin with scissors 1–2 mm higher than the level of the "armpit" (with paw extended at 90°) 2 mm away from the left sternal border (*see* **Note 7**, **Fig. 7.3A**).
3. Separate two layers of thoracic muscle (*see* **Note 8**).
4. Separate intercostal muscles in the 2d intercostal space (*see* **Note 8**).

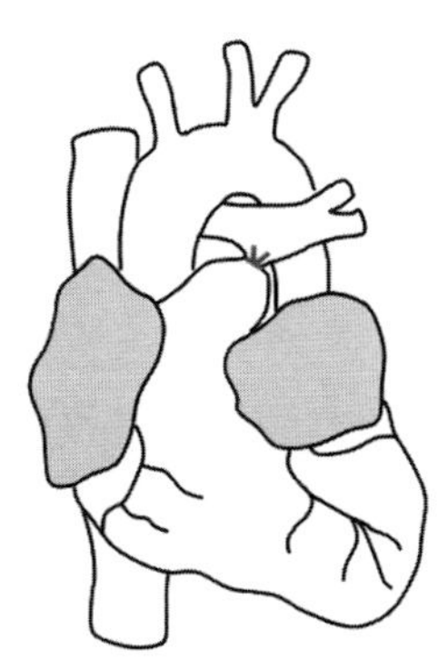

Fig. 7.5. Position of the band in pulmonary artery banding.

5. Insert the chest retractor to facilitate the view (*see* **Note 9**).
6. Pull the thymus and surrounding fat behind the left arm of the retractor. Gently pull pericardial sac and attach it to both arms of the retractor (*see* **Note 9**).
7. After mobilization of the pericardium, the pulmonary trunk (partially covered by the left atrium) should be observable (*see* **Note 15**).
8. With Foerster curved forceps, bluntly dissect the pulmonary trunk from the aorta (on the left) and left atrium (on the right) (*see* **Note 16**).
9. With blunted needle create a passageway under the pulmonary trunk (*see* **Note 16**).
10. With the help of the special "wire and snare" device move the 7-0 silk suture underneath the pulmonary trunk (*see* **Notes 13 and 16**).
11. Tie a loose double knot to create a loop 7–10 mm in diameter (*see* **Note 16**).
12. Position a needle of proper size into the loop (*see* **Note 16**).
13. Tie the loop around the aorta and needle and secure with the second knot; immediately remove the needle to provide a lumen with a stenotic pulmonary artery. Make another knot to secure the tie (*see* **Note 16**).
14. As the heart rate significantly slows down, stimulate the mouse (*see* **Note 17**).
15. Remove the chest retractor and re-inflate the lungs (*see* **Note 18**).
16. Close the chest wound layer-by-layer (*see* **Note 18**).
17. For performing a sham surgery *see* **Note 19**.

3.4. Myocardial Infarction and Ischemia–Reperfusion Models

The major technical steps are very similar. The surgical access is the same (**Fig. 7.6**).

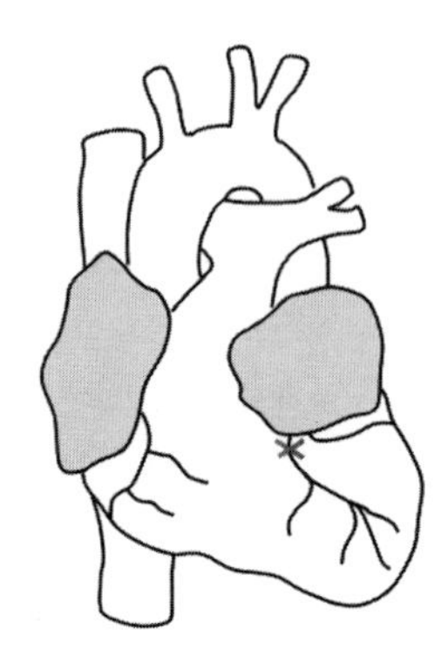

Fig. 7.6. The position of the suture on the LAD coronary artery (illustrated by the cross).

3.4.1. Myocardial Infarction Model (Permanent Ligation of the Artery)

1. To supplement the animal with local anesthesia, inject 0.1 ml of 0.2% lidocaine subcutaneously at the surgical site (*see* below).
2. Make an oblique 8-mm incision of the skin (parallel to ribs) 2 mm away from the left sternal border toward the left armpit (1–2 mm below it; **Fig. 7.3B**).
3. Separate two layers of thoracic muscle (*see* **Note 8**).
4. Separate intercostal muscles in the fourth intercostal space (*see* **Note 20**).
5. Insert the chest retractor to facilitate the view (*see* **Note 20**).
6. Gently pull apart pericardial sac and attach it to both arms of the retractor (*see* **Note 20**).
7. Locate the left anterior descending (LAD) coronary artery (*see* **Note 21**).
8. Pass the tapered needle with 7-0 silk suture underneath the LAD (*see* **Note 21**).
9. Tie the suture with one double and then with one single knot (*see* **Note 21**).
10. Remove the chest retractor and close the wound layer-by-layer (*see* **Notes 11 and 22**).
11. For performing a sham surgery *see* **Note 23**.

3.4.2. Ischemia–Reperfusion Model

Ischemia–reperfusion model is a modification of the permanently ligated MI model described above. Though technically very similar, certain conditions should be addressed (*see* **Note 24**). The procedure may be acute (non-survival, mouse stays on the ventilator for the whole duration of the procedure, or survival), when the animal needs to be reperfused for longer periods of time.

1. The access to the artery and placement of the suture are exactly the same as in the permanent MI model (**Section 3.4.1**, Steps 1–8).
2. To create ischemia, tie the temporary suture around the LAD (alternative ways, *see* **Note 25**). Cover the incision with wet gauze to prevent drying.
3. At the end of ischemia release the tie (do not remove the suture) and observe the reperfusion (*see* **Notes 24 and 25**).
4. If the procedure is survival, leave the suture underneath the LAD and close the chest wound layer-by-layer (*see* **Notes 11** and **22**).
5. At the end of reperfusion, retie the suture to demarcate the area at risk by infusion of blue dye (in the survival model, the mouse needs to be re-anesthetized, intubated, and the chest opened, *see* **Section 3.1**, Steps 1–5).

3.5. Post-operative Care

Post-operative care for all surgeries described above is fairly universal. The measures taken are directed to alleviate pain, provide supplementary heat to prevent hypothermia, and control respiratory depression (17).

1. Administer the first dose of analgesic (Buprenex 0.1 mg/kg) intraperitoneally immediately at the completion of the surgery (*see* **Note 26**).
2. Move the animal to another ventilator in a designated recovery area with 100% oxygen loosely connected to its inflow.
3. Provide heat by either a 60-W lamp or a regular pharmacy heating pad at a low setting (*see* **Note 27**).
4. Once the mouse makes the attempts to breathe spontaneously (normally after 45–60 min), disconnect the intubation tube from the ventilator.
5. Keep the intubation tube in the trachea for another 10–15 min until the mouse resumes the normal breathing pattern. Give the mouse a supplementary oxygen (the mouse is placed next to the source of oxygen).
6. Extubate the mouse and place it into the clean warm cage (*see* **Note 28**).

4. Notes

1. Estrogens could play a protective role in certain cardiac disease conditions and, therefore, introduce an additional confounding variable that needs to be considered. However, modern gender medicine requires the investigation of both males and females (18).
2. Since the surgical procedures described in this chapter are relatively short, it is not necessary to withhold food and water from mice prior to surgery. Pentobarbital sodium at 70 mg/kg (intraperitoneal, i.p.) is the anesthetic of choice since it provides an adequate depth of anesthesia for 30–40 min. Compared to the inhaled anesthetics (i.e., isoflurane), pentobarbital gives certain advantages since it slows down the heart rate to 200–300 beats/min and therefore facilitates manipulations on the beating heart. For the longer procedures (ischemia–reperfusion) an additional dose of 30 mg/kg (i.p) is required. Anesthetic is injected intraperitoneally with a ½-in. 27-ga needle. The animal is then placed into the prewarmed clean empty cage for 5–7 min while the anesthesia takes effect. It is important to not disturb or agitate the animal while the anesthetic is taking effect to ensure smooth induction and facilitate subsequent procedures. The depth of anesthesia may be assessed by pinching the toe or the tail. If the animal responds to the stimuli, then an additional 10–20% of the initial dose of anesthetic should be given.

   Presurgical analgesia (buprenorphine) is not recommended since narcotic analgesics are known to depress the respiratory center (19) and may interfere with the survival after the open-chest surgery. The first dose of buprenorphine (Buprenex 0.1 mg/kg) is given subcutaneously immediately after the completion of surgery. As the animal recovers, additional post-surgical doses are administered if there are signs of abnormal animal condition and behavior (*see* **Note 24**).
3. Any plastic platform works well (simple clip board) as an operating stage. Using a styrofoam base (cover from a commercial styrofoam freezer box) about an inch in thickness may be helpful to lift the operating field above the level of the table. The mouse is secured to the platform by the tail with a piece of tape. A 3–0 silk string is then threaded behind the upper lower incisors, pulled taut, and fixed with tape near the animal's nose. Pulling the animal taut is very important to ensure the proper plane of trachea for the smooth passage of the intubation tube. The forelimbs are then secured in place to the sides of the body with 5-mm wide strands of tape.

Taping of legs is not symmetrical (watch the arrows in the **Fig. 7.2A**) to ensure the proper positioning of the animal for the subsequent surgery. It is important that the front limbs are not overstretched as this can compromise respiration. The mouse is now ready for intubation for respiratory support.

4. There are numerous methods for performing endotracheal intubation described in the literature by both invasive and non-invasive approaches. The invasive or surgical method involves tracheotomy and is routinely used for non-survival procedures, yet it is still used by some investigators for operations which require recovery and long-term survival. It is not advisable to use this method because it subjects the animal to an extra surgical procedure, is time consuming, and may cause complications in the post-operative period. The non-invasive method, described by Brown et al. in 1995 (20), is preferable. The method described here requires no more than proper restraint of the mouse as described in **Note 3**, curved forceps (used as a laryngoscope), and a light source for transillumination of the neck. The intubation tube is made out of a 20-ga i/v catheter (Catheter #Ref 381134), cut at exactly 25 mm (1 in.) in length, attached to a connector (10-mm piece of PE-190 tubing (VWR, cat #63018-769) with one beveled end). The latter is attached to an extender (60-mm piece of PE-240 tubing (VWR, cat #63018-827) used as a handle) (**Fig. 7.2B**). This intubation tube construct is designed for repeated uses (10–20 times).

The platform with the mouse is brought to the very edge of the table with the head of the mouse directed toward the surgeon. The curved forceps (Roboz, cat #RS-5228) and intubation tube with the extender should be kept handy on the platform (tubes are kept in alcohol between surgeries for disinfection). The tube should be shaken vigorously to remove any remaining alcohol trapped inside to avoid aspiration by the mouse. A drop of 1% lidocaine may be placed on the tip of the tube to numb the throat and facilitate the passage of the tube. A power light with flexible horns (MVI, cat #MVI-DGN) may be used to illuminate the neck of the mouse. To enhance the view during the intubation and operation, binocular lenses with 1.75 magnification (SCS Limited, cat #H10L1.75 ×) can be used. Position yourself such that the head of the mouse is at your eye level. It can be achieved either by using the stage recommended by Brown (20) or by simply kneeling down. The tongue of the mouse is grasped with the curved forceps held in the operator's right hand, it should be moved to the left side of the incisors, received with thumb and index finger of the left hand and moved up slightly. The forceps (still held in right hand) should be put under the tongue to hold it firmly to the lower jaw (the forceps should be kept strictly horizontal with the ends turned up and opened to 1–2 mm). The vocal cords and trachea (as a light hole closing like a valve) should then be visible. It is now necessary to move the forceps to the left hand preserving the same position of the tongue. With free right hand the catheter of the intubation tube is gently inserted into the trachea until the connector starts going into the oral cavity (**Fig. 7.2C**). The extender is then detached using forceps and the connector should stay on the tube. With practice the fixation and intubation of the mouse should not take longer than 5 min.

5. For artificial ventilation a mouse volume-controlled ventilator 687 series (Harvard Apparatus) may be used. The tidal volume and ventilation rate are calculated by formulas provided by the company:

$$Vt = 0.0062 \times Mb^{1.01}$$

Where Vt is tidal volume and Mb is animal mass in kg.

$$\text{Rate}(\text{min}^{-1}) = 53.5 \times Mb^{-0.26}$$

Breaths/min

A table of tidal volumes and rates for particular body weights is shown below:

Mass	Vt*	Rate
20 g	0.12 cc	148
30 g	0.18 cc	133
40 g	0.24 cc	123.5
50 g	0.30 cc	116.6
60 g	0.36 cc	111.2

*Does not account for system dead space.

The described intubation construct creates a dead space of about 0.03–0.04 cc.

A modified Y-shaped connector (Fisher, cat #15-320-10A) is used to attach the mouse to the ventilator. The lower edge of the connector should be cut off to reduce the dead space. A 10-mm piece of silicone tube (Masterflex 96420-14) is put on its end to further adapt the Y-shaped connector to the diameter of intubation tube (its PE-190 tube portion). The flexible silicone tube easily adjusts to different diameters. The Y-shaped connector is attached to the surgical platform by a piece of 5-mm wide tape. All restraints except those on the upper legs should be removed and the mouse positioned in proximity to the Y-shaped connector. The mouse is positioned and secured on its right side with the chest rotated and exposed to the operator at a 45° angle to the plane of the table (**Fig. 7.3**) (provided by the proper taping described previously, *see* **Fig. 7.2A**). The PE-190 connector of the intubation tube (beveled) should then be inserted into the silicone adapter of the Y-shaped connector (this connection should ensure the minimal dead space, while not blocking the airway completely). The operator should now visually confirm rhythmic movements of the chest synchronized with the ventilator. The ventilator is set to 133 breaths/min and a tidal volume of 0.20–0.23 ml (accounts for the dead space) for a 25–30 g mouse (*see* table above); 100% oxygen at a slow flow of 0.5–1.0 l/min is loosely connected to the inflow of the ventilator. (Periodic maintenance of the ventilator is important because the plunger in the pump gets worn out with use resulting in a decreased tidal volume.) The hind legs and tail should be fixed to the platform with 5-mm wide strands of tape and then secured with wider tapes (**Fig. 7.3**). To avoid an accidental extubation during surgery, it is wise to secure the nose of the animal with a narrow piece of surgical tape (3 M, 1527-1). Should an accidental extubation occur, close the wound immediately (one superficial suture on the skin is enough), and then re-intubate the mouse.

6. Since the described surgical procedures are relatively short (15–20 min), it is not critical to strictly regulate the core temperature of the mouse. Performing surgery at room temperature 20–25°C is optimal. The body temperature of the mouse may go down to 33–34°C after anesthesia, but it should not interfere with the survival. However, for longer procedures (and specifically for ischemia–reperfusion model), it is crucial to maintain the body temperature at the physiological level to ensure uniformity (it is known that fluctuations in temperature may affect the infarct size at the time of ischemia and reperfusion (21)).
7. Operations are carried out under aseptic conditions. Autoclave the surgical tools a day prior to surgery (sterile pack can be used for five to ten series surgeries) and keep them in a hot bead sterilizer in between surgeries. Rinse them with cold sterile saline or water before use. Cover the operating table with sterile bench underpad. Scrub the operating field of the animal with betadine three times in succession with a water rinse between each scrub, then apply chlorhexidine or betadine (2% solution) to the surgical site as a final step. Drape the animal properly with sterile material (e.g., autoclaved paper

towel) with a 2 × 2 cm window over the surgical site. Wear scrubs, a mask, and a head cover, wash hands with chlorhexidine before surgery, and use fresh, sterile gloves when switching between animals.

For hemostasis, it is usually sufficient to use a piece of gauze or cotton applicators, although cautery is necessary in some instances, as indicated in the text.

8. A transverse 5-mm incision of the skin is made with scissors 2 mm away from the left sternal border, 1–2 mm higher than the level of the armpit (**Fig. 7.3A**). There are two layers of thoracic muscle in this area. At the lateral corner of the incision, the prominent vein on the superficial muscle layer should be visualized. Both muscle layers should be cut avoiding this vein, otherwise cautery may be used. The chest wall is thin and rather transparent, so the moving lung should be easily observable; its tip lies in the second intercostal space and it is good mark for the right incision. In some instances, the tip of the lung is not readily visible, as it gets partially collapsed during intubation and start of artificial ventilation. The lung is easily re-inflated by shutting off the outflow on the ventilator with the finger for two to three breathing cycles. Intercostal muscles should be separated with 5-mm incision, while caution is taken to avoid damaging the lung. Extra care should be taken while approaching the sternal (medial) corner of the incision not to cross the internal thoracic artery; it is invisible, but the bleeding may be rather extensive requiring immediate cauterization.
9. The chest retractor is gently inserted to spread the wound to 4–5 mm in width, taking care to avoid injuring the lungs with its teeth. Pull away the thymus and surrounding fat with forceps to the left arm of the retractor. The pericardial sac should be gently pulled apart with two forceps and attached to both arms of the retractor. The pericardial sac and thymus are intimately connected to the left superior vena cava which runs immediately next to the left side of the heart (mice and rats have two superior venae cavae versus larger animals and humans). The pericardial sac should not be pulled apart with too much force since it may rupture the wall of the left superior vena cava and cause bleeding that can be profuse and fatal.
10. The ascending portion of the aorta should be bluntly dissected on its lateral side from the pulmonary trunk (extreme caution should be taken, since bleeding from any of the great vessels is quickly fatal). From the medial side, the curved forceps should be placed under the ascending aorta so that the tips of the forceps appear between the aorta and the pulmonary trunk. A gentle prodding movement is required

to go through the connective tissue between the aorta and pulmonary trunk. A 7-0 silk is grasped by forceps and moved underneath the aorta. A loose double knot is made to create a loop about 7–10 mm in diameter. For aortic constriction in normal size mice, a 25-ga needle blunted in advance and bent to make an L-shape is recommended (OD 0.51 mm). The size of the needle depends on the degree of stenosis and level of hypertrophy/failure desired. To hold the needle in position, a so-called third hand construct can be used. The needle is attached to a 1-ml syringe which is held off a lab stand by means of two metal rods connected by a two-ball joint connector (**Fig. 7.1**). The needle is delivered through the loose double knot from the left side and placed directly above and parallel to the aorta. The loop is then tied around the aorta and needle and secured with a second knot (this should be done very quickly to minimize ischemia and a build-up of pressure). The needle is immediately removed to provide a lumen with a stenotic aorta. Another knot is made to secure the tie.

11. The chest retractor is then removed and the thymus is moved back to its anatomical position. The lungs, which have likely been partially collapsed by the chest retractor, should now be re-inflated by shutting off the outflow on the ventilator for two to three cycles. The chest cavity should then be closed by bringing together the second and third ribs with one 6-0 absorbable suture, taking care not to grasp the lung in the process. While making a knot, slight pressure is applied on the chest with a needle holder to reduce the volume of free air remaining in the chest cavity. All layers of muscle and skin are closed with two continuous 6-0 absorbable and nylon sutures, respectively. With practice, the whole procedure should normally take about 15 min.
12. An important control for aortic banding as well as the subsequent operations is the sham surgery. For the sham operation, the mice undergo a similar procedure, but the intervention stops when the curved forceps are moved underneath the ascending aorta without placing a ligature. Then the lungs are re-inflated and the chest is closed in the way described above.
13. The intimate link between the thymus and pericardial sac is bluntly separated with curved forceps to demonstrate the aortic arch. It should be further bluntly dissected from the surrounding tissues and vessels with certain caution as all structures here are fragile and vital. By means of the L-shaped blunted needle (1-ml insulin syringe with 28-ga needle) a passageway is created underneath the aortic arch. The needle is passed between the innominate and left carotid arteries to emerge on the opposite side of the aortic arch right above the

pulmonary trunk. The needle can be now removed and a special device called a "wire and snare" (6) can be moved through the passageway to bring the 7-0 silk suture underneath the aorta. This "wire and snare" device is made of 24-ga iv catheter (Catheter #Ref 381212), the needle of which is blunted and bent to L-shape. After releasing the catheter 4–5 mm from the needle, the suture is threaded into it, and snagged when the catheter is moved back. After delivering to the opposite side of the arch, the suture is grasped with forceps and released from the construct. (Some investigators use small curved forceps to move the suture, but the area is usually too tight to accommodate any tool. In this case the described construct appears very handy.) The suture is tied into the loose loop. The whole surgical stage is now temporarily moved 90° clockwise to change the plane of the loop; 27-ga L-shaped needle held by the construct (*see* **Note 10**) is moved into the loop from the left, the loop is tied with two knots and the needle is quickly removed to yield the stenotic aorta. The stage is rotated back to its initial position, another knot is made to secure the tie, lungs are re-inflated, and the wound is closed in the manner described in **Note 11**. In experienced hands, the entire procedure should not take longer than 20 min.

14. The major challenges that arise during this surgery are due to the extremely thin and fragile walls of the pulmonary trunk and the inability of the right ventricle to withstand stress while the pulmonary artery is being manipulated. The blood pressure in the right ventricle is very low compared to that in the left. Thus, any dissection underneath the pulmonary trunk using forceps leads to complete blockage of blood flow to the lungs, which results in immediate respiratory and cardiac distress. Consequently, recovery and survival after surgery is very poor. To overcome this complication, special techniques are required (*see* below).
15. The position of the atrium creates an additional complication as its location and fragile nature make it very vulnerable to damage during the dissection.
16. The pulmonary trunk should be bluntly dissected with curved forceps from the aorta (on the left) and left atrium (on the right). Great care must be taken while performing this dissection; it should be relatively superficial and no attempt should be made to go underneath the pulmonary trunk at this stage. To create a passageway underneath the pulmonary trunk, use an L-shaped 28-ga blunted needle (an insulin syringe with the needle attached serves as a convenient tool). The needle should be placed from the right side of the pulmonary trunk closest to the left atrium,

and gently pushed underneath the pulmonary trunk so that the end of the needle appears between the pulmonary and aortic trunks. A slight prodding movement is required to rupture the connective tissue between the trunks. The shape of the needle allows a passageway to be created underneath the pulmonary trunk without compromising the pulmonary blood flow, which would not be possible with the use of any type of forceps (*see* **Note 14**). Throughout the dissection, special attention should be paid to the atrium and the walls of the pulmonary artery, as these structures are very fragile and bleeding is invariably fatal.

A 7-0 silk suture is threaded through the passageway with the help of the "wire and snare" device described in **Note 13**. The device with suture is moved behind the pulmonary trunk from the right side and the suture is then caught with forceps between the pulmonary and aortic trunks (the suture easily detaches from the catheter). A loose double knot is made (7–10-mm diameter loop). For moderate stenosis in the average-sized mouse (23–25 g) the suture is tied against a 25-ga needle (0.51-mm OD); for more severe stenosis, a 26-ga needle (0.46-mm OD) (9) can be used. For easier access, the platform with the mouse is temporarily rotated 90° in a clockwise direction. The needle, held in position using the "needle holder construct" (*see* **Note 10**), is introduced into the loop above and parallel to the pulmonary trunk. The loop is then tied around the pulmonary trunk and needle and secured with a second knot (this should be done very quickly so as not to severely compromise pulmonary blood flow). The needle is immediately removed to provide a lumen with a stenotic pulmonary artery.

17. At this point, the heart rate noticeably slows down as the right ventricle has difficulty withstanding the increased pressure. To override this condition and improve the heart rate, the animal is stimulated by pinching its tail with forceps. This step is very important and the heart rate usually gets accelerated instantly (consider that even slight overdosing the animal with anesthetic may not make this step successful). If the right ventricle cannot compensate for the increase in pressure, the heart rate may not recover, resulting in loss of the animal. This is an unavoidable complication of this surgery. After the needle is removed, a third knot is made to secure the tie.
18. The chest retractor is removed and the thymus is moved back to its anatomical position. The lungs are re-inflated, the chest cavity is closed with 6-0 nylon suture, and the muscles and

skin are closed layer by layer with 6-0 absorbable and 6-0 nylon sutures, respectively. The entire procedure takes approximately 15–20 min.

19. For the sham operation, dissection of the pulmonary trunk is performed using the L-shaped needle but a ligature is not placed or tied. The lungs are then re-inflated and the wound is closed as previously described.

20. The rib cage and left lung should now be visible. The fourth intercostal space represents the area between those ribs where the lowest part of the lung is observed (the lung should be fully inflated by shutting of the outflow of the ventilator for two to three cycles). The chest cavity is then opened between the fourth and fifth ribs taking care not to damage the lung. The chest retractor is inserted and opened to spread the wound to 8–10 mm in width (to avoid breaking ribs, the wound may be spread in two efforts). The heart, partially covered by the lung should now be visible. The pericardium is gently picked up with two curved forceps (RS-5136 and RS-5101), pulled apart, and placed behind the arms of the retractor. This maneuver pulls up the heart slightly and at the same time "screens away" the lung, providing better exposure of the heart.

21. The left anterior descending (LAD) coronary artery should be observable as a pulsating bright red spike running in the midst of the heart wall from underneath the left atrium toward the apex. Proper lighting (pinpoint but not too bright) is essential to discern the vein from the pulsating artery. If LAD is not visible, its origin from the aorta can be located by lifting the left atrium. Positioning of the ligature is dictated by the volume of infarction desired. For most studies, the LAD artery is ligated 1–2 mm below the tip of the left atrium in its normal position (**Fig. 7.6**), which induces roughly 30–50% ischemia of the left ventricle (this may vary depending on the mouse strain and individual anatomy of the left coronary artery). Once the site of ligation has been determined, curved forceps can be used to gently press on the artery a little below the intended site to enhance the view of the artery and stabilize the heart. Next, with a tapered needle, a 7-0 silk ligature is passed underneath the LAD artery. For easier and smoother passage the needle should be bent in advance to make the curvature rounder. When placing the ligature, it is important not to go too deep as the needle will enter the cavity of the ventricle, but at the same time not to be too superficial as the ligature will cut through the artery wall. The ligature is secured with one double and then one single knot. Occlusion is confirmed by characteristic pallor of the anterior wall of the left ventricle. To prevent arrhythmia, a

drop of 1% lidocaine may be placed on the apex of the heart (this step is debatable and may be omitted if it conflicts with the study design).

22. The retractor is then removed and the lungs are re-inflated by shutting off the ventilator outflow as previously described in **Note 11**. The chest cavity is closed by bringing together the fourth and fifth ribs with one 6-0 nylon suture while slight pressure is applied to the chest wall to reduce the amount of the free air in the chest cavity. The muscles and skin are closed layer by layer with continuous 6-0 absorbable and nylon sutures, respectively. The duration of the whole surgical procedure should take about 12–15 min.
23. The sham-operated mice undergo the same procedure without tying the suture but moving it behind the LAD artery. The chest is closed as described above.
24. The body temperature should be strictly monitored and kept at 37°C throughout the whole surgical procedure, since the occlusion of the artery is transient and it is known that changes in temperature may affect the amount of cardiac damage (21). The animal is placed on the heating pad connected to the controlling device. The rectal probe is inserted after the mouse has been intubated and ventilated.

    Duration of ischemia and subsequent reperfusion are dictated by the goals of the study. Ischemia is typically induced for 15–60 min and the length of reperfusion ranges from minutes to days. An immediate complication of this model is cardiac arrhythmia, which may occur after the restoration of the blood flow (reperfusion). To prevent this condition and improve survival, two to three boluses of lidocaine (6 mg/kg) may be given to the animal intraperitoneally during the procedure if this intervention is not expected to influence the study measurements.

    The goal of the surgical procedure is to completely occlude the LAD without causing severe vascular damage. The artery should be clearly visible and the ischemia and reperfusion events readily observable for inclusion of the animal in the study.
25. Demarcation of the area at risk is a final step of the procedure, so the untied suture should be left in place for the period of reperfusion. There are several approaches for creating a temporary occlusion of the artery. A 7-0 silk suture is passed underneath the LAD in the same manner as described in **Note 21**. The suture may be tied into a slipknot around a piece of PE-10 tubing to minimize damage to the artery. To do this, a loose double knot (5-mm loop) is crafted around the artery. Then a short 2–3-mm piece of PE-10 tubing is placed with forceps into the loop and the double knot is tied

firmly around it. Next, a slipknot (shoe lace tie) is tied on top of it. Alternatively, both ends of the suture can be threaded through a 10-mm piece of propylene (PE-50) tubing to form a snare. The snare is occluded by clamping the tube together with suture using a mosquito hemostat. The successful occlusion is confirmed by the characteristic pallor of the heart region distal to the suture. At the end of ischemia the suture is untied or the snare is released and the suture is left in place for the period of reperfusion. Reperfusion is verified by the appearance of hyperemia of the previously pale area. At the end of reperfusion, to demarcate the area at risk, the suture is retied and the heart is perfused with the blue dye or pigment.

26. Analgesia (Buprenex 0.1 mg/kg) should be continued every 8–12 h for the next 48 h. Nylon sutures should be removed from the skin 10 days after surgery.
27. The temperature of the heating pad should not exceed 40°C to prevent possible burns to the animal.
28. After the mouse resumes a normal breathing pattern, it can be extubated and placed in a clean cage. To prevent accidental aspiration, the mouse should not be placed on the bedding but rather on a soft paper towel at the bottom of the cage. The cage is placed on a regular heating pad on a low setting. After the mouse starts moving around, it can be returned to its regular housing cage.

References

1. Lin, MC, Rockman, HA, Chien KR. (1995) Heart and lung disease in engineered mice. *Nat Med* **1**, 749–751.
2. Braunwald, E, Zipes, DP, Libby, P. (2001) *Heart Disease* (6th ed.). Saunders, Philadelphia, pp. 1–18, 1114–1219, 1955–1976.
3. Ding, B, Price, RL, Borg, TK, et al. (1999) Pressure overload induces severe hypertrophy in mice treated with cyclosporine, an inhibitor of calcineurin. *Circ Res* **84**, 729–734.
4. Fard, A, Wang, CY, Takuma, S, et al. (2000) Noninvasive assessment and necropsy validation of changes in left ventricular mass in ascending aortic banded mice. *J Am Soc Echocardiogr* **13**, 582–587.
5. Hamawaki M, Coffman TM, Lashus A, et al. (1998) Pressure-overload hypertrophy is unabated in mice devoid of AT_{1A} receptors. *Am J Physiol Heart Circ Physiol* **274**, H868–H873.
6. Hu P, Zhang D, Swenson L, Chakrabarti G, Abel ED, and Litwin SE. (2003) Minimally invasive aortic banding in mice: effects of altered cardiomyocyte insulin signaling during pressure overload. *Am J Physiol Heart Circ Physiol* **285**, H1261–H1269. First published May 8, 2003; 10.1152/ajpheart.00108.2003.
7. Rockman HA, Ross RS, Harris AN, et al. (1991) Segregation of atrial-specific and inducible expression of an atrial natriuretic factor transgene in an in vivo murine model of cardiac hypertrophy. *Proc Natl Acad Sci USA* **88,** 8277–8281.
8. Fuster, V, Alexander, RW, O'Rourke, RA. (2001) *Hurst's the Heart* (10th ed.). McGraw-Hill, New York, pp. 3–17.
9. Rockman HA, Ono S, Ross RS, et al. (1994) Molecular and physiological alterations in murine ventricular dysfunction. *Proc Natl Acad Sci USA* **91**, 2694–2698.
10. Katz AM. (2000) *Heart Failure.* Lippincott Williams and Wilkins, Philadelphia, PA.
11. Bayat H, Swaney JS, Ander AN, et al. (2002) Progressive heart failure after myocardial infarction in mice. *Basic Res Cardiol* **97**, 206–213.

12. Eberli, FR, Sam, F, Ngoy, S, et al. (1998) Left-ventricular structural and functional remodeling in the mouse after myocardial infarction: assessment with the isovolumetrically-contracting Langendorff heart. *J Mol Cell Cardiol* **30**, 1443–1447.
13. Gao, XM, Dilley, RJ, Samuel, CS, et al. (2002) Lower risk of postinfarct rupture in mouse heart overexpressing beta 2-adrenergic receptors: importance of collagen content. *J Cardiovasc Pharmacol* **40**, 632–640.
14. Kanno S, Lerner DL, Schuessler RB, et al. (2002) Echocardiographic evaluation of ventricular remodeling in a mouse model of myocardial infarction. *J Am Soc Echocardiogr* **15**, 601–609.
15. Michael LH, Entman ML, Hartley CJ, et al. (1995) Myocardial ischemia and reperfusion: a murine model. *Am J Physiol Heart Circ Physiol* **269**, H2147–H2154.
16. Woldbaek PR, Hoen IB, Christensen G, et al. (2002) Gene expression of colony-stimulating factors and stem cell factor after myocardial infarction in the mouse. *Acta Physiol Scand* **175**, 173–181.
17. Waynforth HB, Flecknell PA. (2001) *Experimental and Surgical Technique in the Rat.* Academic, San Diego, CA
18. Klocke R, Tian W, Michael T, et al. (2007) Surgical animal models of heart failure related to coronary heart disease. *Cardiovasc Res* April 1; **74**(1), 29–38. Epub 2006 November 23.
19. Davis JA. (2008) Current Protocols in Neuroscience, Appendix 4B, Mouse and Rat, Anesthesia and Analgesia, Copyright © 2008 John Wiley & Sons, Inc.
20. Brown RH, Walters DM, Greenberg RS, et al. (1999) A method of endotracheal intubation and pulmonary functional assessment for repeated studies in mice. *J Appl Physiol* **87**, 2362–2365.
21. Hale SL, Kloner RA. (1998) Myocardial temperature reduction attenuates necrosis after prolonged ischemia in rabbits. *Cardiovasc Res* **40**, 502–507.
22. Tarnavski O, McMullen JR, Schinke M, et al. (2004) Mouse cardiac surgery: comprehensive techniques for the generation of mouse models of human diseases and their application for genomic studies. *Physiol Genomics* **16**, 349–360.

Chapter 8

Echocardiographic Examination in Rats and Mice

Jing Liu and Dean F. Rigel

Abstract

Rats and mice are the predominant experimental species in cardiovascular research due to the widespread availability of genetic and transgenic rodent models of heart disease. Phenotyping of these models requires reliable and reproducible methods to noninvasively and serially assess cardiovascular structure and function. However, the small size of rodents has presented a challenge. Many of these challenges have been overcome in recent years due to significant technological advances in echocardiographic capabilities. For example, improved spatial resolution and increased frame rates have allowed more precise and accurate quantification of diminutive structures, myocardial function, and blood flow in mice. Consequently, transthoracic echocardiography (TTE) has emerged as a popular and powerful tool for cardiac phenotypic characterization in rodents. This chapter will focus on the use of TTE in rodents for evaluating (1) left ventricular (LV) chamber dimensions and wall thickness, (2) LV mass, (3) global LV systolic and diastolic function, (4) regional LV systolic function by newly developed tissue Doppler imaging (TDI), and (5) hemodynamic parameters. Reliability of these measurements depends on various factors such as the skill and experience of the sonographer and the image analyzer, the type, depth, and duration of anesthesia, and animal characteristics. These topics will also be discussed.

Key words: Echocardiography, mouse, rat, systolic function, diastolic function, strain rate, left ventricular mass, anesthesia.

1. Introduction

Genetic and surgical models of cardiovascular disease in rats and mice have greatly contributed to our understanding of molecular mechanisms underlying both normal and pathophysiological cardiovascular function in humans. Complete cardiac evaluation of phenotypic changes in rodents requires reliable and reproducible methods to noninvasively and serially assess cardiovascular structure and function. This has been a challenge due to the small size and rapid heart rate (HR) of these species. Traditionally,

K. DiPetrillo (ed.), *Cardiovascular Genomics*, Methods in Molecular Biology 573,
DOI 10.1007/978-1-60761-247-6_8, © Humana Press, a part of Springer Science+Business Media, LLC 2009

hemodynamic studies in small animals have been performed using invasive and terminal techniques thereby precluding longitudinal assessments.

Transthoracic echocardiography (TTE) has emerged as a powerful noninvasive tool for serially characterizing LV geometry and function in rodents. Accurate and reproducible TTE images and measurements are now attainable (*see* **Note 1**) with the currently available high-frequency ultrasound transducers, which can achieve high spatial and temporal resolutions as well as near-field imaging (1, 2). Also, the introduction of fully digital machines has greatly facilitated post-acquisition image processing.

The goal of this chapter is to describe the basic step-by-step procedures for acquiring and analyzing TTE images and for deriving relevant cardiac structural and functional parameters in laboratory rats and mice. Furthermore, conditions that are critical for obtaining high-quality TTE measurements in these species will be identified (*see* **Note 2**).

2. Materials

1. Echocardiographic instrument (Vivid 7, GE Healthcare, Milwaukee, WI, USA) and M12L and i13L probes (*see* **Note 3**).
2. Isoflurane, vaporizer, oxygen supply, induction chamber, and nose cones for anesthetizing rats and mice.
3. Avertin® (2,2,2-tribromoethanol, product T48402, Sigma-Aldrich, St. Louis, MO, USA) and 1-cc syringe with attached 26-ga needle for anesthetizing mice (*see* **Note 4**).
4. Deltaphase™ Isothermal Pad (model 39DP, Braintree Scientific, Inc., Braintree, MA, USA) for maintaining normothermia (*see* **Note 5**).
5. Rectal thermometer (model BAT-12, Physitemp Instruments, Inc., Clifton, NJ, USA).
6. Hair clipper with size 40 blade (Oster, McMinnville, TN, USA).
7. Electrocardiogram (ECG) needle electrodes (model E2-48, Grass Technologies, West Warwick, RI, USA).
8. Ultrasound transmission gel (Aquasonic® 100) and gel warmer (Thermasonic® model 82-01) (Parker Laboratories, Inc., Fairfield, NJ, USA).
9. EchoPAC™ Dimension analysis software (GE Healthcare).

3. Methods

3.1. Anesthesia

1. For isoflurane induction in mice or rats, place the rodent for approximately 2 min in an isolation chamber filled with isoflurane (5% in 100% oxygen) at a flow rate of 0.5–1 L/min.
2. Maintain anesthesia by applying isoflurane (2% in 100% oxygen at a flow rate exceeding the minute ventilation of the rodent) via a nose cone during spontaneous breathing.
3. For Avertin® anesthesia in mice, administer i.p. 0.3 mg/g body weight of the prepared tribromoethanol cocktail (*see* **Note 4**).

3.2. Animal Preparation

1. After anesthesia, place the animal in the left-lateral decubitus position on the warming pad (*see* **Note 5**).
2. Shave the animal from the left sternal border to the left axillary line. Clean and wet the imaging area with alcohol or water to improve probe-to-skin coupling.
3. Attach ECG needle electrodes to the animal's legs.
4. Apply a thin layer of pre-warmed ultrasound gel to the chest wall.

3.3. Image Acquisition

Our flowchart for conducting a comprehensive TTE examination in rodents is outlined in **Fig. 8.1**. Various images are first acquired from each of the four indicated views, the images are analyzed offline, and

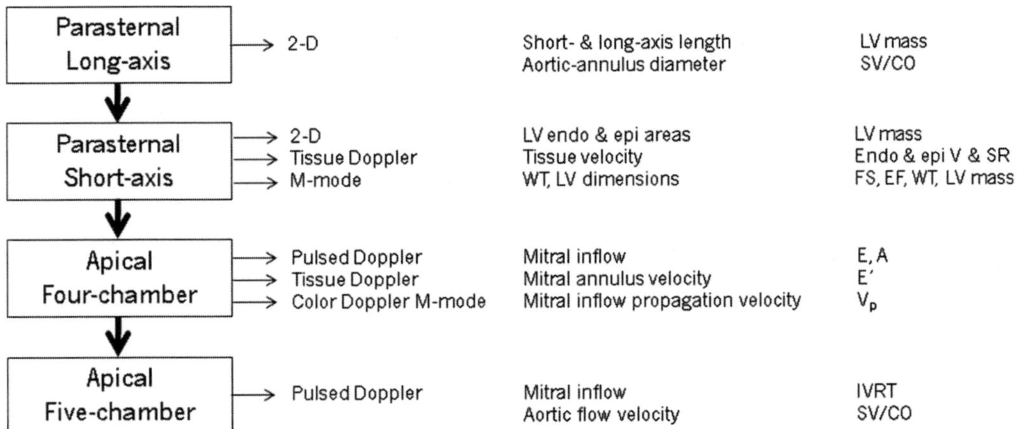

Fig. 8.1. Flowchart for conducting a comprehensive TTE examination in rodents. Images are first acquired in the indicated sequence from the various modes within each of the four TTE views. The images are subsequently analyzed offline, and the appropriate parameters derived from these measurements. For consistency and to minimize variability, it is essential to maintain the predetermined order and timing of these acquisitions between subjects and for longitudinal measurements within a subject. TTE, transthoracic echocardiography; 2D, two-dimensional; LV, left ventricular; SV, stroke volume; CO, cardiac output; endo, endocardial; epi, epicardial; *V*, velocity; SR, strain rate; WT, wall thickness; FS, fractional shortening; EF, ejection fraction; *E*, early; *A*, atrial; V_p, mitral inflow propagation velocity; IVRT, isovolumic relaxation time.

appropriate parameters derived from these measurements. Based on personal preferences and the priorities of acquired data, investigators may elect to alter this sequence of image acquisition. Nevertheless, for consistency and to minimize variability, it is essential to maintain the chosen order and timing of these acquisitions between subjects and for longitudinal measurements within a subject. As indicated in **Fig. 8.1**, a full examination will include two-dimensional (2D) imaging, Doppler flow imaging, M-mode imaging, and tissue Doppler imaging (TDI) if available. Some newer instruments also provide 2D speckle strain capabilities.

The following description of steps assumes that the sonographer: (1) has a basic understanding of cardiac anatomy and echocardiographic views, (2) is familiar with the specific echocardiographic equipment being used, (3) adjusts the gain settings, gray scale, sweep speed, frame rate, depth, focus, and zoom factors in order to optimize image quality for the specific measurements, (4) acquires all 2D images as a cineloop, which includes at least three to five cardiac cycles, and (5) uses the smallest sample volume available on the machine for recording all pulsed-wave (PW) Doppler flow images. Furthermore, these descriptions are based on the use of a General Electric Vivid 7 machine; however, they would also generally apply to equipment from other vendors.

3.3.1. Parasternal Long-Axis View

1. Acquire a parasternal long-axis view by placing the transducer (*see* **Note 6**) over the left third or fourth intercostal space and orienting its notch toward the animal's right shoulder (*see* **Note 7**).
2. Capture a 2D image (for later measurement of LV chamber dimensions and aortic-annulus diameter).

3.3.2. Parasternal Short-Axis View

1. With the transducer still in its previous position, rotate it 90° clockwise so that the transducer's notch is directed toward the animal's left shoulder to obtain a short-axis view (*see* **Note 8**).
2. Capture a 2D image (for later measurement of LV cross-sectional area).
3. Turn on the tissue Doppler feature and capture a 2D image (for later assessment of tissue velocity and strain rate; *see* **Note 9**).
4. Turn on 2D-guided M-mode. Place the M-mode cursor at the mid-papillary level and perpendicular to the interventricular septum and posterior wall of the LV.
5. Freeze and then record an M-mode image (for later assessment of chamber dimensions and wall thickness; *see* **Note 10**).

3.3.3. Apical Four-Chamber View

1. Place the transducer at the cardiac apex and orient it toward the animal's right scapula such that the transducer's notch faces the animal's left axilla in order to obtain a 2D apical

four-chamber view. When properly adjusted, this image includes the four chambers, both atrioventricular valves, and the interventricular and interatrial septa.

2. Turn on color Doppler and select the PW Doppler setting.
3. Place the PW sample volume between the tips of the mitral valve (*see* **Note 11**), adjust baseline and gain, and record a flow image (for later measurement of the early (E) and atrial (A) phases of mitral flow or fused E and A; *see* **Note 12**).
4. Turn off color Doppler and turn on the TDI feature.
5. Place the sample volume at the septal corner of the mitral annulus (i.e., at the junction between the anterior mitral leaflet and ventricular septum) and adjust gains and filters to acquire a clear mitral tissue signal without background noise (for later assessment of E', i.e., peak early mitral annulus velocity).
6. Turn off TDI and return to color Doppler mode.
7. Place the M-mode cursor in the middle of the mitral flow jet.
8. Freeze and capture an M-mode image (for later measurement of mitral inflow propagation velocity (V_p)).

3.3.4. Apical Five-Chamber View

1. From the apical four-chamber view, tilt the transducer into a shallower angle relative to the chest wall to simultaneously visualize the left ventricular outflow tract (LVOT), aortic valve, and aortic root (*see* **Note 13**).
2. Turn on color Doppler and select the PW Doppler setting.
3. Place the sample volume within the LVOT, but in proximity to the anterior mitral valve leaflet, to simultaneously record LV inflow and outflow signals (for later measurement of isovolumic relaxation time (IVRT)).
4. Maintain the PW Doppler setting and place the sample volume at the level of the aortic annulus to record the maximum aortic flow (*see* **Note 14**).

3.4. Image Analysis

Digital images from the Vivid 7 are analyzed offline in any sequence on a personal computer workstation with GE Echo-PAC™ software. The measurements derived from each image are summarized in **Fig. 8.1**.

3.4.1. Parasternal Long-Axis Images

1. Measure LV length from the LV apex to the middle of the mitral annulus at end-diastole (*see* **Note 15**; for 2D area-length LV mass measurement (3, 4)).
2. Measure aortic annulus diameter (*see* **Note 16**).

3.4.2. Parasternal Short-Axis Images

1. On the 2D image, trace the epicardium and endocardium to obtain the total (A_1) and the cavity (A_2) areas, respectively, during end-diastole (*see* **Note 15**). Exclude the papillary muscle area from the measurement.
2. On the TDI image, place sample volumes (i.e., regions of interest, ROI) in the subendocardium and subepicardium within the posterior wall (*see* **Fig. 8.2**). From the automatically generated subendocardial and subepicardial velocity–time plots, select the peak systolic velocities for each beat (*see* **Note 17**).

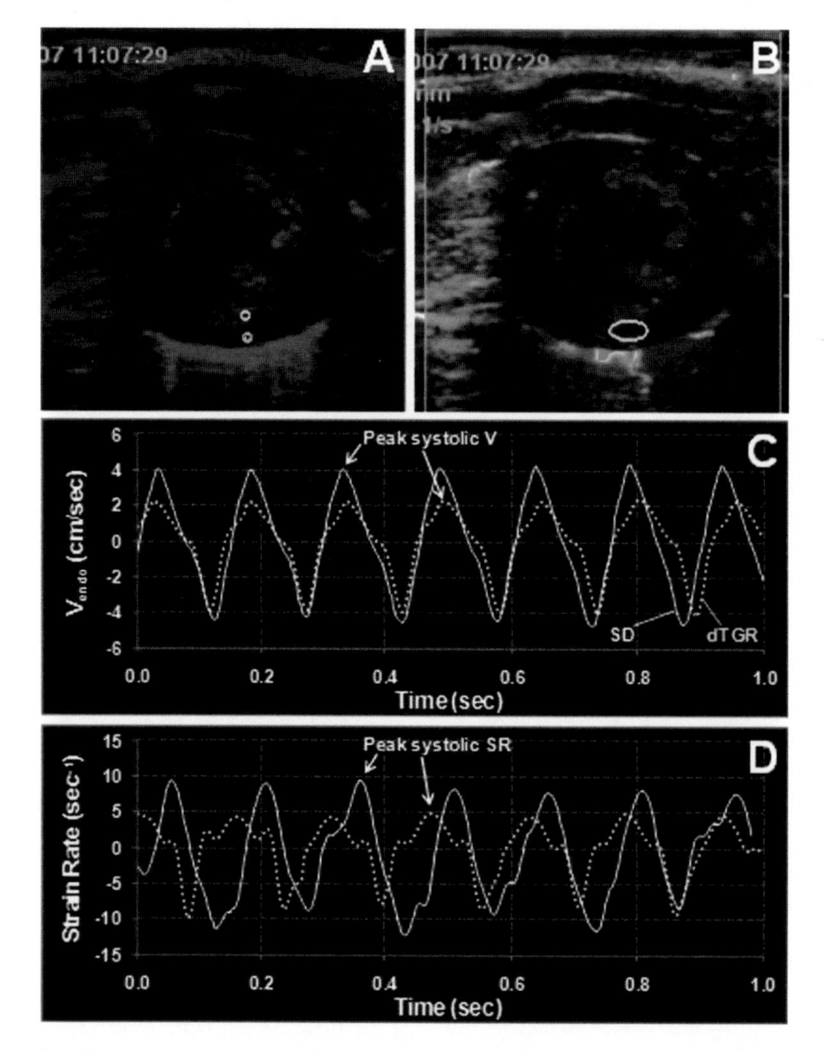

Fig. 8.2. Tissue Doppler assessment of endocardial and epicardial *V* and SR in a 7-week-old normal SD rat and an age-matched dTGR with LV dysfunction. Representative 2D TDI images (**parasternal short-axis view**) of a dTGR heart at end-diastole depict the placement of the two *V* sample volumes in the subendocardium and subepicardium of the posterior wall (**Panel A**) and the SR sample volume spanning the subendocardium to subepicardium (**Panel B**). V_{endo} (**Panel C**) and SR (**Panel D**) time-course plots generated for seven cardiac cycles (1 s) indicate that peak systolic function is depressed by ~50% in the transgenic rats relative to the SD rats. (Please refer to the online figure for clearer color images.) *V*, velocity; SR, strain rate; SD, Sprague-Dawley; dTGRs, double-transgenic rats overexpressing human renin and angiotensinogen; LV, left ventricular; 2D, two-dimensional; TDI, tissue Doppler imaging; endo, endocardial.

3. Switch to SR mode and place a single ROI that is within the posterior wall and spans the subendocardium to subepicardium. Select peak systolic SR values for each beat as described above for velocities.
4. Measure LV end-diastolic (*see* **Note 15**) and end-systolic internal dimensions ($LVID_d$ and $LVID_s$) and septal and posterior wall thicknesses (IVS and LVPW) using 2D-guided M-mode (*see* **Note 18**).

3.4.3. Apical Four-Chamber Images

1. On the PW Doppler flow recording, measure peak E and A mitral flow velocities if possible (*see* **Note 12**).
2. On the tissue Doppler image, measure peak E' and A' mitral annulus velocities (*see* **Note 19**).
3. On the color Doppler M-mode image, measure V_p (*see* **Note 20**).

3.4.4. Apical Five-Chamber Images

1. On the PW Doppler recording of simultaneous mitral and LVOT flows, measure IVRT (*see* **Note 21**).
2. On the PW Doppler aortic flow recording, trace the outer edge of the aortic flow wave to generate the time–velocity integral (TVI; *see* **Note 22**).

3.5. Derivation of Parameters

Figure 8.1 lists the functional and structural cardiac parameters that are derived from the measurements in **Section 3.4**. Since many of these calculations incorporate values measured from multiple images, the parameter descriptions below are grouped topically and not according to the source images.

3.5.1. Indexes of Systolic Function

Fractional shortening (FS) and ejection fraction (EF) are the most commonly used indexes of global LV systolic function because they are easily measured and exhibit low intra- and inter-observer variabilities (4). In spite of this high reproducibility, both parameters are subject to potential errors when there are marked differences in regional cardiac function or asymmetric geometry (e.g., in ischemia/infarct models). This is because both estimates are derived from LV dimensions measured along a single M-mode interrogation line. In these cases, FS and EF can either be overestimated or underestimated depending on the location of the regional dysfunction relative to the M-mode cursor placement. Alternatively, both FS and EF have been estimated from the 2D fractional area (short axis) changes; however, this method is likewise subject to the aforementioned errors. FS and EF are also limited by their rate dependence and load dependence.

Both FS and EF are derived from the 2D-guided M-mode short-axis measurements:

$$\text{FS}(\%) = [(\text{LVID}_\text{d} - \text{LVID}_\text{s})/\text{LVID}_\text{d}] \times 100$$

$$\text{EF}(\%) = [(\text{LVID}_\text{d}^3 - \text{LVID}_\text{s}^3)/\text{LVID}_\text{d}^3] \times 100$$

where LVID_d and LVID_s are the respective diastolic and systolic LV internal dimensions.

More recently, TDI has been used to assess local tissue velocity and SR, which are indexes of regional myocardial function and are relatively load-independent (*see* **Note 9**). Myocardial subendocardial and subepicardial velocities and SR are directly measured from the respective parameter–time plots as already outlined in **Section 3.4.2**.

3.5.2. Indexes of Diastolic Function

We routinely assess LV diastolic function in rodents based on three Doppler-derived parameters: E/E', V_p, and IVRT/RR (*see* **Note 23**). Another index is E/A; however, distinct E and A peaks are generally not discernible in anesthetized rodents (*see* **Note 12**). All of these have been described in earlier sections and are measured directly from the echocardiographic images or derived as the quotient of the two indicated parameters (*see* **Notes 19** and **24**).

3.5.3. Hemodynamic Parameters

Stroke volume (SV) and cardiac output (CO) are derived from the aortic flow TVI, the aortic-annulus diameter (D), and HR (*see* **Note 25**).

$$\text{SV} = \text{TVI} \times \text{CSA}$$

where CSA is the aortic-annulus cross-sectional area, which is estimated assuming that the aortic annulus is circular:

$$\text{CSA} = D^2 \times \pi/4 = D^2 \times 0.785$$

$$\text{CO} = \text{SV} \times \text{HR}$$

3.5.4. LV Mass Estimation

Two approaches are generally used to estimate LV mass in rodents – the M-mode “cubed” and 2D “area-length” methods (*see* **Note 26)**. Although the former is more commonly used because it is convenient, our experience (*see* **Fig. 8.3**) and reports by others (4, 5, 6) indicate that it usually overestimates the gravimetric LV mass. This method relies on unidimensional M-mode measurements, which are often overestimated because the ultrasound beam is not perpendicular to the plane of the heart. Therefore, these errors are further amplified by the cubing of these measurements:

$$\text{LV mass(mg)} = 1.05[(\text{IVS} + \text{LVID} + \text{LVPW})^3 - \text{LVID}^3]$$

where IVS and LVPW are the thicknesses of the interventricular septum and LV posterior wall, respectively, LVID is the LV internal diameter (*see* **Section 3.4.2**), and 1.05 mg/mm^3 is the density of myocardium (5). Additional errors arise because the rodent LV may not fit the ellipsoid geometry of this model.

In contrast, the area-length method relies on a half-ellipsoid/cylinder model that more closely resembles the LV geometry under varied conditions:

$$\text{LV mass(mg)} = 1.05[5/6(A_1(L+t)) - 5/6(A_2 \times L)]$$

where A_1 and A_2 are the epicardial (total) and endocardial (cavitary) parasternal short-axis areas at end-diastole (*see* **Section 3.4.1**), L is the LV length from the parasternal long-axis view (*see* **Section 3.4.2**), and t is the wall thickness, which is calculated as the difference of the A_1 and A_2 radii assuming that the areas are circular (i.e., $\sqrt{(A_1/\pi)} - \sqrt{(A_2/\pi)}$). Accordingly, LV masses estimated by this method correlate well with gravimetric measures (*see* **Fig. 8.3**).

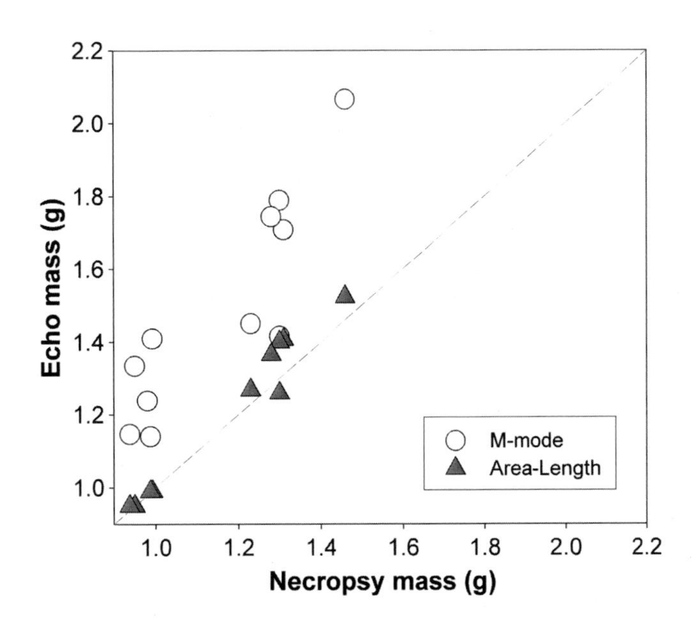

Fig. 8.3. Relationships between LV masses in rats measured gravimetrically ("Necropsy mass") and estimated by two TTE approaches ("Echo mass"): (1) the M-mode "cubed" and (2) the 2D "area-length" methods. The former method is more commonly used because it is convenient; however, it tends to overestimate the LV mass ($y = 1.37\ x - 0.10$; $R^2 = 0.76$). In contrast, the area-length method relies on a model that more closely resembles the LV geometry and thereby better estimates gravimetric LV masses ($y = 1.15\ x - 0.14$; $R^2 = 0.97$). LV, left ventricular; TTE, transthoracic echocardiography; 2D, two-dimensional.

4. Notes

1. TTE is highly operator-dependent and susceptible to significant variability in rodents (2). Therefore, considerable experience is required to acquire the ability to generate reproducible and high-quality images; consistent effort is needed to maintain this skill level (7).

2. TTE in rodents is quantitatively and qualitatively dependent on the selection of an appropriate anesthetic, dosing regimen, and timing of echocardiography image acquisition relative to anesthesia induction. For example, **Fig. 8.4** compares the TTE

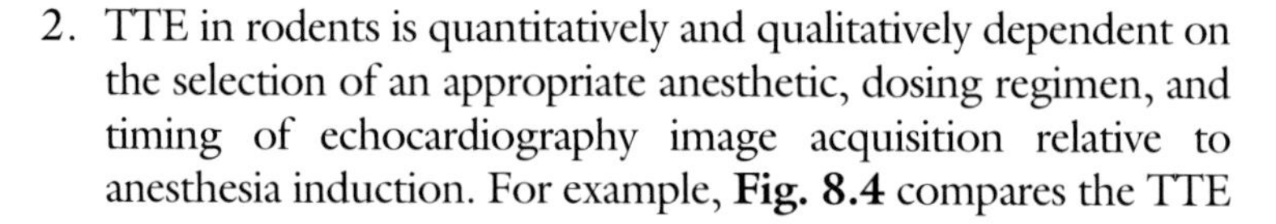

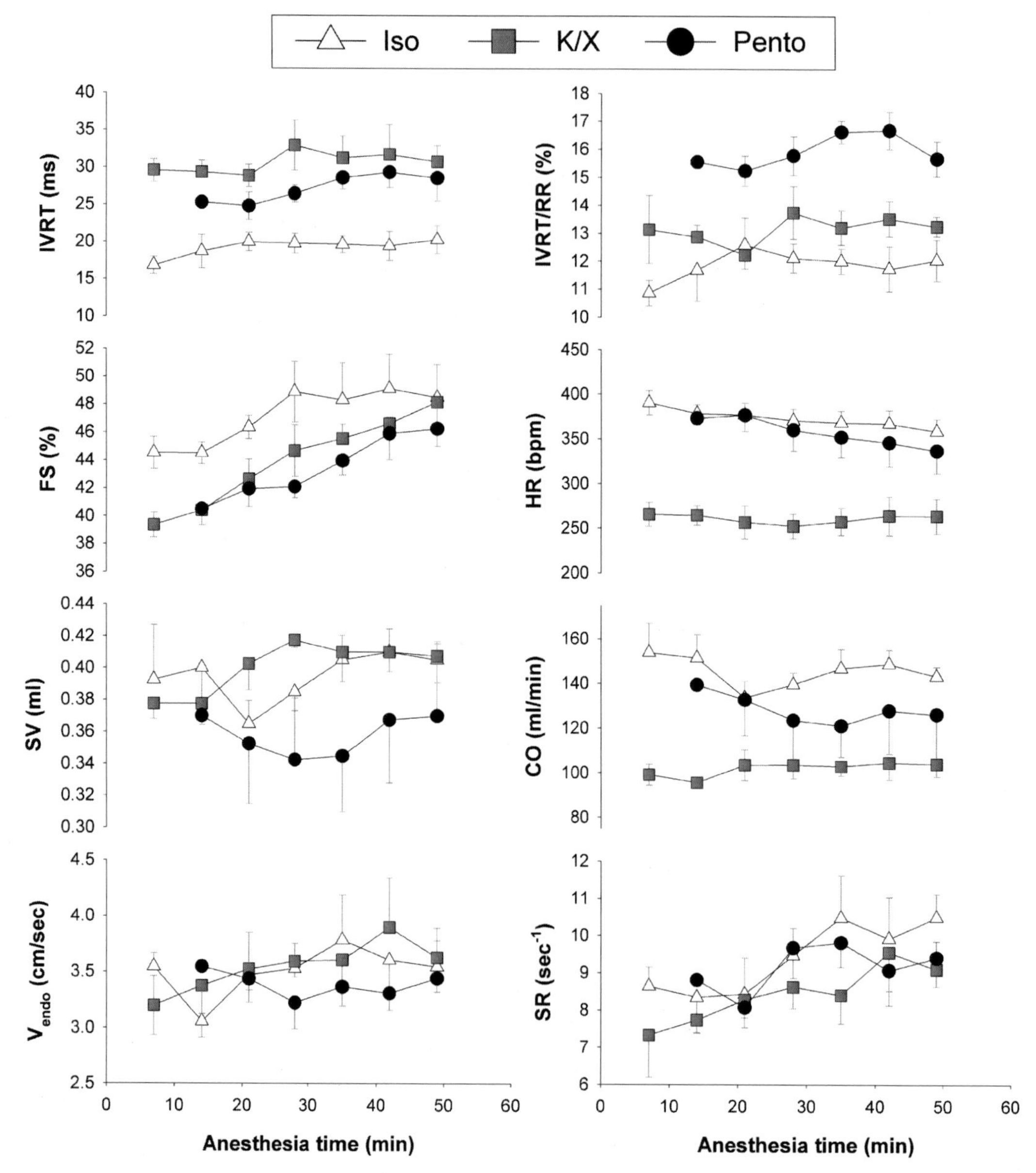

Fig. 8.4. Effects of time after anesthesia induction on several TTE and hemodynamic parameters in SD rats ($n = 4$ rats/group). Rats were evaluated under either iso (2%), K/X (50/10 mg/kg), or pento (50 mg/kg) anesthesia. In each TTE session, seven examinations were conducted sequentially every 7 min between 7 and 49 min after anesthesia induction (except in the pento group in which none and two of the four rats were sufficiently deep at 7 and 14 min, respectively). Anesthetic- and temporal-dependent differences in values are evident. TTE, transthoracic echocardiography; SD, Sprague-Dawley; Iso, isoflurane; K/X, ketamine/xylazine; Pento, pentobarbital; IVRT, isovolumic relaxation time; RR, cardiac cycle length determined by time between successive R waves on the ECG; FS, fractional shortening; HR, heart rate; SV, stroke volume; CO, cardiac output; V_{endo}, endocardial velocity; SR, strain rate.

parameters derived from Sprague-Dawley rats examined under either isoflurane (2%), ketamine/xylazine (50/10 mg/kg), or pentobarbital (50 mg/kg) anesthesia. In each TTE session, seven examinations were conducted sequentially every 7 min between 7 and 49 min after anesthesia induction. Anesthetic- and temporal-dependent differences in values are evident. Likewise, others have reported that pentobarbital and ketamine/xylazine anesthesia markedly depress LV function and cardiac output in rodents (1, 8, 9) whereas isoflurane may yield the most stable and reproducible measurements in repeated studies (1, 8). The short acting, alcohol-based anesthetic Avertin® (tribromoethanol) produces only a modest (12%) reduction in LV fractional shortening (10) and is commonly used in mouse echocardiography. The reported doses of all of these agents vary considerably and can therefore variably alter cardiac function or hemodynamics due to their pharmacologic actions. For example, xylazine is an alpha-2 receptor agonist that can induce hyperglycemia and hypoinsulinemia. Consequently, some investigators conduct TTE examinations in fully conscious mice (9). However, this approach also has drawbacks.

3. We routinely use the GE M12L transducer for rats and the i13L transducer for mice. The former is a multi-frequency (9–14 MHz) linear probe with high frame rate (103.2–210.3 frames per second (fps) in 2D mode and 319.3–365.6 fps in TDI mode) and that provides gray scale, color, and PW Doppler imaging. The i13L probe has similar features as the M12L with high frequency (10–14 MHz) and higher frame rate (186.1–570.6 fps in 2D mode and 312.8–540.3 fps in TDI mode). Its small size and light weight render it suitable for mice. Other state-of-the-art clinical echocardiographic platforms (e.g., from Phillips (Bother, WA, USA) and Acuson/Siemens (Malvern, PA, USA)) are equipped with high-frequency transducers and can also be used. Furthermore, dedicated research echocardiographic systems such as VisualSonics' Vevo 2100 now support ultra-high frequency (up to 70 MHz) linear array probes that can achieve frame rates up to 1,000 fps and 30 μm spatial resolution (http://www.visualsonics.com/).
4. Prepare a tribromoethanol stock solution by dissolving 1 g of Avertin® powder in 1 mL of tertiary amyl alcohol (2-methyl-2-butanol, product 240486, Sigma--Aldrich). Tribromoethanol undergoes photo-degradation in solution and should be prepared freshly or stored in the dark at 4°C (stable for a week). For preparing the anesthesia solution, add 25 μL of the stock solution to 1 mL of sterile water. For anesthetic induction in mice, administer 12 μL/g (12 mL/kg) body weight of this 2.5% (25 mg/mL) solution i.p. for achieving a dose of ∼300 mg/kg. This dose should maintain a stable

plane of anesthesia for at least 20 min. If additional anesthesia time is required to complete the TTE examination, an additional 100 mg/kg (4 μL/g) can be administered.

5. Maintenance of normothermia during the TTE examination is critical since cardiac function and heart rate are highly dependent on core temperature (11). Small rodents (e.g., mice and young rats) or animals with transgenes that affect thermoregulation are especially susceptible to rapid hypothermia after anesthesia. Therefore, when investigating a novel rodent strain, rectal temperature should be monitored initially to assure that the means for maintaining core temperature is adequate; thereafter, temperature monitoring can be abandoned. The Deltaphase® Isothermal Pads are especially useful for safely stabilizing body temperature without the use of electrical devices. These pads are filled with a polyethylene glycol formulation, which remains isothermic at 39°C during the material's prolonged (hours) liquid-to-solid phase transition period (http://www.braintreesci.com/PDF-Files/Deltaphase%20Pads.pdf). The pad also serves as a convenient support for securing the animal's position during the TTE examination.
6. Avoid placing excessive pressure on the chest wall with the transducer since the weight of the transducer alone may cause bradycardia and hypotension. A slight upward lifting of the transducer while continuing contact with the chest wall is recommended.
7. Acquiring a parasternal long-axis view is usually the starting point (**Fig. 8.1**). It is critical that this view includes the full length of the LV to avoid LV foreshortening. To this end, while adjusting the image depth, width, and gain, slightly move, angulate, and tilt the transducer to simultaneously visualize the LV apex, mitral valve, aortic valve, left atrium, and right ventricle.
8. A proper short-axis view will be acquired at the mid-papillary muscle level. Rotate or tilt the probe as needed to attain a circular LV geometry.
9. TDI allows quantification of myocardial tissue velocity from which tissue strain and strain rate (SR) can be derived (12, 13) (**Fig. 8.2**). SR is the rate of fractional tissue deformation in response to an applied force. SR is most commonly measured in small animals from the parasternal short-axis view at the papillary muscle level since apical views are difficult to obtain and poorly reproducible in mice and rats (14). SR has the advantage over traditional indexes of systolic function in that it can assess regional myocardial performance, exhibits low intra- and inter-observer variabilities, and is more sensitive and relatively load-independent (15, 16, 17). However, TDI in mice and rats is challenging because of their small size and rapid HR, thereby requiring a fine

balance between high frame-rate setting and spatial resolution. We typically use a frame rate of 365 fps in rats although rates as low as 203 fps have been reported (16). Higher frame rates are required in mice (e.g., 483 fps) due to the higher heart rates (15).

10. Although it is common in clinical echocardiography to measure cardiac dimensions by M-mode from the parasternal long-axis view, in our experience it is rarely possible to align the M-mode cursor perpendicular to the long axis of the LV in rodents. Since this alignment is mandatory to obtain an accurate minor-axis dimension measurement, we routinely skip this step in the parasternal long-axis view and rely solely on the measurements from the parasternal short-axis images. The GE Vivid 7 EchoPAC™ software is also equipped with "anatomical M-mode" (18), which allows the M-mode cursor to be arranged in any direction on the digital 2D image. Even this feature rarely allows proper alignment of the cursor on the long-axis image but can be quite beneficial for improving the reliability of measurements from the short-axis images.
11. Mitral inflow is recorded by placing the smallest available sample volume at the tip of the mitral valve and with the Doppler beam parallel to mitral flow. This is the point at which the mitral flow velocities are maximal. If the sample volume is placed closer to the annulus, the measured velocities will be lower because of the relatively larger cross-sectional area for flow.
12. We measure E and A from three consecutive cardiac cycles. In reality, however, E and A can usually not be measured in anesthetized rats and mice because the high HR causes fusion of the two peaks. However, anesthetics such as ketamine/xylazine that lower HR will generally yield distinct E and A peaks. Also, some investigators resort to unphysiological maneuvers such as increasing the isoflurane concentration well above the MAC (minimum alveolar concentration) in order to briefly slow HR sufficiently to separate the E and A. For us, this practice is unacceptable. In any event, we still measure the fused peak "E" velocity, which will be used for determining E/E' (*see* **Section 3.5.2**). Although in this case, the true peak E and peak E' velocities cannot be discerned, we and others (19) have found that this still provides a useful index of diastolic function when HR is high.
13. This is frequently referred to as the "five-chamber view" since it places both the LV inflow and outflow roughly parallel to the ultrasound beam. By simultaneously recording LVOT and mitral flows, the timing of the aortic valve closing and mitral valve opening can be quantified for estimating the IVRT. Furthermore, aortic flow can be measured for estimating stroke volume and cardiac output.

14. Although it can be a challenge in rodents, it is essential to angle the probe such that the Doppler beam is parallel to the aortic flow. Otherwise, the aortic flow measurement will be underestimated. Based on our experience, this can be achieved in most rodents only by slightly angling the probe's face upward toward the sternum or right clavicle. Likewise, for an accurate estimate of SV, it is imperative to place the sample volume at the level of the aortic annulus. The closing click of the aortic valve is often seen when the sample volume is correctly positioned.
15. We determine the timing of end diastole as the peak of the R wave. In most rodent hearts, this point also corresponds to the maximal LV diameter. Nevertheless, the maximum LV diameter is taken as the end-diastolic diameter.
16. Aortic-annulus diameter should be measured at its maximum, which occurs during the early ejection phase of the cardiac cycle (i.e., after the QRS complex), and can be determined by manually interrogating the cine loop frames during that interval.

    Aortic-annulus diameter is measured from the junction of the aortic leaflets with the septal endocardium to the junction of the leaflet with the mitral valve posteriorly, using inner edge to inner edge (20). If the aortic diameter measurements from the first three cardiac cycles are similar, the average of these values is acceptable. If there is a discrepancy between these values, additional cardiac cycles are measured and the average of the three largest diameter readings is used.
17. We typically record velocity and SR measurements from seven consecutive cardiac cycles, which is a sufficient number to provide a robust estimate of these parameters and is also within the reported range (five to ten cardiac cycles) (15, 16). It is critical to verify that the sample volumes remain within the respective subendocardial and subepicardial segments throughout each cardiac cycle by manually tracking the ROIs frame by frame (21). For tissue velocity measurements, we use a sample volume size of 0.4×0.4 mm in rats and 0.2×0.2 mm in mice (15). For SR measurements, we set the strain length to 2 mm in rats (16) and 1 mm in mice (15) and the ROI to 1×2 mm in rats.
18. Chamber dimensions and wall thickness should be assessed according to current recommendations (4, 22, 23), i.e., by directly measuring the distance between the actual tissue–blood interfaces instead of between the leading-edge echoes. If the M-mode cursor cannot be properly placed then anatomical M-mode can be used as an alternative (*see* **Note 10**). If there are still uncertainties about the reliability of the M-mode measurements (e.g., off-axis cursor or interference

from intra-ventricular structures such as papillary muscle or chordae tendineae), the dimensions can alternatively be acquired using direct 2D measurement.

19. E' measures the velocity of myocardial displacement as the LV expands during diastole and can, therefore, be considered a surrogate marker for tau (the time constant of relaxation) (24). The same discussion of E and A measurements in **Note 12** also apply to E' and A'.
20. V_p is measured as the slope of the first aliasing isovelocity line during early diastolic filling (25).
21. IVRT is defined as the time between the closing of the aortic valve (outer edge of the aortic flow zero-crossing point) and the opening of the mitral valve (outer edge of the onset of mitral flow). We measure IVRT from three consecutive cardiac cycles. RR intervals should also be measured on the same three beats for later normalization of IVRT (*see* **Section 3.5.2**).
22. We measure aortic flow from three consecutive cardiac cycles. HR should also be measured on the same three beats for later calculation of CO.
23. IVRT is generally reported as absolute time (ms). However, since IVRT is frequency-dependent, the directly measured value is often "corrected" for HR as either IVRT/HR, IVRT/RR, or IVRT/$\sqrt{RR}$. We have found that scaling IVRT to the cardiac cycle length (i.e., as a duty cycle or percent of RR interval) reduces the interanimal variability and provides a more reliable and intuitively understandable parameter (*see* **Fig. 8.4**).
24. Mitral inflow indexes (E, A, and E/A) are complex and subject to misinterpretation because these parameters are dependent on not only intrinsic relaxation properties of the LV but also on other factors such as HR, the volume loading conditions, left atrial–LV pressure gradients, and LV compliance. Although the ratio of peak early transmitral flow velocity (E) to the peak early myocardial tissue velocity (E') is a more robust indicator of LV diastolic dysfunction, it too has its shortcomings. Likewise, V_p is relatively independent of changes in preload and heart rate and also closely reflects tau (26). A more detailed description of the impact of various conditions on these indexes is available in reference (24).
25. Alternatively, SV can also be estimated as the difference between the LV end-systolic and end-diastolic volumes from the 2D images (1, 3, 4, 27, 28).
26. Other approaches recommended by clinical guidelines (3, 4) such as the truncated ellipsoid model and Simpson's rule method have also been used to estimate LV mass in rodents (1, 29). The latter may more accurately estimate LV mass

when cardiac geometry is irregular (29). Also, use of contrast agents to more clearly define blood–endocardial interfaces can improve LV mass estimates independent of the geometric model applied (30, 31).

References

1. Collins, KA, Korcarz, CE, Lang, RM. (2003) Use of echocardiography for the phenotypic assessment of genetically altered mice. *Physiol Genomics* **13:** 227–239.
2. Wasmeier, GH, Melnychenko, I, Voigt, JU, et al. (2007) Reproducibility of transthoracic echocardiography in small animals using clinical equipment. *Coron Artery Dis* **18:** 283–291.
3. Schiller, NB, Shah, PM, Crawford, M, et al. (1989) Recommendations for quantitation of the left ventricle by two-dimensional echocardiography. *J Am Soc Echocardiogr* **2:** 358–367.
4. Lang, RM, Bierig, M, Devereux, RB, et al. (2006) Recommendations for chamber quantification. *Eur J Echocardiogr* **7:** 79–108.
5. Devereux, RB, Alonso, DR, Lutas, EM, et al. (1986) Echocardiographic assessment of left ventricular hypertrophy: comparison to necropsy findings. *Am J Cardiol* **57:** 450–458.
6. Collins, KA, Korcarz, CE, Shroff, SG, et al. (2001) Accuracy of echocardiographic estimates of left ventricular mass in mice. *Am J Physiol Heart Circ Physiol* **280:** H1954–H1962.
7. Syed, F, Diwan, A, Hahn, HS. (2005) Murine echocardiography: a practical approach for phenotyping genetically manipulated and surgically modeled mice. *J Am Soc Echocardiogr* **18:** 982–990.
8. Roth, DM, Swaney, JS, Dalton, ND, et al. (2002) Impact of anesthesia on cardiac function during echocardiography in mice. *Am J Physiol Heart Circ Physiol* **282:** H2134–H2140.
9. Yang, XP, Liu, YH, Rhaleb, NE, et al. (1999) Echocardiographic assessment of cardiac function in conscious and anesthetized mice. *Am J Physiol Heart Circ Physiol* **277:** H1967–H1974.
10. Hoit, BD. (2001) New approaches to phenotypic analysis in adult mice. *J Mol Cell Cardiol* **33:** 27–35.
11. Arras, M, Autenried, P, Rettich, A, et al. (2001) Optimization of intraperitoneal injection anesthesia in mice: drugs, dosages, adverse effects, and anesthesia depth. *Comp Med* **51:** 443–456.
12. Voigt, JU, Flachskampf, FA. (2004) Strain and strain rate. New and clinically relevant echo parameters of regional myocardial function. *Z Kardiol* **93:**249–258.
13. Pislaru, C, Abraham, TP, Belohlavek, M. (2002) Strain and strain rate echocardiography. *Curr Opin Cardiol* **17:** 443–454.
14. Pollick, C, Hale, SL, Kloner, RA. (1995) Echocardiographic and cardiac Doppler assessment of mice. *J Am Soc Echocardiogr* **8:** 602–610.
15. Sebag, IA, Handschumacher, MD, Ichinose, F, et al. (2005) Quantitative assessment of regional myocardial function in mice by tissue Doppler imaging: comparison with hemodynamics and sonomicrometry. *Circulation* **111:** 2611–2616.
16. Hirano, T, Asanuma, T, Azakami, R, et al. (2005) Noninvasive quantification of regional ventricular function in rats: assessment of serial change and spatial distribution using ultrasound strain analysis. *J Am Soc Echocardiogr* **18:** 907–912.
17. Weytjens, C, Franken, PR, D'hooge, J, et al. (2008) Doppler myocardial imaging in the diagnosis of early systolic left ventricular dysfunction in diabetic rats. *Eur J Echocardiogr* **9:** 326–333.
18. Donal, E, Coisne, D, Pham, B, et al. (2004) Anatomic M-mode, a pertinent tool for the daily practice of transthoracic echocardiography *J Am Soc Echocardiogr* **17:** 962–967.
19. Sohn, DW, Kim, YJ, Kim, HC, et al. (1999) Evaluation of left ventricular diastolic function when mitral E and A waves are completely fused: role of assessing mitral annulus velocity. *J Am Soc Echocardiogr* **12:** 203–208.
20. Quinones, MA, Otto, CM, Stoddard, M, et al. (2001) Recommendations for quantification of Doppler echocardiography: a report from the Doppler Quantification Task Force of the Nomenclature and Standards Committee of the American Society of Echocardiography. *J Am Soc Echocardiogr* **15:** 167–184.
21. Maclaren, G, Kluger, R, Prior, D, et al. (2006) Tissue Doppler, strain, and strain rate echocardiography: principles and potential perioperative applications. *J Cardiothorac Vasc Anesth* **20:** 583–593.

22. Sahn, DJ, Demaria A, Kisslo, J. (1978) Recommendations regarding quantitation in M-Mode echocardiography: results of a survey of echocardiographic measurements. *Circulation* **58:** 1072–1083.
23. Feigenbaum, H, Armstrong, WF, Ayan, T. (2005) Evaluation of systolic and diastolic function of the left ventricle, in *Feigenbaum's Echocardiography*. Lippincott Williams & Wilkins, Philadephia, PA, pp. 138–180.
24. Maurer, MS, Spevack, D, Burkhoff, D, et al. (2004) Diastolic dysfunction: can it be diagnosed by Doppler echocardiography. *J Am Coll Cardiol* **44:** 1543–1549.
25. Tsujita, Y, Kato, T, Sussman, MA. (2005) Evaluation of left ventricular function in cardiomyopathic mice by tissue Doppler and color M-mode Doppler echocardiography. *Echocardiography* **22:** 245–253.
26. Garcia, MJ, Smedira, NG, Greenberg, NL, et al. (2000) Color M-mode Doppler flow propagation velocity is a preload insensitive index of left ventricular relaxation: animal and human validation. *J Am Coll Cardiol* **35:** 201–208.
27. Janssen, B, Debets, J, Leenders, P, et al. (2002) Chronic measurement of cardiac output in conscious mice. *Am J Physiol Regul Integr Comp Physiol* **282:** R928–R935.
28. Ding, B, Price, RL, Goldsmith, EC, et al. (2000) Left ventricular hypertrophy in ascending aortic stenosis mice: anoikis and the progression to early failure. *Circulation* **101:** 2854–2862.
29. Kanno, S, Lerner, DL, Schuessler, RB, et al. (2002) Echocardiographic evaluation of ventricular remodeling in a mouse model of myocardial infarction. *J Am Soc Echocardiogr* **15:** 601–609.
30. Mor-Avi, V, Korcarz, C, Fentzke, RC, et al. (1999) Quantitative evaluation of left ventricular function in a Transgenic Mouse model of dilated cardiomyopathy with 2-dimensional contrast echocardiography. *J Am Soc Echocardiogr* **12:** 209–214.
31. Scherrer-Crosbie, M, Steudel, W, Ullrich, R, et al. (1999) Echocardiographic determination of risk area size in a murine model of myocardial ischemia. *Am J Physiol Heart Circ Physiol* **277:** H986–H992.

Chapter 9

QTL Mapping in Intercross and Backcross Populations

Fei Zou

Abstract

In the past two decades, various statistical approaches have been developed to identify quantitative trait locus with experimental organisms. In this chapter, we introduce several commonly used QTL mapping methods for intercross and backcross populations. Important issues related to QTL mapping, such as threshold and confidence interval calculations are also discussed. We list and describe five public domain QTL software packages commonly used by biologists.

Key words: Bootstrap, genome-wide significance, interval mapping, multiple QTL mapping, permutation.

1. Introduction

Traits showing continuous pattern of variation are called quantitative traits. Quantitative traits usually do not have corresponding distinct and non-overlapping classes for different genotypes and are often controlled by both genetic and non-genetic factors, which complicate the statistical inference of quantitative traits. In the last two decades, quantitative genetics has developed rapidly and various statistical approaches have been developed to identify quantitative trait locus (QTL) using molecular markers. Sax (1) proposed a single marker t-test to detect genetic markers that are close to a QTL. However, the single marker t-test cannot estimate the QTL position. Furthermore, the estimate of QTL effect is often biased. To overcome these problems, Lander and Botstein (2) proposed an interval mapping (IM) method that detects and localizes QTL simultaneously. Haley and Knott (3) developed a regression mapping approach by combining the regression analysis and the IM method. The regression approach approximates the

K. DiPetrillo (ed.), *Cardiovascular Genomics,* Methods in Molecular Biology 573,
DOI 10.1007/978-1-60761-247-6_9, © Humana Press, a part of Springer Science+Business Media, LLC 2009

IM approach quite well with far less computation. Later, the IM approach was extended to multiple interval mapping (MIM) (4, 5) and to composite interval mapping (CIM) (6–8) where effects of other QTL outside the interval of the putative QTL are adjusted for improved mapping power and precision. Bayesian methods have also enjoyed popularity in QTL community, particularly in multiple QTL mapping due to its flexibility in handling a large number of QTL, missing data (an un-avoidable phenomena in QTL data), and prior information (9–15).

Threshold calculation is an important practical issue in the design and analysis of QTL. Due to the multiple tests on the whole genome, the usual point-wise significance level based on the chi-square approximation is inappropriate, since the entire genome is tested for the presence of a QTL. Theoretical approximation under simplified assumptions (such as dense-map with no missing genotypes and large sample size) is available (2). A shuffling permutation approach (16) is widely used in QTL community. An efficient resampling method has been proposed by Zou et al. (17) to quickly assess the genome-wide significance for QTL mapping.

In this chapter, we first introduce the backcross (BC) and F2 intercross experimental mapping populations and review several mainstream QTL mapping methods. Some other important issues related to QTL mapping, such as threshold and confidence interval (CI) calculations and QTL software, will be discussed as well.

2. Experimental Design and Data Structure

Quantitative traits are often controlled by both genetic and non-genetic factors, such as environment. When studying natural populations, like humans, it is difficult to separate environmental and genetic effects. With experimental organisms, uniform genetic backgrounds and controlled breeding schemes can avoid the variability which often confounds genetic effects. It is considerably easier to map quantitative traits with experimental populations than with natural populations. For this reason, crosses between completely inbred lines are often used for detecting QTL and have been widely applied in plant science. It has also been used successfully in a number of animal species, such as mice and rats (18, 19). Because of the homology between humans, mice, and rats, rodent models are extremely useful in helping us to understand human diseases.

The BC and F2 intercross are two of the most popular mapping populations in QTL studies. Suppose two inbred parents (P1 and P2) differ in some quantitative traits. At each locus, let us label the allele of parent P1 as m while that of P2 as M. An F1 generation is completely heterozygous with genotype Mm at all loci, receiving one allele from each parent. Thus, there is no segregation in F1 individuals. F1 can be then crossed with P1 or P2 to generate BC1 or BC2. At each locus, every backcross individual has probability of $1/2$ to be Mm, and $1/2$ to be mm (for BC1) or MM (for BC2), respectively. In the sequel, BC always refers to BC2 without further explanation. By crossing F1 individuals, we generate an F2 population in which each individual has probability $1/4$, $1/2$, and $1/4$ of being mm, Mm, and MM, respectively. Thus there is segregation in BC (and F2) since BC (and F2) individuals are no longer genetically identical at each locus.

For each individual i ($i = 1, 2, \ldots, n$, n = the total number of observations), we observe genotype M_{il} located at d_l ($l = 1, \ldots, L$, L=the number of markers) along the genome and the quantitative trait y_i. Therefore, QTL data consist of two parts: the marker data $\{\{M_{il}\}, \{d_l\}\}$ and the phenotype data $\{y_i\}$. In addition to the marker and phenotype data, there often exist non-genetic covariate data, such as age, gender, and body weight.

For BC, we code M_{il} as

$$M_{il} = \begin{cases} 0 & \text{if individual } i\text{'s } l\text{th genotype is } Mm. \\ 1 & \text{if individual } i\text{'s } l\text{th genotype is } MM \end{cases}$$

while for F2, M_{il} is often coded as

$$M_{il} = \begin{cases} -1 & \text{if individual } i\text{'s } l\text{th genotype is } mm \\ 0 & \text{if individual } i\text{'s } l\text{th genotype is } Mm. \\ 1 & \text{if individual } i\text{'s } l\text{th genotype is } MM \end{cases}$$

Marker data therefore contain information about segregation at various positions. Putative QTL genotypes are often expressed as $\{qq, Qq, QQ\}$ instead of $\{mm, Mm, MM\}$ to be distinguished from marker genotypes.

Before QTL analysis, the marker genetic locations $\{d_l\}$ should be estimated either directly from their physical locations, or more commonly, from the observed marker genotypes, based on the concept of recombination frequency. Crossover between non-sister chromatids of two homologs within a segment between two loci (genes or markers) is the basis of estimating the genetic distance between the two loci. Gametes resulting from the odd number of crossovers between two loci are called recombinants. Recombination frequency, or recombination rate, r, between two loci is the probability that recombination event takes place during meiosis and measures how closely two loci are linked. $r = 0$ if two loci are linked completely and no recombinants would be observed during meiosis; while $r = \frac{1}{2}$ if the loci are unlinked and roughly equal number of recombinants and non-recombinant would be observed during meiosis. Recombination frequency is non-additive. A mapping function, such as Haldane's mapping function, can be used to transform the recombination frequencies into additive map distances (20). Haldane's mapping function assumes no interference, that is, the crossover events occur independently in adjacent intervals. Other mapping functions take interference into consideration and employ different assumptions on the degree of the dependency (*see* **Chapter 14** of Lynch and Walsh (21)). Because of its simplicity, Haldane's mapping function is widely used in QTL community.

3. Single QTL Mapping

3.1. Single Marker Analysis

Among all available methods for single QTL mapping, single marker analysis is the simplest one but conveys the key idea on how the phenotypic difference at one QTL is linked to genetic markers, the basis of QTL mapping.

Let us take the BC population as an example. Suppose QTL Q is linked to marker A with recombination rate r (**Fig. 9.1**) and the phenotypic means corresponding to the two QTL genotypes are y_{QQ} and y_{Qq}, respectively. Then the phenotypic means of the two marker classes, y_{MM} and y_{Mm} are $(1-r)y_{QQ}+ry_{Qq}$ and $ry_{QQ}+(1-r)y_{Qq}$, respectively (*see* **Chapter 5** of Liu (22)). Therefore, $y_{MM}-y_{Mm}=(1-r)(y_{QQ}-y_{Qq})$, which is a composite function of the QTL effect, $y_{QQ}-y_{Qq}$, and the recombination rate, r, between the marker and the QTL. The closer the marker to the QTL, the smaller r is, and therefore the bigger the phenotypic difference is at the marker. The single marker analysis is built on this fact. In single marker analysis, the association between the quantitative trait and the genotypes at each marker is tested separately. That is, at each marker, individuals are grouped according to their genotypes, and then whether the phenotypic means between the two marker genotypes are the same is tested by:

$$t=\frac{\hat{y}_{MM}-\hat{y}_{Mm}}{\sqrt{\hat{s}^2\left(\frac{1}{n_1}+\frac{1}{n_2}\right)}}$$

where $\hat{y}_{MM}$ and $\hat{y}_{Mm}$ are the observed phenotypic means of the two marker genotypes MM and Mm, while $\hat{s}^2$ is the sample variance of the pooled data. A significant t-test indicates that there exists a

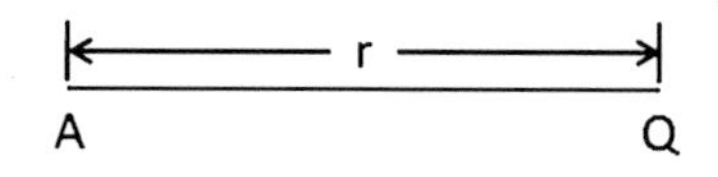

Fig. 9.1. Single marker analysis.

QTL and the market is linked to the QTL. For F2 or other crosses where there are more than two genotypes at each marker, ANOVA approach can be used.

Though simple, the single marker t-test cannot provide QTL position estimate. Furthermore, the estimate of QTL effect is often biased due to the confounding of the QTL effect and the recombination frequency r.

3.2. Interval Mapping

To overcome the drawbacks of the single marker analysis, Lander and Botstein (2) proposed the following interval mapping (IM) method, where pairs of markers are used together to search for QTL (**Fig. 9.2**).

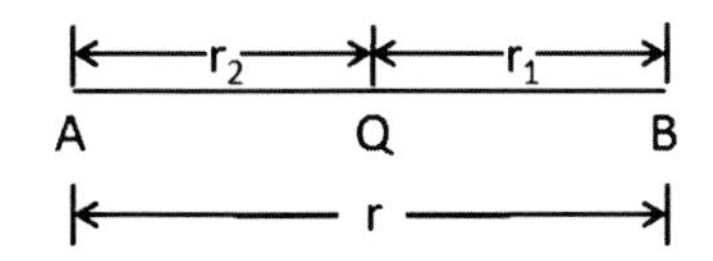

Fig. 9.2. Interval mapping analysis.

The IM is now a standard approach for QTL mapping. It extends the single marker approach by taking flanking markers into the modeling procedure. The IM model assumes that there is a QTL affecting the trait according to the following linear regression model:

$$y_i = \mu + ax_i + e_i, i = 1, 2, \ldots, n,$$

where μ is the grand phenotypic mean, a is the QTL effect, $x_i (= 0$ or 1 if the QTL genotype is Qq or QQ, respectively) is an indicator of the QTL genotype, and e_i is a random error, usually assumed to be normally distributed with mean 0 and unknown variance σ^2. The null hypothesis is $H_0 : a = 0$.

If the QTL genotypes are observed, QTL mapping is simply a regression problem. However, in practice, the QTL position is rarely known and the genotypes are typically unobserved, resulting in all missing x_i s. The idea of the interval mapping is that at each fixed location d of the genome, the conditional probabilities of the unobserved (putative QTL) genotype are calculated given the flanking markers. In intervals between known loci, the conditional probabilities of the unobserved QTL genotypes can be inferred using flanking markers (*see* **Table 9.1**). The distribution of the quantitative trait given the markers thus follows a discrete mixture model. Specifically, for a given locus d, the log likelihood for $\theta = (\mu, a, \sigma^2)$ takes the form

$$l(\theta; d) = \sum_{i=1}^{n} l_i(\theta; d)$$

Table 9.1
Probabilities of putative QTL genotypes given flanking markers

Marker		**QTL**	
A	*B*	Qq($x=0$)	QQ($x=1$)
BC			
Mm	*Mm*	$\frac{(1-r_1)(1-r_2)}{1-r}$	$\frac{r_1 r_2}{1-r}$
Mm	*MM*	$\frac{(1-r_1)r_2}{r}$	$\frac{r_1(1-r_2)}{r}$
MM	*Mm*	$\frac{r_1(1-r_2)}{r}$	$\frac{(1-r_1)r_2}{r}$
MM	*MM*	$\frac{r_1 r_2}{1-r}$	$\frac{(1-r_1)(1-r_2)}{1-r}$

Marker			**QTL**	
A	*B*	qq($x=-1$)	Qq($x=0$)	QQ($x=1$)
F2				
mm	*mm*	$\frac{(1-r_1)^2(1-r_2)^2}{(1-r)^2}$	$\frac{2r_1r_2(1-r_1)(1-r_2)}{(1-r)^2}$	$\frac{r_1^2r_2^2}{(1-r)^2}$
mm	*Mm*	$\frac{(1-r_1)^2r_2(1-r_2)}{r(1-r)}$	$\frac{r_1(1-r_1)(r_2^2+(1-r_2)^2)}{r(1-r)}$	$\frac{r_1^2r_2(1-r_2)}{r(1-r)}$
mm	*MM*	$\frac{(1-r_1)^2r_2^2}{r^2}$	$\frac{2r_1r_2(1-r_1)(1-r_2)}{r^2}$	$\frac{r_1^2(1-r_2)^2}{r^2}$
Mm	*mm*	$\frac{r_1(1-r_1)(1-r_2)^2}{r(1-r)}$	$\frac{(r_1^2+(1-r_1)^2)r_2(1-r_2)}{r(1-r)}$	$\frac{r_1(1-r_1)r_2^2}{r(1-r)}$
Mm	*Mm*	$\frac{2r_1r_2(1-r_1)(1-r_2)}{r^2+(1-r)^2}$	$\frac{((1-r_1)^2+r_1^2)((1-r_2)^2+r_2^2)}{r^2+(1-r)^2}$	$\frac{2r_1r_2(1-r_1)(1-r_2)}{r^2+(1-r)^2}$
Mm	*MM*	$\frac{r_1(1-r_1)r_2^2}{r(1-r)}$	$\frac{(r_1^2+(1-r_1)^2)r_2(1-r_2)}{r(1-r)}$	$\frac{r_1(1-r_1)(1-r_2)^2}{r(1-r)}$
MM	*mm*	$\frac{r_1^2(1-r_2)^2}{r^2}$	$\frac{2r_1r_2(1-r_1)(1-r_2)}{r^2}$	$\frac{(1-r_1)^2r_2^2}{r^2}$
MM	*Mm*	$\frac{r_1^2r_2(1-r_2)}{r(1-r)}$	$\frac{r_1(1-r_1)(r_2^2+(1-r_2)^2)}{r(1-r)}$	$\frac{(1-r_1)^2r_2(1-r_2)}{r(1-r)}$
MM	*MM*	$\frac{r_1^2r_2^2}{(1-r)^2}$	$\frac{2r_1r_2(1-r_1)(1-r_2)}{(1-r)^2}$	$\frac{(1-r_1)^2(1-r_2)^2}{(1-r)^2}$

where $l_i(\theta; d) = \log[\pi_i(0; d)\phi(\frac{y_i-\mu}{\sigma}) + \pi_i(1; d)\phi(\frac{y_i-\mu-a}{\sigma})]$, $\phi(x)$ is the density of a standard normal random variable, and for $i = 1, \ldots, n$ and $k = 0$ and 1, $\pi_i(k; d) = Pr(x_i = k$ at locus $d|$subject i's flanking markers at locus $d)$.

Note that $\pi_i(k; d)$ depends on the flanking marker genotypes, the distance between the two flanking markers as well as the distances between the putative QTL locus d and the right and left markers (*see* **Table 9.1** for details).

The maximum likelihood estimator $\hat{\theta} \equiv (\hat{\mu}, \hat{a}, \hat{\sigma}^2)$ is often obtained by using the EM algorithm (23) in which the unknown QTL genotypes are naturally dealt as missing data. Let the maximum likelihood estimator of θ under $H_0 : a = 0$ be denoted by $\tilde{\theta} = (\tilde{\mu}, 0, \tilde{\sigma}^2)$. Then the likelihood ratio test statistic for testing $H_0 : a = 0$ against $H_1 : a \neq 0$ at location d takes the form

$$LRT(d) = 2[l(\hat{\theta}; d) - l(\tilde{\theta}; d)],$$

which is approximately chi-square distributed with one degree of freedom. Traditionally, an LOD score, which is calculated directly from the likelihood ratio test statistic as $LRT/2log(10)$, is reported in linkage analysis.

In traditional IM, a single QTL is fixed on a grid of putative positions in the genome and an LOD score for the QTL is maximized at each position, resulting in the so-called profile LOD. In any region where the profile exceeds a (genome-wide) significance threshold, a QTL is declared at the position with the highest LOD score.

To improve computation efficiency, Haley and Knott (3) combined the interval idea and the regression method to develop regression (interval) mapping (in contrast, Lander and Botstein's method is called maximum likelihood (ML) interval mapping). Knapp et al. (24) and Martinez and Curnow (25) proposed similar methods independently. In regression mapping, the unknown QTL genotype xi used for interval mapping is replaced by its conditional expectation given the flanking markers. That is,

$$y_i = \mu + ac_i + e_i$$

where $c_i = E(x_i|\text{flanking markers})$.

The corresponding LOD score is calculated as $\frac{n}{2}\log_{10}\left(\frac{RSS_0}{RSS_1}\right)$ where RSS_0 and RSS_1 are the residual sum of squares under the null (no QTL) and alternative hypotheses, respectively.

An advantage of regression mapping is that the statistical analysis is straightforward and existing statistical software, such as SAS (SAS Institute, Inc.) and R (http://www.r-project.org/) can be used fairly easily here. The regression interval mapping has been shown to be a good approximation to the ML interval mapping (3). Xu (26) proposed a weighted least squares analysis to improve the efficiency of the regression mapping.

3.3. Overall Significance Level and Confidence Interval

Determining the threshold of the test statistic is complicated and important in QTL mapping. Many factors, such as genome size, map density and the proportion of missing data, may affect the null distribution of the test statistic. A great deal of effort has been spent to understand the appropriate LOD threshold to use. The usual point-wise significance level based on the chi-square approximation is inadequate because the entire genome is tested for the presence of a QTL. Theoretical approximations based on the Ornstein–Uhlenbeck diffusion process have been developed to determine threshold and power (2, 27–32) in some simple experimental crosses. However, the theoretical approximation is not readily available for any study design and hard to obtain for complicated models.

Churchill and Doerge (16) proposed a permutation procedure to obtain an empirical threshold. Permutation is a method of establishing significance without making assumptions about the data. Permutation (along with other randomization tests such as bootstrapping and jackknifing) has been around for many years. Instead of assuming an obviously incorrect test statistic distribution, a Monte Carlo sampling process is a reasonable framework in which one is able to decide appropriately among the test statistics. The idea is to replicate the original analysis many times on data sets generated by randomly reshuffling the original trait data while leaving the marker data unchanged. This approach accounts for missing marker data, actual marker densities, and non-random segregation of marker alleles. However, the permutation procedure is time consuming. Furthermore, permutation testing is limited to situations in which there is complete exchangeability under the null hypothesis. It is this exchangeability that ensures the validity of inference based on the permutation distribution. Naive application of permutation testing may lead to inflated error rates (33, 34).

Recently, Zou et al. (17) has proposed a new resampling procedure to assess the genome-wide significance for QTL mapping. At each genome position, a score statistic, which is equivalent to the LOD score, can be computed. Under the null hypothesis, the score statistics are functions of certain zero-mean Gaussian processes over the genome positions and the realizations from the Gaussian processes can be generated by Monte Carlo simulations efficiently. The resampling procedure is far more efficient than the permutation procedure since it needs only to maximize the likelihood of the observed data once with no need to maximize the likelihood in each resampling iteration. The resampling method is also more accurate than theoretical approximations when rigid requirements of theoretical approximations are not satisfied and also avoids the derivation of parameters in the Ornstein–Uhlenbeck diffusion approximations, which can be a difficult task when the model is complicated.

Another important issue in QTL mapping is to construct a confidence interval (CI) for the QTL position with proper coverage, which would provide us directions about future experiments and fine mapping strategies. However, none of the QTL mapping methods, whether based on maximum likelihood or regression, lead to a direct calculation of a CI for the QTL position. The LOD drop-off method of Lander and Botstein (2) is widely used to construct CIs. The total width corresponding to a 1.5 and 1.8 LOD drop-off from the peak of the LOD curve can be taken as the confidence interval with approximately 95% CI for the BC and F2 intercross (35). However, Mangin et al. (36) observed that the coverage of LOD support CIs depends on the QTL effect, and therefore do not behave as true CIs.

Visscher et al. (37) proposed use of a bootstrap method to construct CIs. They showed that the bootstrap method has reasonable coverage but in general were slightly conservatively biased. The bootstrap CIs are generally less biased if only significant replicates are used for constructing CIs. However, Manichaikul et al. (35) have shown that the coverage of bootstrap CIs for QTL location depends critically upon the location of the QTL relative to the typed genetic markers. They alternatively recommended an approximate Bayes credible interval and showed that the Bayes credible intervals provide consistent coverage, irrespective of the QTL effect, marker density, and sample sizes.

4. Multiple QTL Mapping

The methods described thus far are designed to detect one QTL at a time. However, complex traits are often affected by the joint action of multiple genes. Lander and Botstein (2) have shown that the interval mapping tends to identify a "ghost" QTL if there exist, for example, two closely linked QTL, a phenomena that can only be solved via multiple QTL mapping.

4.1. Composite Interval Mapping (CIM)

Composite interval mapping proposed by Zeng (7) and Jansen and Stam (8) aims to remove the confounding effects of other QTL outside the current interval to be mapped by introducing additional markers as covariates into the model. Specifically, the CIM model is

$$y_i = \mu + ax_i + \sum_j \beta_j m_{ij} + e_i, i = 1, \ldots, n,$$

where x_i is the genotype at the putative QTL and m_{ij} is the genotype of the jth selected marker for controlling confounding effects of other QTL. Selecting appropriate markers as co-factors is an important issue of CIM. Zeng (38) provided some guidelines for marker selection. Though much simplified, the CIM model lacks the ability to map multiple interacting QTL and is inefficient in studying the genetics architecture of complex traits under the control of multiple (potentially) interacting QTL.

4.2. Multiple Interval Mapping (MIM)

To model genetic architecture of complex traits, MIM straightforwardly extends interval mapping where multiple QTL with gene–gene interaction (also referred as epistasis) are modeled (4, 5). Specifically, for k putative QTL, the MIM model is

$$y_i = \mu + \sum_{j=1}^{k} a_j x_{ij} + \sum_{1 \le j < r \le k} b_{jr} x_{ij} x_{ir} + e_i$$

where $x_{ij} = 0$ or 1 if the QTL genotype of individual i at the jth QTL is Qq or QQ, respectively. Therefore, a_j refers to the main effect of the jth QTL and b_{jr} is the epistasis between the jth and rth QTL. Similar to the IM, all QTL positions are unknown and therefore, all x_{ij} s are missing. Following the idea of the IM, the MIM again uses flanking markers to calculate the conditional probabilities of the QTL genotypes and EM algorithm to maximize the likelihood and obtain the maximum likelihood estimate of parameters. Model selection techniques, such as the forward selection and step-wise selection, have been applied to the MIM in searching for the best genetic model (4, 5, 39). Since the number of parameters increases exponentially with the number of QTL in the MIM model, the implementation of the MIM model becomes computationally intensive. Worse, we face a saturation problem when the number of covariates gets larger than the number of samples if a large number of QTL are fitted in the model. Alternatively, Bayesian QTL mapping methods provide a partial solution to the problem.

4.3. Bayesian Multiple QTL Mapping

It is quite beneficial to view the multiple QTL mapping problem as a large variable selection problem: for example, for a BC population with m markers (or m potential QTL) where m is in the hundreds or thousands, there are 2^m possible main effect models. Variable selection methods are needed that are capable of selecting variables that are not necessarily all individually important but rather together important. By treating multiple QTL mapping as a model/variable selection problem (40), forward and step-wise selection procedures have been proposed in searching for multiple QTL. Though simple, these methods have their limitations, such as the uncertainty of number of QTL and the sequential model building that makes it unclear how to assess the significance of the associated tests. Bayesian QTL mapping (9–14) has been developed, in particular, for detection of multiple QTL by treating the number of QTL as a random variable and specifically modeling it using reversible jump Markov chain Monte Carlo (MCMC) (41). Due to the change of dimensionality, care must be taken in determining the acceptance probability for such dimension change, which in practice, may not be handled correctly (42). To avoid such a problem by the uncertain dimensionality of parameter space, another leading approach to variable selection in QTL analysis implemented by Markov chain Monte Carlo (MCMC) and based on the composite model space framework (43, 44) has been introduced to genetic mapping by Yi (15). Reversible jump MCMC and stochastic search variable selection (SSVS) (45) are special cases. SSVS is a variable selection method that keeps all possible variables in the model and limits the posterior distribution of non-significant variables in a small neighborhood of 0 and therefore eliminates the need to remove non-significant variables from the model. The Bayesian method of Yi (15) has the ability to control the genetic variances of a large number of QTL where each has small effect (46). The Bayesian mapping methods enjoy their flexibility in handling the uncertainty of number of QTL, their locations, and missing QTL genotypes. The prior knowledge can be easily reflected in the prior distributions of the parameters. For more about the Bayesian QTL mapping methods, we suggest readers to consult the review paper by Yi and Shriner (47).

5. Miscellaneous

In previous sections, we covered the most important issues on the complex trait mapping, there are several remaining issues that deserve some discussions, which we summarize briefly in this section.

5.1. Combined Crosses

All the methods discussed above have been developed for experimental designs with a single cross from two inbred parents. Multiple crosses from two or multiple inbred lines are common in plant and animal breeding programs, such as a study using both BC and F2 populations. Different crosses might have different genetic backgrounds due to genetic factors other than the major QTL considered. In practice, one may separately analyze data for each population, then compare and combine the results in some fashion. However, power to detect the QTL may be reduced by doing this way.

Some methods have been proposed to analyze all data simultaneously. Bernardo (48) used Wright's relationship matrix A to accommodate differential correlations when analyzing diallel crosses. However, it is more reasonable to treat the cross effects as fixed when closely related crosses are from a small number of inbred lines. Rebai et al. (28) extended the regression method of Haley and Knott (3) to several F2s with all effects fixed. Most recently, Liu and Zeng (49) extended the composite interval mapping to crosses derived from multiple inbred lines. Their model can be applied to various designs, such as diallel cross, factorial cross, and cyclic cross originating in two or more parental lines. Careful attention need be paid when applying the permutation procedure to combined crosses, since naive application of the permutation procedure to combined crosses may result in grossly inflated type I error (33).

5.2. Beyond Normality Assumption

Most QTL mapping methods share a common assumption that the phenotypic trait follows a normal distribution for each QTL genotype. The normality assumption of the underlying distributions greatly simplifies the form of the likelihood function. Many traits, such as tumor counts, however, are not normally distributed. Using standard QTL interval mapping under normality assumption for such traits may lead to low power or unacceptably high false positive rates. When the underlying distributions are suspected to be non-normal, one strategy is to use a parametric likelihood approach by transforming the original data to normally distributed data using, for example, the Box–Cox transformation (50) or applying an appropriate distribution for the original data. Several investigators have studied non-normal parametric phenotype models, including binomial and threshold models (12, 51–55), Poisson (56, 57), and negative binomial (19). Hackett and Weller (58) considered ordinal threshold models. Broman (55) proposed a two-part parametric model for phenotype with a spike at one value, including structural zeros and type I censoring. Weibull model was examined by Diao et al. (59).

Alternatively, one can apply model-free Wilcoxon rank-sum statistics to test the significance of the genetic effects. For BC, Kruglyak and Lander (60) proposed rank-based interval mapping through the use of linear rank statistics defined as

$$Z = \frac{\sum_{i=1}^{n}(n + 1 - 2 * rank(y_i))c_i}{\sqrt{\frac{n^3-n}{3} v(r_1, r_2, r)}}$$

where $rank(y_i)$ is the rank of y_i among $y_1, \ldots, y_n$; $v(r_1, r_2, r) = \left\{\frac{(r_1-r_2)^2}{r} + \frac{[1-(r_1+r_2)^2]}{1-r}\right\}$ (*see* **Fig. 9.2** for r_1, r_2 and r). Z is a generalization of the Wilcoxon rank-sum test to QTL data and is asymptotically standard normal under H_0. When the putative QTL is located exactly on a genetic marker, Z reduces to the Wilcoxon rank-sum test.

Generalization of Z to other crosses, such as F2, has been discussed by Kruglyak and Lander (60) and Poole and Drinkwater (61). It was suggested by Kruglyak and Lander (60) that QTL data be analyzed both parametrically and non-parametrically. If the results differ between the two approaches, the experiment should be interpreted with considerable caution.

Other semi-parametric and non-parametric mapping methods have been developed in the last decade. Zou and others (62, 63) developed semi-parametric QTL mapping using the exponential tilt. Many parametric models are special cases of this semi-parametric model, including normal, Poisson, and binomial. Fine et al. (64) and Huang et al. (65) provided non-parametric estimators to mixture models, which can aid in selecting a parametric mapping model. Lange and Whittaker (66) investigated QTL mapping using generalized estimating equations. Symons et al. (67), Diao and Lin (68), and Epstein et al. (69) considered a semi-parametric Cox proportional hazards model and a Tobit model, respectively, for censored survival data.

5.3. Software

Some QTL mapping methods, such as single marker analysis and regression interval mapping, can be fairly easily implemented using statistical software packages. However, most of the QTL mapping methods are often complicated and specialized software packages are needed for implementing them. Most available QTL software for experimental crosses are based on: single marker analysis; IM of Lander and Botstein (2); regression interval mapping of Haley and Knott (3); and composite interval mapping of Zeng (6, 7). Packages usually handle commonly used mapping populations, such as BC or F2 intercross. For the same data set, all should provide similar if not the same results. The differences among those packages are data format, computer platforms, user interface, graphic output, etc. Below we discuss three public domain software packages (*MapMaker/QTL*, *QTL Cartographer*, and *MapManager*) which are commonly used by biologists and two newly emerging software, *R/QTL* and *R/QTLbim*.

The original software, *Mapmaker/QTL*, is one of the first available QTL mapping software (http://hpcio.cit.nih.gov/lserver/MAPMAKER_QTL.html) (70). *Mapmaker/QTL* is a companion program to *MapMaker*, which creates genetic maps from marker genotypes. *Mapmaker/QTL* runs on most platforms and implements the interval mapping approach. It uses command-driven user interface and provides no graphic interface, however, the output graphs can be saved as a postscript file.

QTL Cartographer (http://statgen.ncsu.edu/qtlcart/index.php) (71) implements CIM, MIM with epistasis, multiple trait analysis, and also maps categorical traits. The original software was command-driven, not user-friendly. The new Windows-version, *WinQTlCart*, is user-friendly with a powerful graphic interface.

MapManager QT/QTX (http://mapmanager.org/mmQTX.html) (72, 73) allows for the single marker analysis, IM, CIM, and interacting QTL search. The unique features of *MapManager QT/QTX* are its ability in supporting advanced crosses and recombinant inbred intercross (RIX) and its rich database on common recombinant inbred (RI) mouse resources. *MapManager QT/QTX* is also user-friendly with nice graphic interface.

One drawback of all the above packages is their restricted implementation on specific mapping methods and concentration on computing instead of model building. Thus they are limited in handling mapping data with complex models, such as modeling non-genetic covariate effects. In practice, regression analysis is done first to adjust the effects of non-genetic covariates. The adjusted phenotypes are then used for further QTL mapping. This two-stage analysis may not be efficient. Simultaneously modeling both genetic and non-genetic factors will be statistically more powerful. *R/QTL* (http://www.rqtl.org/) (74) is a new QTL mapping package with emphasis on improving modeling rather than computing. *R/QTL* is implemented as an add-on package for the freely available and widely used statistical language/software R (http://www.r-project.org/). *R/QTL* can perform single-QTL, two-dimensional, two-QTL genome scans with the possible inclusion of covariates. *R/qtl* requires some basic R programming skill. To help users without programming skill to analyze their data, a Java GUI for *R/qtl*, called *J/qtl* has been developed. *R/qtlbim* (75) (http://www.qtlbim.org/) is a software for Bayesian Interval Mapping. It provides a Bayesian model selection approach to map multiple interacting QTL. The package can handle continuous, binary, and ordinal traits. It allows for epistasis and interacting covariates. *R/qtlbim* is built upon *R/qtl* and requires package *R/qtl* version 1.03 or later.

WinQTlCart, MapManager QT, and *R/QTL* all calculate empirical threshold by permutation procedure (16) and estimate the confidence interval for QTL position by bootstrap (37). *R/QTL* and *MapMaker/QTL* also permit non-parametric mapping methods described in **Section 5** for non-normal phenotypes.

All the software assume a known linkage map in their analysis. *Mapmaker/QTL, MapManager QT*, and *R/QTL* have their own companion packages to construct linkage map and check genotyping errors before QTL analysis. QTL Cartographer can directly incorporate the linkage results from *MapMaker* into its analysis.

6. Conclusions

In this chapter, we described several standard QTL methods and their extensions for both single QTL and multiple QTL mapping. Major statistical issues on QTL mapping, such as threshold, confidence interval calculations were also discussed. Some important topics, such as expression QTL (eQTL) analysis, are left to the following chapters.

References

1. Sax, K. (1923) The association of size differences with seed-coat pattern and pigmentation *Phaseolus Vulgaris. Genetics* **8**: 552–560.
2. Lander, ES, Botstein, D. (1989) Mapping Mendelian factors underlying quantitative traits using RFLP linkage maps. *Genetics* **121**: 185–199.

3. Haley, CS, Knott, SA. (1992) A simple regression method for mapping quantitative trait in line crosses using flanking markers. *Heredity* **69**: 315–324.
4. Kao, CH, Zeng, ZB. (1997) General formulas for obtaining the MLEs and the asymptotic variance-covariance matrix in mapping quantitative trait loci when using the EM algorithm. *Biometrics* **53**, 653–665.
5. Kao, CH, Zeng, ZB, Teasdale, RD. (1999) Multiple interval mapping for quantitative trait loci. *Genetics* **152**: 1203–1216.
6. Zeng, ZB. (1993) Theoretical basis of separation of multiple linked gene effects on mapping quantitative trait loci. *Proc Nat Acad Sci USA* **90**: 10972–10976.
7. Zeng, ZB. (1994) Precision mapping of quantitative traits loci. *Genetics* **136**: 1457–1468.
8. Jansen, RC, Stam, P. (1994) High resolution of quantitative traits into multiple quantitative trait in line crosses using flanking markers. *Heredity* **69**: 315–324.
9. Satagopan, JM, Yandell, BS, Newton, MA, et al. (1996) A Bayesian approach to detect quantitative trait loci using Markov chain Monte Carlo. *Genetics* **144**: 805–816.
10. Sillanpää, MJ, Arjas, E. (1998) Bayesian mapping of multiple quantitative trait loci from incomplete inbred line cross data. *Genetics* **148**: 1373–1388.
11. Stephens, DA, Fisch, RD. (1998) Bayesian analysis of quantitative trait locus data using reversible jump Markov chain Monte Carlo. *Biometrics* **54**: 1334–1347.
12. Yi, N, Xu, S. (2000) Bayesian mapping of quantitative trait loci for complex binary traits. *Genetics* **155**: 1391–1403.
13. Yi, N, Xu, S. (2001) Bayesian mapping of quantitative trait loci under complicated mating designs. *Genetics* **157**: 1759–1771.
14. Hoeschele, I. (2001) Mapping quantitative trait loci in outbred pedigrees. In: Balding, D, Bishop, M, Cannings, O, (eds) Handbook of Statistical Genetics, Wiley and Sons, New York, pp. 599–644.
15. Yi, N. (2004) A unified Markov chain Monte Carlo framework for mapping multiple quantitative trait loci. *Genetics* **167**: 967–975.
16. Churchill, GA, Doerge, RW. (1994) Empirical threshold values for quantitative trait mapping. *Genetics* **138**: 963–971.
17. Zou, F, Fine, JP, Hu, J, et al. (2004) An efficient resampling method for assessing genome-wide statistical significance in mapping quantitative trait loci. *Genetics* **168**: 2307–2316.
18. Stoehr, JP, Nadler, ST, Schueler, KL, et al. (2000) Genetic obesity unmasks nonlinear interactions between murine type 2 diabetes susceptibility loci. *Diabetes* **49**: 1946–1954.
19. Lan, H, Kendziorski, CM, Haag, JD, et al. (2001) Genetic loci controlling breast cancer susceptibility in the Wistar-Kyoto rat. *Genetics* **157**: 331–339.
20. Ott, J (1999) *Analysis of Human Genetic Linkage*. The Johns Hopkins University Press, Baltimore, MD.
21. Lynch, M, Walsh, B. (1998) Genetics and Analysis of Quantitative Traits. Sinauer Associates, Sunderland, MA.
22. Liu, BH. (1998) Statistical Genomics: Linkage, Mapping, and QTL Analysis. CRC Press, Boca Raton, FL
23. Dempster, AP, Laird, NM, Rubin, DB. (1997) Maximum likelihood from incomplete data via the EM algorithm. *J R Stat. Soc. Series B* **39**: 1–38.
24. Knapp, SJ, Bridges, WC, Birkes, D. (1990) Mapping quantitative trait loci using molecular marker linkage maps. Theor. *Appl. Genet.* **79**: 583–592.
25. Martinez, O, Curnow, RN. (1992) Estimating the locations and sizes of the effects of quantitative trait loci using flanking markers. *Theor. Appl. Genet.* **85**: 480–488.
26. Xu, SZ. (1998) Iteratively reweighted least squares mapping of quantitative trait loci. *Behav. Genet.* **28**: 341–355.
27. Dupuis, J, Siegmund, D. (1999) Statistical methods for mapping quantitative trait loci from a dense set of markers. *Genetics* **151**: 373–386.
28. Rebai, A, Goffinet, B, Mangin, B, et al. (1994) Detecting QTLs with diallel schemes. In: van Ooijen JW, Jansen, J. (eds) Biometrics in plant breeding: applications of molecular markers. 9th meeting of the EUCARPIA, Wageningen, The Netherlands.
29. Rebai, A, Goffinet, B, Mangin, B. (1995) Comparing power of different methods for QTL detection. *Biometrics* **51**: 87–99.
30. Piepho, HP. (2001) A quick method for computing approximate thresholds for quantitative trait loci detection. *Genetics* **157**: 425–432.
31. Zou, F, Yandell, BS, Fine, JP. (2001) Statistical issues in the analysis of quantitative traits in combined crosses. *Genetics* **158**: 1339–1346.
32. Zou, F, Fine, JP. (2002) Note on a partial empirical likelihood. Biometrika **89**: 958–961.
33. Zou, F, Xu, ZL, Vision, TJ. (2006) Assessing the significance of quantitative trait loci

in replicated mapping populations. *Genetics* **174**: 1063–1068.

34. Churchill, GA, Doerge, RW. (2008) Naive application of permutation testing leads to inflated type I error rates. *Genetics.* **178**: 609–610.
35. Manichaikul, A, Dupuis, J, Sen, S, et al. (2006) Poor performance of bootstrap confidence intervals for the location of a quantitative trait locus. *Genetics* **174**: 481–489.
36. Mangin, B, Goffinet, B, Rebai, A. (1994) Constructing confidence intervals for QTL location. *Genetics* **138**: 1301–1308.
37. Visscher, PM, Thompson, R, Haley, CS, (1996) Confidence intervals in QTL mapping by bootstrapping. *Genetics* **143**: 1013–1020.
38. Zeng, ZB. (2000). Unpublished notes on Statistical model for mapping Quantitative trait loci. North Carolina State University, Raleigh, NC.
39. Zeng, ZB, Kao, CH, Basten, CJ. (1999) Estimating the genetic architecture of quantitative traits. Genet Res **74**: 279–289.
40. Broman, KW, Speed, TP. (2002) A model selection approach for the identification of quantitative trait loci in experimental crosses. *J. R. Stat. Soc. Series B* **64**: 641–656.
41. Green, PJ. (1995) Reversible jump Markov Chain Monte Carlo computation and Bayesian model determination. *Biometrika* **82**: 711–732.
42. Ven, RV. (2004) Reversible-Jump Markov Chain Monte Carlo for quantitative trait loci mapping. *Genetics* **167**: 1033–1035.
43. Godsill, SJ. (2001) On the relationship between Markov chain Monte Carlo methods for model uncertainty. *J. Comput. Graph. Stat.* **10**: 230–248.
44. Godsill, SJ. (2003) Proposal densities, and product space methods. In: Green, PJ, Hjort, NL, Richardson, S, (eds) *Highly Structured Stochastic Systems.* Oxford University Press, London/New York/Oxford.
45. George, EI, McCulloch, RE, (1993) Variable selection via gibbs sampling. *J. Am. Stat. Assoc.* **88**: 881–889.
46. Wang, H, Zhang, YM, Li, X, et al. (2005) Bayesian shrinkage estimation of quantitative trait loci parameters. *Genetics* **170**: 465–480.
47. Yi, N, Shriner, D. (2008) Advances in Bayesian multiple QTL mapping in experimental 11 designs. *Heredity* **100**: 240–252.
48. Bernardo, R. (1994) Prediction of maize single-cross performance using RFLPs and information from related hybrids. *Crop Science* **34**: 20–25.
49. Liu, Y, Zeng, ZB. (2000) A general mixture model approach for mapping quantitative trait loci from diverse cross designs involving multiple inbred lines. *Genet. Res.* **75**: 345–355.
50. Draper, NR, Smith, H. (1998) *Applied Regression Analysis.* John Wiley & Sons, New York.
51. Visscher, PM, Haley, CS, Knott, SA. (1996) Mapping QTLs for binary traits in backcross and F-2 populations. *Genet. Res.* **68**: 55–63.
52. Xu, S, Atchley, WR. (1995) A random model approach to interval mapping of quantitative genes. *Genetics* **141**: 1189–1197.
53. McIntyre, LM, Coffman, C, Doerge, RW. (2000) Detection and location of a single binary trait locus in experimental populations. *Genet. Res.* **78**: 79–92.
54. Rao, SQ, Li, X. (2000) Strategies for genetic mapping of categorical traits. *Genetica* **109**: 183–197.
55. Broman, KW. (2003) Quantitative trait locus mapping in the case of a spike in the phenotype distribution. *Genetics* **163**: 1169–1175.
56. Mackay, TF, Fry, JD. (1996) Polygenic mutation in Drosophila melanogaster: genetic interactions between selection lines and candidate quantitative trait loci. *Genetics* 144: 671–688.
57. Shepel, LA, Lan, H, Haag, JD, et al. (1998) Genetic identification of multiple loci that control breast cancer susceptibility in the rat. *Genetics* **149**: 289–299.
58. Hackett, CA, Weller, JI. (1995) Genetic mapping of quantitative trait loci for traits with ordinal distributions. *Biometrics* **51**: 1254–1263.
59. Diao, G, Lin, DY, Zou, F. (2004) Mapping quantitative trait loci with censored observations. *Genetics* **168**: 1689–1698.
60. Kruglyak, L, Lander, ES, (1995) A nonparametric approach for mapping quantitative trait loci. *Genetics* **139**: 1421–1428.
61. Poole, TM, Drinkwater, NR, (1996) Two genes abrogate the inhibition of murine hepatocarcinogenesis by ovarian hormones. *Proc Nat Acad Sci USA* **93**: 5848–5853.
62. Zou, F, Fine, JP, Yandell, BS. (2002) On empirical likelihood for a semiparametric mixture model. *Biometrika* **89**: 61–75.
63. Jin, C, Fine, JP, Yandell, B. (2007) A unified semiparametric framework for QTL mapping, with application to spike phenotypes. *J Am Stat Assoc* **102**: 56–57.

64. Fine, JP, Zou, F, Yandell, BS. (2004) Non-parametric estimation of mixture models, with application to quantitative trait loci. *Biostatistics* **5**: 501–513.
65. Huang, C, Qin, J, Zou, F. (2007) Empirical likelihood-based inference for genetic mixture models. *Can J Stat* **35**: 563–574.
66. Lange, C., Whittaker, J. C. (2001). Mapping quantitative trait loci using generalized estimating equations. *Genetics* **159**: 1325–1337.
67. Symons, RCA, Daly, MJ, Fridlyand, J, et al. (2002) Multiple genetic loci modify susceptibility to plasmacytoma-related morbidity in E5-v-abl transgenic mice. *Proc Nat Acad Sci USA* **99**: 11299–11304.
68. Diao, G, Lin, DY. (2005) Semiparametric methods for mapping quantitative rait loci with censored data. *Biometrics* **61**: 789–798.
69. Epstein, MP, Lin, X, Boehnke, M, (2003) A Tobit variance-component method for linkage analysis of censored trait data. *Am J Hum Genet* **72**: 611–620.
70. Lincoln, SE, Daly, MJ, Lander, ES. (1993) A tutorial and reference manual. 3rd edn. Technical Report Whitehead Institute for Biomedical Research.
71. Basten, CJ, Weir, BS, Zeng, ZB. (1997) *QTL Cartographer: A Reference Manual and Tutorial for QTL Mapping*. North Carolina State University, Raleigh, NC.
72. Manly, KF, Cudmore, JRH, Meer, JM. (2001) Map Manager QTX, cross-platform software for genetic mapping. *Mamm Genome* **12**: 930–932.
73. Manly, KF, Olson, JM, (1999) Overview of QTL mapping software and introduction to Map Manager QT. *Mamm Genome* **10**: 327–334.
74. Broman, KW, Wu, H, Sen, S, et al. (2003) R/qtl: QTL mapping in experimental crosses. *Bioinformatics* **19**: 889–890.
75. Yandell, BS, Mehta, T, Banerjee, S, et al. (2007) R/qtlbim: QTL with Bayesian Interval Mapping in experimental crosses. *Bioinformatics* **23**: 641–643.

Chapter 10

Quantitative Trait Locus Analysis Using J/qtl

Randy Smith, Keith Sheppard, Keith DiPetrillo, and Gary Churchill

Abstract

Quantitative trait locus (QTL) analysis is a statistical method to link phenotypes with regions of the genome that affect the phenotypes in a mapping population. R/qtl is a powerful statistical program commonly used for analyzing rodent QTL crosses, but R/qtl is a command line program that can be difficult for novice users to run. J/qtl was developed as an R/qtl graphical user interface that enables even novice users to utilize R/qtl for QTL analyses. In this chapter, we describe the process for analyzing rodent cross data with J/qtl, including data formatting, data quality control, main scan QTL analysis, pair scan QTL analysis, and multiple regression modeling; this information should enable new users to identify QTL affecting phenotypes of interest within their rodent cross datasets.

Key words: Quantitative trait locus analysis, QTL.

1. Introduction

Mouse and rat inbred strains exhibit wide variations in many phenotypes, including some phenotypes that model chronic human diseases. Quantitative trait locus (QTL) analysis is a method to link regions of the genome with phenotypes that vary amongst inbred strains (*see* (1, 2) for review), a first step toward identifying the causal genes underlying the phenotypes. The QTL analysis process typically involves generating a cross between inbred strains that differ in one or more phenotypes of interest (i.e., backcross, intercross, advanced intercross, etc.), phenotyping the animals generated, genotyping the animals with genetic markers evenly spaced across the genome, and then linking the genotypes at those markers with phenotypes measured in the population. R/qtl is a program commonly used for the statistical analysis of rodent QTL crosses (3). However, R/qtl is a command

K. DiPetrillo (ed.), *Cardiovascular Genomics,* Methods in Molecular Biology 573,
DOI 10.1007/978-1-60761-247-6_10, © Humana Press, a part of Springer Science+Business Media, LLC 2009

line program that can be difficult for novice users to run, so J/qtl was developed as an R/qtl graphical user interface that enables even novice users to utilize R/qtl for QTL analyses.

In this chapter, we describe the process for analyzing rodent cross data with J/qtl to identify QTL affecting phenotypes of interest. As users choose analysis parameters throughout the analysis process, the corresponding R/qtl commands are shown in the preview window to allow users to see the commands and insert notations; all commands and notations can be exported or saved for future reference (*see* **Note 1**). In addition to the procedure described here, J/qtl has an extensive help section (*see* **Note 2**) that provides more detailed information about the theory and procedures employed in the program.

2. Materials

2.1. Hardware

1. Operating system – Windows XP, Windows Vista, Mac OS X 10.4 or greater.
2. A minimum of 512 MB RAM is recommended. Additional memory may be required for large datasets and will increase analysis speed.

2.2. Software

1. Java – http://java.com
2. R – http://cran.r-project.org
3. R/qtl – http://www.rqtl.org/ (*see* **Note** 3)
4. J/qtl – http://research.jax.org/faculty/churchill/software/Jqtl/index.html

3. Methods

The software packages listed above should be downloaded and installed prior to performing QTL analysis. The data analysis process starts by combining the genotype and phenotype data into an appropriately formatted dataset, then checking the quality of the dataset. Missing genotypes, abnormal recombination patterns, and non-normal phenotype distributions should be corrected and the data quality confirmed by repeating the quality analysis steps. After ensuring a high-quality dataset, the data are analyzed for main effect QTL and permuted to calculate significance thresholds (4). The dataset is further analyzed for epistasis by identifying interacting QTL pairs linked to the phenotype(s) of interest. Finally, these main effect and interacting QTL are fit into a multiple regression model.

3.1. Data Format

1. Combine the genotype and phenotype data into a single Excel spreadsheet (**Fig. 10.1**) in a comma-delimited format (.csv).

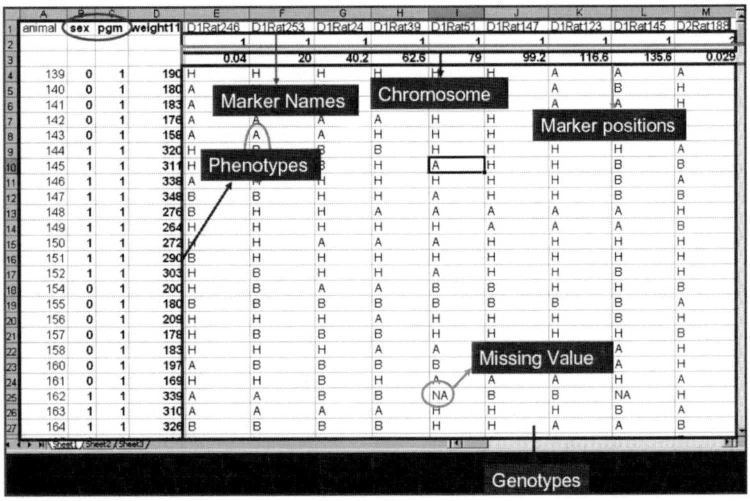

Fig. 10.1. Example of dataset format required for J/qtl.

2. Row 1 should contain the column headings, including the phenotype designations and the names of the genotyping markers.
3. Row 2 should be empty in the columns containing phenotype data and contain the chromosome numbers of the genotype marker listed in that column.
4. Row 3 should be empty in the columns containing phenotype data and contain the centimorgan (cM) position (*see* **Note 4**) of the genotype marker listed in that column.
5. Rows 4 and higher should contain the phenotype data and genotype information for the animal listed in that row. Phenotype information should be numerical or numerically coded, covariates may be quantitative or categorical, and genotype data can be either numerically coded or alphabetical (genotype codes will be specified within the program later on).
6. To include Chromosome X data in the analysis, you must have a phenotype for sex coded as female = 0 and male = 1. When analyzing Chromosome X for an intercross, you additionally need a pgm phenotype (paternal grandmother) to indicate the direction of the cross. For example, pgm should

be coded as 0 or 1, with 0 indicating crosses (A × B) × (A × B) or (B × A) × (A × B) and 1 indicating crosses (A × B) × (B × A) or (B × A) × (B × A).

7. Save the file in a comma-delimited format (.csv).

3.2. Loading the Data

1. Start J/qtl, choose "File" → "New QTL Project" within the main window (**Fig. 10.2**), then name the new project.

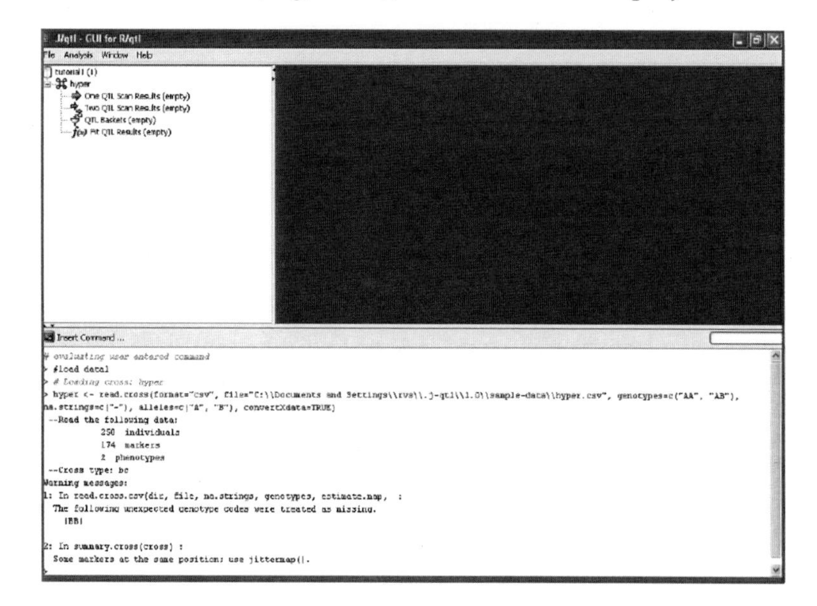

Fig. 10.2. J/qtl main window enables users to load datasets, open analysis wizards, view analysis results, and save/export results.

2. To load the dataset, choose "File" → "Load Cross Data" to open the Load Cross Data Wizard (**Fig. 10.3**).
3. Click the "Browse Files" tab to find and select the .csv file containing the dataset.
4. Choose the file format (i.e., comma-delimited) from the drop-down menu. This will load data into the Cross Data Preview window.
5. Type a Cross Name into the text box and choose a cross type from the drop-down menu.
6. Type genotype and allele codes into the available text boxes.
7. Click the "Load Cross" tab (*see* **Note 5**).

3.3. Data Quality Control

1. To examine normality of the phenotypes, choose "Analysis" → "Phenotype Checking" → "Histograms."
2. Review the histograms to confirm that the phenotype data are normally distributed.
3. Transform phenotype data that are not normally distributed (*see* **Note 6**).

Load Cross Data

Cross Data File: C:\Documents and Settings\rvs\.j-qtl\1.0\sample-data\hyper.csv
Browse Files ...
File Format: Comma-Delimited
Cross Data Preview (First 10 Lines):

Cross Name: hyper Cross Type: Backcross
Genotype Codes:
AA: AA AB: AB BB: B Not AA: D Not BB: C Missing: -
Allele Codes:
First Allele: A Second Allele: B
Convert X Chromosome Data to Internal Format (Recommended)

Command Preview:

```
hyper <- read.cross(format="csv", file="C:\\Documents and
Settings\\rvs\\.j-qtl\\1.0\\sample-data\\hyper.csv", genotypes=c("AA", "AB"),
na.strings=c("-"), alleles=c("A", "B"), convertXdata=TRUE)
```

Load Cross Cancel Help...

Fig. 10.3. Load Cross Data Wizard allows users to import a dataset in J/qtl for analysis and input the cross and genotype information.

4. Choose "Analysis" → "Genotype Checking" → "Genotype Plot" to open the corresponding wizard (**Fig. 10.4**). Choose "Input Order" from the Sort By drop-down menu and "Genotype" from the Things to plot drop-down menu. Type "all" into the Choose Individuals text box.

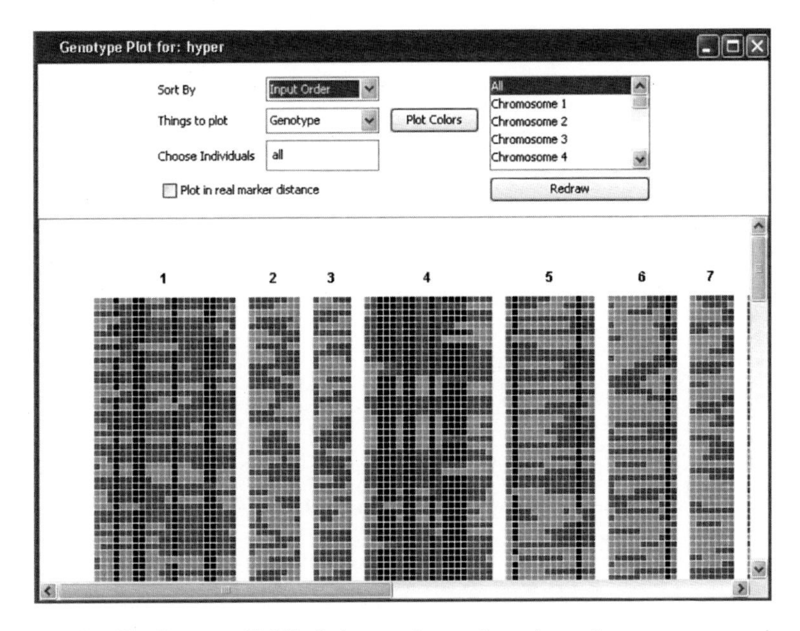

Fig. 10.4. The Genotype Plot illustrates genotype patterns for each mouse across each chromosome, including missing genotypes shown in *black*, and serves as an important quality control of the genotype data.

5. Click on an individual box to retrieve detailed spot information if needed.
6. Choose "Analysis" → "Genotype Checking" → "Genetic Map" → "Display Genetic Map" to show the genetic map for the cross and illustrate the genome coverage of the genetic markers tested on the animals (**Fig. 10.5**).

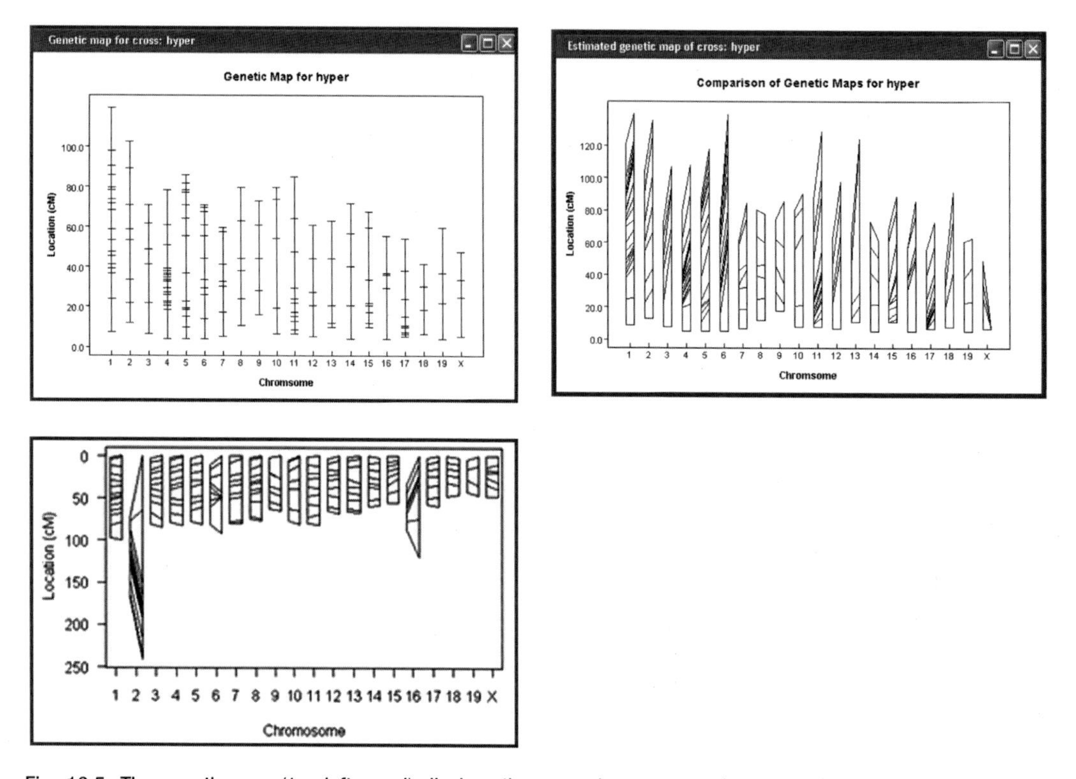

Fig. 10.5. The genetic map (*top left panel*) displays the genomic coverage of genotyped markers, as well as the intermarker interval. The estimated genetic map (*top right panel*) compares the known genetic length of each chromosome with the genetic length estimated from the dataset. The expanded lengths of chromosomes 2 and 16 (*bottom panel*) indicate misplaced marker order or high genotyping error rate on those chromosomes.

7. Choose "Analysis" → "Genotype Checking" → "Genetic Map" → "Estimate Genetic Map" to compare the known genome map with the one estimated from the experimentally derived recombination within the dataset. Large discrepancies between the known and estimated genetic maps, indicated by one or more estimated chromosomes that are much longer than the rest (**Fig. 10.5**), signify a genotyping problem that should be fixed before starting the QTL analysis.
8. Choose "Analysis" → "Genotype Checking" → "RF Plot" to generate a graph illustrating pairwise recombination fractions across the genome (calculated from the genotype data in the

dataset). Linkage of a marker with a chromosome other than the one on which it is thought to reside, indicated by red zones off the diagonal of the graph (*see* **Note 7** and **Fig. 10.6**), points to a genotyping error that should be fixed before starting the QTL analysis.

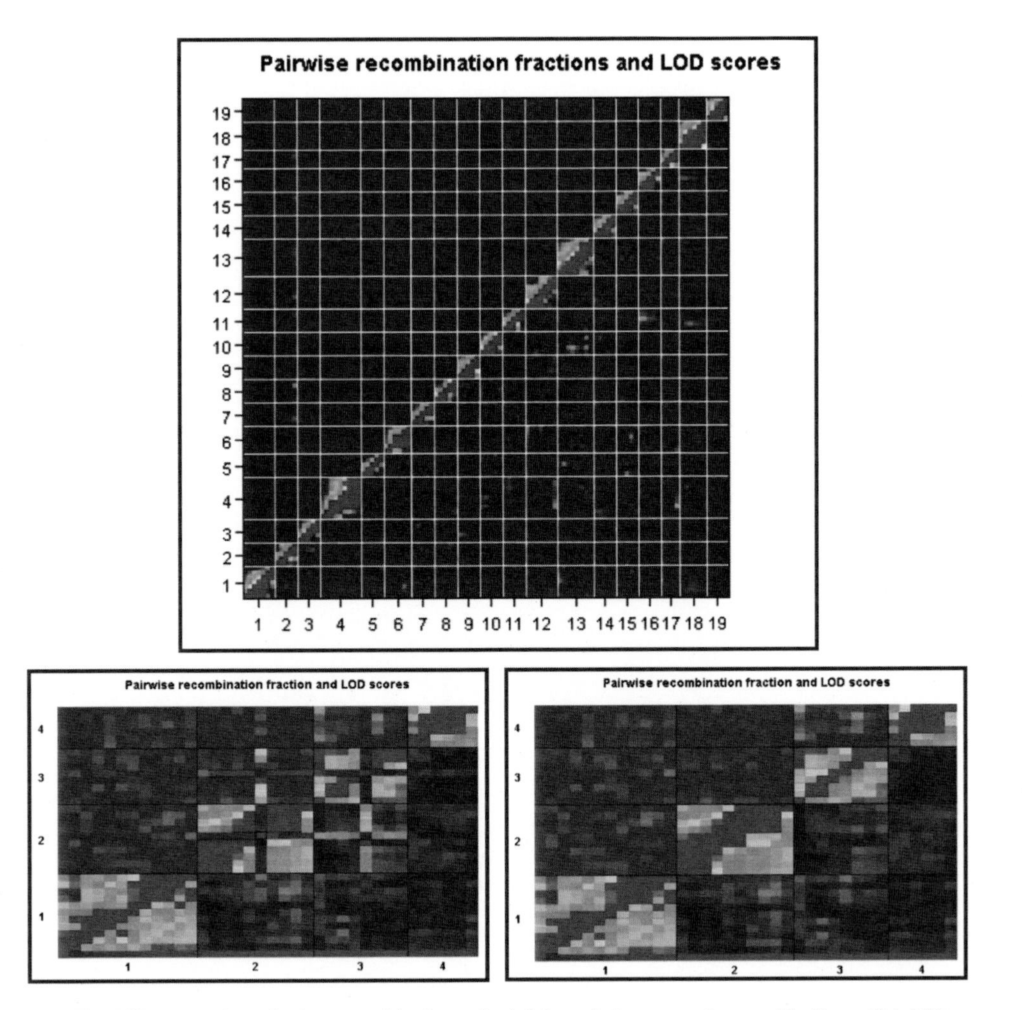

Fig. 10.6. The LOD scores for pairwise recombination reflect linkage between marker combinations. High LOD scores along the diagonal (*top panel*) indicate linkage between markers on the same chromosome, as expected. Linkage between markers on different chromosomes (*bottom left panel*) indicates genotyping errors or marker misplacement. Identifying and correcting the genotyping errors will improve the recombination plot (*bottom right panel*).

3.4. QTL Analysis – Main Scan

1. To open the One QTL Genome Scan Wizard (**Fig. 10.7**), choose "Analysis" → "Main Scan" → "Run One QTL Genome Scan."
2. Specify the cross name, phenotype for analysis, and phenotype distribution from the drop-down menus.

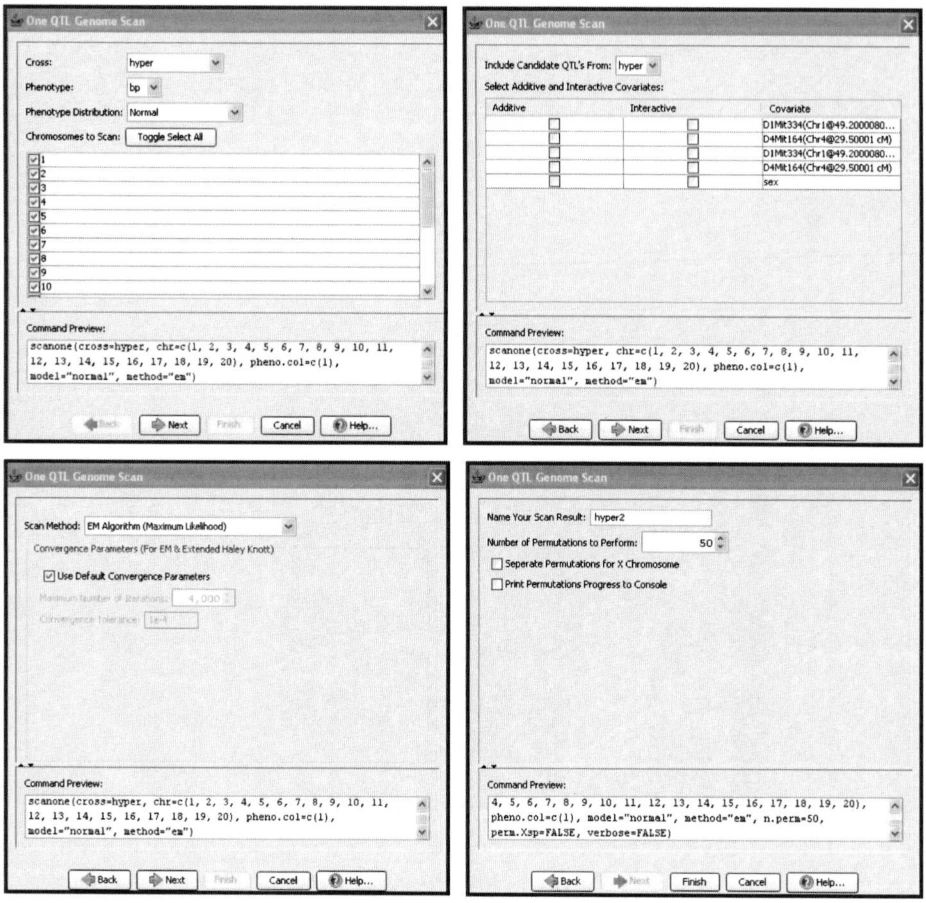

Fig. 10.7. Four windows within the One QTL Genome Scan Wizard.

3. Select the chromosomes to scan by clicking the "Toggle Select All" tab or by individually clicking the check-boxes for each chromosome.
4. Click the "Next" tab to continue.
5. Choose interactive or additive covariates to include in the main scan by clicking the check-boxes for QTL or phenotypes.
6. Click the "Next" tab to continue.
7. Select the scan method from the drop-down menu (*see* **Table 10.1** for choosing an appropriate scan method) and either click the check-box to use default convergence parameters or enter chosen parameters into the text boxes (*see* **Note 8**).
8. Click the "Next" tab to continue.

Table 10.1
Dataset characteristics determine the appropriate main scan method

Phenotype distribution	Model	Method	Allow covariates
Continuous	Normal	EM, IMP, MR, HK	Yes
	Non-parametric	EM	No
	Two-part	EM, MR	Yes
Binary	Binary model	EM, MR only	Yes

EM: maximum likelihood is performed via the EM algorithm (5).
IMP: multiple imputation is used, as described by (6).
HK: Haley–Knott regression of the phenotypes on the multipoint QTL genotype probabilities is used (7).
MR: Marker regression is used. Analysis is performed only at the genetic markers and individuals with missing genotypes are discarded (8).

9. Enter a name for the analysis results in the text box, select a number of permutations to perform to compute significance thresholds (*see* **Note 9**), and click the check-boxes for the permutation options desired.
10. Click the "Finish" tab to start the analysis. When the analysis is complete, the results will populate the one QTL scan results section in the main J/qtl window.
11. Open the one QTL scan results to see the LOD score plot for the genome scan (**Fig. 10.8**; *see* **Note 10**).
12. Open the scan summary (**Fig. 10.8**) and choose the appropriate cross, phenotype, threshold, and chromosome peaks from the drop-down menus to see a summary of the QTL

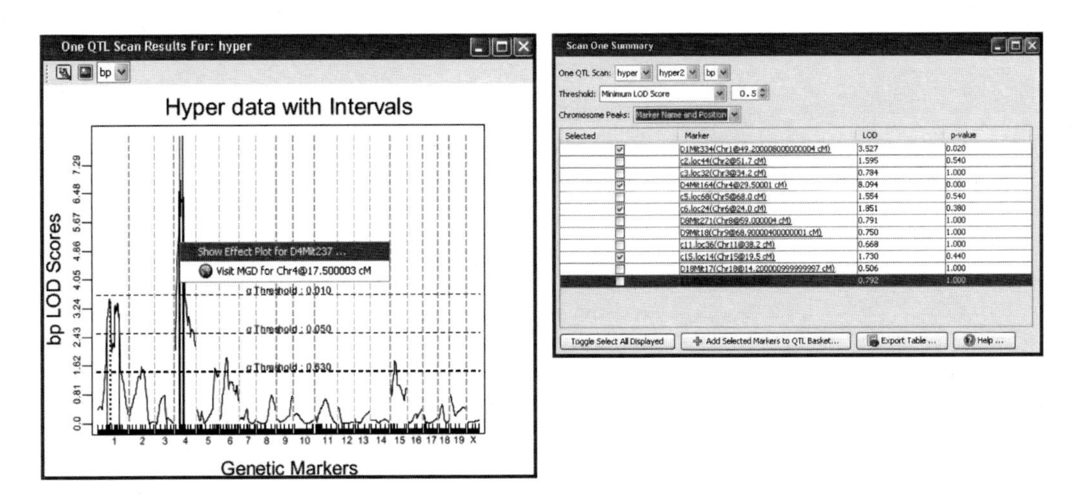

Fig. 10.8. Results from the One QTL Genome Scan are displayed as a genome-wide LOD score plot (*left panel*) and summary table (*right panel*).

fitting the criteria. To send some QTL to the QTL Basket for subsequent multiple regression analysis, click the check-box for the appropriate QTL then click the "Add Selected Markers to QTL Basket" button.

3.5. QTL Analysis – Pair Scan

1. To open the One QTL Genome Scan Wizard, choose "Analysis" → "Main Scan" → "Run Two QTL Genome Scan."
2. Specify the cross name, phenotype for analysis, and phenotype distribution from the drop-down menus.
3. Select the chromosomes to scan by clicking the "Toggle Select All" tab or by individually clicking the check-boxes for each chromosome.
4. Click the "Next" tab to continue.
5. Choose interactive or additive covariates to include in the main scan by clicking the check-boxes for QTL or phenotypes.
6. Click the "Next" tab to continue.
7. Select the scan method from the drop-down menu and either click the check-box to use default convergence parameters or enter chosen parameters into the text boxes.
8. Click the "Next" tab to continue.
9. Enter a name for the analysis results in the text box, select a number of permutations to perform to compute significance thresholds (permutations can take a long time for pair scans), and click the check-boxes for the permutation options desired. Pair scans can take a long time to complete, thus it is useful to perform and time a preliminary analysis with only a few chromosomes and no permutations to estimate how long the full scan with permutations will take.
10. Click the "Finish" tab to start the analysis. When the analysis is complete, the results will populate the two QTL scan results section in the main J/qtl window.
11. Open the two QTL scan results to see the interaction plot for the pair scan (**Fig. 10.9**).
12. Open the scan summary (**Fig. 10.9**) and choose the appropriate scan, model to optimize, threshold type and levels, and chromosome pair peaks from the drop-down menus to see a summary of the QTL fitting the criteria. To send some QTL to the QTL Basket (**Fig. 10.10**) for subsequent multiple regression analysis, click the check-box for the appropriate QTL, then click the "Add Selected Markers to QTL Basket" tab.

3.6. Multiple Regression Analysis

1. Choose "Analysis" → "Fit QTL Model" → "Fit QTL" to open the window to input multiple regression analysis options (**Fig. 10.10**).

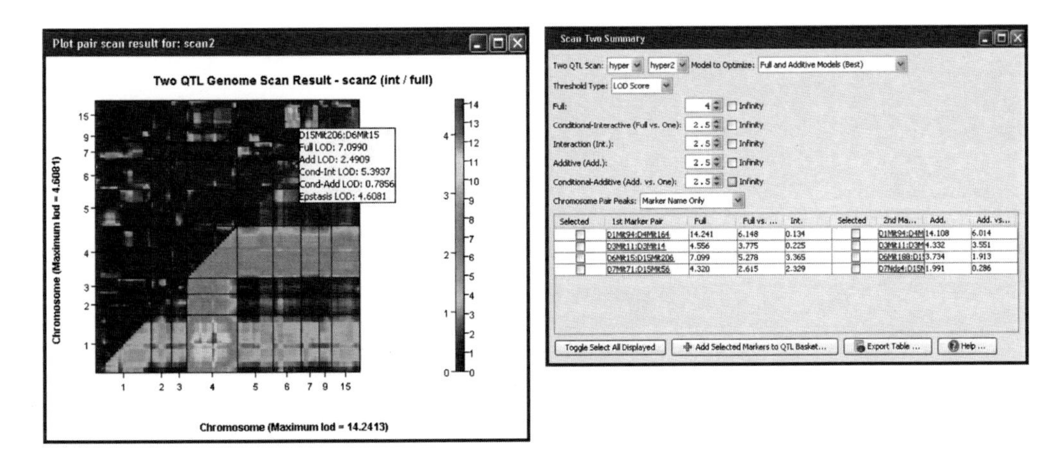

Fig. 10.9. Results from the two QTL genome scan are displayed as an interaction plot (*left panel*) and summary table (*right panel*).

Fig. 10.10. QTL basket (*left panel*) contains main effect and interacting QTL that can be included in the multiple regression analysis (i.e., fit analysis; *right panel*).

2. From the drop-down menus, specify the cross name, phenotype to fit, and QTL basket containing the results on which to perform the multiple regression analysis.
3. Add or remove terms from the analysis using the appropriate tabs.
4. Click the appropriate check-boxes to perform drop-one-term analysis or estimate QTL effects.
5. Enter a name for the fit analysis in the text box.
6. Click the "OK" tab to start the analysis.
7. The multiple regression analysis results are shown in a separate window. Click "Export Table" to save the results.

4. Notes

1. J/qtl allows users to annotate and save their QTL analyses to facilitate reproducible research. Users can add comment lines into the R command windows to specify why particular options, for example, interacting covariates in the main scan wizard, were chosen for that analysis. Choose "File" → "Save Project" to save the object definitions and the R data file in a zip file for archiving. Choose "File" → "Export R Commands" to save an R script containing all of the R commands used throughout the analysis; this is valuable for archiving the analysis and as a template for future analysis of similar datasets. To save all of the commands, comments, and output in the R window, click inside the R window, highlight all, copy, and then paste into a separate text document.
2. The J/qtl help section provides detailed information explaining the rationale for steps in the QTL analysis process and guides users through how to perform various steps within J/qtl. The help section can be accessed from the "Help" menu in the main window or by clicking the "Help" present in most wizards within J/qtl.
3. J/qtl will search for a compatible version of R/qtl and attempt to download and install the latest version of R/qtl if it does not detect one.
4. With the availability of full genome sequences for many species, most genotyping markers available for genetic analysis have been physically mapped to base pair positions in the genome, but not to genetic map positions (i.e., cM). J/qtl requires cM map positions for genotyping markers, so physical positions must be converted to genetic positions; the mouse map conversion tool at http://cgd.jax.org/mousemapconverter can be used for this purpose. It is also possible to estimate the genetic positions of the markers by using recombination frequencies calculated from the genotyping within the dataset, but the accuracy will be limited by the number of animals in the cross. A very rough approximation of genetic positions in the mouse genome is to assume a genome-wide conversion of 1 cM for every 2 Mb and divide the physical position (in Mb) of each marker by 2 to estimate the cM position. While this method fails to account for local differences in recombination frequency, this method is consistent across different crosses and works because QTL mapping is robust to variation in marker position (but not in marker order).
5. When loading the cross data into J/qtl using the wizard, pay attention to the warning messages in the R command preview window. For example, genotyping codes that are found in the

dataset but not specified in the Load Cross Data window will be treated as missing. Also, R will warn users if multiple markers are listed at the same position and suggest jittermap to resolve the problem. Users can run jittermap by right clicking on the cross name in the work tree within the J/qtl main window.

6. Non-normally distributed phenotype data can often be transformed to a more symmetrical distribution by logarithmic, power, or other transformation. Since non-normally distributed phenotype data can reduce statistical power during QTL analysis, this is an important quality-control step. One alternative to data transformation is to use non-parametric analysis of non-normally distributed data by choosing "Other (non-parametric)" from the phenotype distribution list on the first menu of the One QTL Genome Scan Wizard. Only phenotypes with approximately normal distributions should be included in the multiple regression analysis.
7. The RF plot illustrates the LOD scores for linkage disequilibrium (LD) between every pair of markers in the dataset. A pair of markers located on the same chromosome should be in greater LD than a pair of markers located on different chromosomes, so we expect the highest LOD scores in the RF plot (indicated by red) to fall along the diagonal. The off-diagonal areas of the plot should be blue to indicate low LD between markers on different chromosomes. Yellow, orange, or red off the diagonal indicates high LD between a pair of markers on different chromosomes, which often indicates that one or more markers has been incorrectly mapped. In this case, determine the correct position of the marker or genotype the population with a substitute marker to ensure that all markers are mapped to the correct positions in the genome and rerun the RF plot to confirm that the issue has been resolved. Pattern shifts in the RF plot can also indicate out of order markers and need to be corrected before QTL analysis.
8. If you have not run a genotype probability analysis for this data, the software will remind you of the correct method (e.g., Calculate Genotype Probability for EM methods or Simulate Genotype Probabilities for Imputation). A good initial test is to use the EM method and calculate genotype probabilities using a step size (cM) of 2.0. We recommend using EM for initial scans, but IMP is the only option available for fitting multiple QTL models. With the IMP method, you should use at least 64 (preferably 256) imputations; increasing the number of imputations not only improves mapping accuracy but also increases the computation time.

9. Permutation testing can be a time-consuming process, especially for pair scans. For preliminary QTL analysis of datasets, use approximately 50 permutations to calculate LOD thresholds to minimize the analysis time. However, use 1,000 permutations for final datasets to ensure accurate threshold determination.
10. The options button in the top left corner of the main scan plots allows users to (1) add a graph title and *X*- and *Y*-axis labels under the "Labels" tab; (2) specify the method and chromosomes used for calculating QTL confidence intervals under the "Intervals" tab; and (3) add lines depicting the significance threshold(s) to the graph under the "Thresholds" tab.

Acknowledgments

We are grateful to Sharon Tsaih and Qian Li for preparing the summary table and several figures.

References

1. Doerge, RW. (2002) Mapping and analysis of quantitative trait loci in experimental populations. *Nat Rev Genet* **3**(1): 43–52.
2. Broman, KW. (2001) Review of statistical methods for QTL mapping in experimental crosses. *Lab Anim* **30**(7): 44–52.
3. Broman, KW, Wu, H, Sen, S, et al. (2003) R/qtl: QTL mapping in experimental crosses. *Bioinformatics* **19**(7): 889–890.
4. Churchill, GA, Doerge, RW. (1994) Empirical threshold values for quantitative trait mapping. *Genetics* **138**(3): 963–971.
5. Dempster, AP, Laird, NM, Rubin, DB (1977) Maximum likelihood from incomplete data via Em algorithm. *J R Stat Soc Ser B-Methodol* **39**(1): 1–38.
6. Sen, S, Churchill, GA, (2001) A statistical framework for quantitative trait mapping. *Genetics* **159**(1): 371–387.
7. Haley, CS, Knott, SA (1992) A simple regression method for mapping quantitative trait loci in line crosses using flanking markers. *Heredity* **69**: 315–324.
8. Soller, M, Brody, T, Genizi, A (1976) Power of experimental designs for detection of linkage between marker loci and quantitative loci in crosses between inbred lines. *Theor Appl Gen* **47**(1): 35–39.

Chapter 11

Computing Genetic Imprinting Expressed by Haplotypes

Yun Cheng, Arthur Berg, Song Wu, Yao Li and Rongling Wu

Abstract

Different expression of maternally and paternally inherited alleles at certain genes is called genetic imprinting. Despite its great importance in trait formation, development, and evolution, it remains unclear how genetic imprinting operates in a complex network of interactive genes located throughout the genome. Genetic mapping has proven to be a powerful tool that can estimate the distribution and effects of imprinted genes. While traditional mapping models attempt to detect imprinted quantitative trait loci based on a linkage map constructed from molecular markers, we have developed a statistical model for estimating the imprinting effects of haplotypes composed of multiple sequenced single-nucleotide polymorphisms. The new model provides a characterization of the difference in the effect of maternally and paternally derived haplotypes, which can be used as a tool for genetic association studies at the candidate gene or genome-wide level. The model was used to map imprinted haplotype effects on body mass index in a random sample from a natural human population, leading to the detection of significant imprinted effects at the haplotype level. The new model will be useful for characterizing the genetic architecture of complex quantitative traits at the nucleotide level.

Key words: Genetic imprinting, haplotype, EM algorithm, single-nucleotide polymorphism, risk haplotype, genetic mapping.

1. Introduction

Genetic imprinting, also called the parent-of-origin effect, is an epigenetic phenomenon in which an allele from one parent is expressed while the same from the other parent remains inactive (1). The expression of imprinted genes from only one parental allele results from differential epigenetic marks that are established during gametogenesis. Since its first discovery in the early 1980s, genetic imprinting has been thought to play an important role in embryonic growth and development and disease pathogenesis in a variety of organisms (2–5). Several dozen genes in mice were

K. DiPetrillo (ed.), *Cardiovascular Genomics*, Methods in Molecular Biology 573,
DOI 10.1007/978-1-60761-247-6_11, © Humana Press, a part of Springer Science+Business Media, LLC 2009

detected to express from only one of the two parental chromosomes, some of which have been tested in other mammals including humans (6,7). Many experiments were conducted to investigate the biological mechanisms for imprinting and its influences on fetus development and various disease syndromes (8).

Thus far, no experiment has documented how the imprinting process affects development and disease. To better understand the imprinting regulation, it has been proposed to discover and isolate imprinting genes that may be distributed sporadically or located in clusters forming an imprinted domain (1, 7). One approach to discovering imprinting genes is based on genetic mapping in which individual quantitative trait loci (QTLs) showing parent-of-origin effects are localized with a molecular marker-based linkage map (9–17). Using an outbred strategy appropriate for plants and animals, significant imprinting QTLs were detected for body composition and body weight in pigs (15, 18–21), chickens (22), and sheep (23). Cui et al. (24) proposed an F_2-based strategy to map imprinting QTLs by capitalizing on the difference in the recombination fraction between sexes. More explorations on the development of imprinting models are given in Cui and others (25, 26).

Current models for genetic mapping characterize imprinting genes at the QTL level, but lack an ability to estimate the imprinting effects of DNA sequence variants that encode a complex trait. By altering the biological function of a protein that underlies a phenotype, single-nucleotide polymorphisms (SNPs) residing within a coding sequence open up a gateway to study the detailed genetic architecture of complex traits, such as many human diseases (27, 28). However, current experimental techniques do not allow the separation of maternally and paternally derived haplotypes (i.e., a linear combination of alleles at different SNPs on a single chromosome) from observed genotypes (29). More recently, a battery of computational models has been derived to estimate and test haplotype effects on a complex trait with a random sample drawn from a natural population (30–39). These models implement the population genetic properties of gene segregation into a unifying mixture-model framework for haplotype discovery. They define a so-called risk haplotype, i.e., one that is different from the remaining haplotypes in terms of its genetic effect on a trait.

The motivation of this work is to develop a statistical model for estimating the imprinting effects of SNP-constructed haplotypes on a natural population. Our model is built on Wu et al.'s (39) multiallelic haplotype model in which parent-dependent effects of a haplotype can be estimated and tested within a maximum likelihood framework. We derived closed forms for the EM algorithm to estimate additive, dominance, and imprinted effects of haplotypes. A real example from a pharmacogenetic study of obesity was used to demonstrate the application of the model.

2. Imprinting Model

Consider a pair of SNPs, **A** (with two alleles A and a) and **B** (with two alleles B and b), genotyped from a candidate gene. These two SNPs form nine possible genotypes, each derived from the combination of maternally and paternally inherited haplotypes, symbolized as $\text{Hap}^{\text{M}} = (AB,Ab,aB,ab)$ and $\text{Hap}^{\text{P}} = (AB,Ab,aB,ab)$, respectively (**Table 11.1**). We are interested in the detection of risk haplotype(s) and its imprinting effect on a quantitative trait. Much of our description about the definition and modeling of haplotype effects is derived from (39).

Table 11.1.
Nine genotypes of a pair of SNPs, their diplotypes and diplotype frequencies constructed by maternally and paternally inherited haplotypes, and composite diplotypes (combinations between risk and non-risk haplotypes) under biallelic, triallelic, and quadriallelic models

	Diplotype		Composite diplotype (value) by risk haplotypes			
Genotype	$\text{Hap}^{\text{F}}\vert\text{Hap}^{\text{M}}$	Frequency	Biallelic1	Biallelic2	Triallelic	Quadriallelic
A	$AB\vert AB$	$p^F_{11}p^M_{11}$	$R_1R_1(\mu_{11})$	$R_1R_1(\mu_{11})$	$R_1R_1(\mu_{11})$	$R_1R_1(\mu_{11})$
$AABb$	$AB\vert Ab$	$p^F_{11}p^M_{10}$	$R_1R_0(\mu_{10})$	$R_1R_1(\mu_{11})$	$R_1R_2(\mu_{12})$	$R_1R_2(\mu_{12})$
	$Ab\vert AB$	$p^F_{10}p^M_{11}$	$R_0R_1(\mu_{01})$	$R_1R_1(\mu_{11})$	$R_2R_1(\mu_{21})$	$R_2R_1(\mu_{21})$
$AAbb$	$Ab\vert Ab$	$p^F_{10}p^M_{10}$	$R_0R_0(\mu_{00})$	$R_1R_1(\mu_{11})$	$R_2R_2(\mu_{22})$	$R_2R_2(\mu_{22})$
$AaBB$	$AB\vert aB$	$p^F_{11}p^M_{01}$	$R_1R_0(\mu_{10})$	$R_1R_0(\mu_{10})$	$R_1R_0(\mu_{10})$	$R_1R_3(\mu_{13})$
	$aB\vert AB$	$p^F_{01}p^M_{11}$	$R_0R_1(\mu_{01})$	$R_0R_1(\mu_{01})$	$R_0R_1(\mu_{01})$	$R_3R_1(\mu_{31})$
$AaBb$	$AB\vert ab$	$p^F_{11}p^M_{00}$	$R_1R_0(\mu_{10})$	$R_1R_0(\mu_{10})$	$R_1R_0(\mu_{10})$	$R_1R_0(\mu_{10})$
	$ab\vert AB$	$p^F_{00}p^M_{11}$	$R_0R_1(\mu_{01})$	$R_0R_1(\mu_{01})$	$R_0R_1(\mu_{01})$	$R_0R_1(\mu_{01})$
	$Ab\vert aB$	$p^F_{10}p^M_{01}$	$R_0R_0(\mu_{00})$	$R_1R_0(\mu_{10})$	$R_2R_0(\mu_{20})$	$R_2R_3(\mu_{23})$
	$aB\vert Ab$	$p^F_{01}p^M_{10}$	$R_0R_0(\mu_{00})$	$R_0R_1(\mu_{01})$	$R_0R_2(\mu_{02})$	$R_3R_2(\mu_{32})$
$Aabb$	$Ab\vert ab$	$p^F_{10}p^M_{00}$	$R_0R_0(\mu_{00})$	$R_1R_0(\mu_{10})$	$R_2R_0(\mu_{00})$	$R_2R_0(\mu_{20})$
	$ab\vert Ab$	$p^F_{00}p^M_{10}$	$R_0R_0(\mu_{00})$	$R_0R_1(\mu_{01})$	$R_0R_2(\mu_{12})$	$R_0R_2(\mu_{02})$
$aaBB$	$aB\vert aB$	$p^F_{01}p^M_{01}$	$R_0R_0(\mu_{00})$	$R_0R_0(\mu_{00})$	$R_0R_0(\mu_{00})$	$R_3R_3(\mu_{33})$
$aaBb$	$aB\vert ab$	$p^F_{01}p^M_{00}$	$R_0R_0(\mu_{00})$	$R_0R_0(\mu_{00})$	$R_0R_0(\mu_{00})$	$R_3R_0(\mu_{30})$
	$ab\vert aB$	$p^F_{00}p^M_{01}$	$R_0R_0(\mu_{00})$	$R_0R_0(\mu_{00})$	$R_0R_0(\mu_{00})$	$R_0R_3(\mu_{03})$
$aabb$	$ab\vert ab$	$p^F_{00}p^M_{00}$	$R_0R_0(\mu_{00})$	$R_0R_0(\mu_{00})$	$R_0R_0(\mu_{00})$	$R_0R_0(\mu_{00})$

Risk haplotype(s) is assumed as AB for the first biallelic model, a mixed AB and Ab for the second biallelic model, AB and Ab for the triallelic model, and AB, Ab, and aB for the quadriallelic model.

2.1. Biallelic Model

The four haplotypes (AB,Ab,aB,ab) are sorted into two groups, risk and non-risk, based on their different effects on a given trait. The combination of risk and non-risk haplotypes is defined as a composite diplotype. Let R_1 and R_0 denote the risk and non-risk haplotypes, respectively, which can be thought of as two distinct alleles if the two associated SNPs considered are viewed as a "locus." At such a "biallelic locus," we have three possible composite diplotypes with the corresponding genotypic values specified as

$$\begin{array}{cc} \text{Composite diplotype } (\text{Hap}^{\text{M}}|\text{Hap}^{\text{P}}) & \text{Genotypic value} \\ R_1|R_1 & \mu_{11} = \mu + a \\ R_1|R_0 & \mu_{10} = \mu + d + i \\ R_0|R_1 & \mu_{10} = \mu + d - i \\ R_1|R_1 & \mu_{00} = \mu - a \end{array} \quad [1]$$

where μ is the overall mean, a is the additive effect due to the substitution of the risk haplotype by the non-risk haplotype, d is the dominance effect due to the interaction between the risk and non-risk haplotypes, and i is the imprinting effect of the risk haplotype, with the direction depending on the origin of its parent. If $i = 0$, this means that there is no difference between maternally and paternally inherited haplotypes. A positive or negative value of i reflects that the paternally or maternally inherited risk haplotype is imprinted, respectively.

Among the four haplotypes constructed by two SNPs, there are a total of seven options to choose the risk and non-risk haplotypes. These options include considering any one single risk haplotype or any two combined risk haplotypes from AB, Ab, aB and ab, tabulated as

No.	Risk haplotype	Non-risk haplotype
B_1	AB	Ab, aB, ab
B_2	Ab	AB, aB, ab
B_3	aB	AB, Ab, ab
B_4	ab	AB, Ab, aB
B_5	AB, AB	aB, ab
B_6	AB, aB	Ab, ab
B_7	AB, ab	Ab, aB

[2]

The optimal choice of a risk haplotype for the biallelic model is based on the maximum of the likelihoods calculated for each of the seven options described above.

2.2. Triallelic Model

If there are two distinct risk haplotypes, denoted as R_1 and R_2, which are different from non-risk haplotypes, denoted as R_0, a "triallelic" model can be used to quantify the imprinting effects of haplotypes. The genotypic values of the nine composite diplotypes are expressed as

Composite diplotype ($\text{Hap}^{\text{M}}\|\text{Hap}^{\text{P}}$)	Genotypic value	
$R_1\|R_1$	$\mu_{11} = \mu + a_1$	
$R_2\|R_2$	$\mu_{22} = \mu + a_2$	
$R_0\|R_0$	$\mu_{00} = \mu - a_1 - a_2$	
$R_1\|R_2$	$\mu_{12} = \mu + \frac{1}{2}(a_1 + a_2) + d_{12} + i_{12}$	[3]
$R_2\|R_1$	$\mu_{21} = \mu + \frac{1}{2}(a_1 + a_2) + d_{12} - i_{12}$	
$R_1\|R_0$	$\mu_{10} = \mu - \frac{1}{2}a_2 + d_{10} + i_{10}$	
$R_0\|R_1$	$\mu_{01} = \mu - \frac{1}{2}a_2 + d_{10} - i_{10}$	
$R_2\|R_0$	$\mu_{20} = \mu - \frac{1}{2}a_1 + d_{20} + i_{20}$	
$R_0\|R_2$	$\mu_{02} = \mu - \frac{1}{2}a_1 + d_{20} - i_{20}$	

where μ is the overall mean, a_1 and a_2 are the additive effects due to the substitution of the first and second risk haplotypes by a non-risk haplotype, respectively, d_{12}, d_{10}, and d_{20} are the dominance effects due to the interaction between the first and second risk haplotypes, between the first risk haplotype and the non-risk haplotype and between the second risk haplotype and non-risk haplotype, and i_{12}, i_{10}, and i_{20} are the imprinting effects due to different parental origins between the first and second risk haplotypes, between the first risk haplotype and the non-risk haplotype and between the second risk haplotype and the non-risk haplotype. The sizes and signs of i_{12}, i_{10}, and i_{20} determine the extent and direction of imprinting effects at the haplotype level.

The triallelic model includes a total of six haplotype combinations, which are

No.	Risk haplotype 1	Risk haplotype 2	Non-risk haplotype	
T_1	AB	Ab	aB, ab	
T_2	AB	aB	Ab, ab	
T_3	AB	Ab	Ab, aB	[4]
T_4	Ab	aB	AB, ab	
T_5	Ab	Ab	AB, aB	
T_6	aB	ab	AB, Ab	

The optimal combination of risk haplotypes for the triallelic model corresponds to the maximum of the likelihoods calculated for each of the six possibilities.

2.3. Quadriallelic Model

A quadriallelic genetic model will be used to specify haplotype effects if there are three distinct risk haplotypes. Let R_1, R_2, and R_3 be the first, second, and third risk haplotypes, and R_0 be the non-risk haplotypes, which form 16 composite diplotypes with genotypic values expressed as

$$
\begin{array}{ll}
\text{Composite diplotype } (\text{Hap}^{\text{M}}|\text{Hap}^{\text{P}}) & \text{Genotypic value} \\
R_1|R_1 & \mu_{11} = \mu + a_1 \\
R_2|R_2 & \mu_{22} = \mu + a_2 \\
R_3|R_3 & \mu_{33} = \mu + a_3 \\
R_0|R_0 & \mu_{44} = \mu - (a_1 + a_2 + a_3) \\
R_1|R_2 & \mu_{12} = \mu + \frac{1}{2}(a_1 + a_2) + d_{12} + i_{12} \\
R_2|R_1 & \mu_{21} = \mu + \frac{1}{2}(a_1 + a_2) + d_{12} - i_{12} \\
R_1|R_3 & \mu_{13} = \mu + \frac{1}{2}(a_1 + a_3) + d_{13} + i_{13} \\
R_3|R_1 & \mu_{31} = \mu + \frac{1}{2}(a_1 + a_3) + d_{13} - i_{13} \\
R_2|R_3 & \mu_{23} = \mu + \frac{1}{2}(a_2 + a_3) + d_{23} + i_{23} \\
R_3|R_2 & \mu_{32} = \mu + \frac{1}{2}(a_2 + a_3) + d_{23} - i_{23} \\
R_1|R_0 & \mu_{10} = \mu - \frac{1}{2}(a_2 + a_3) + d_{10} + i_{10} \\
R_0|R_1 & \mu_{01} = \mu - \frac{1}{2}(a_2 + a_3) + d_{10} - i_{10} \\
R_2|R_0 & \mu_{20} = \mu - \frac{1}{2}(a_1 + a_3) + d_{20} + i_{20} \\
R_0|R_2 & \mu_{02} = \mu - \frac{1}{2}(a_1 + a_3) + d_{20} - i_{20} \\
R_3|R_0 & \mu_{30} = \mu - \frac{1}{2}(a_1 + a_2) + d_{30} + i_{30} \\
R_0|R_3 & \mu_{03} = \mu - \frac{1}{2}(a_1 + a_2) + d_{30} - i_{30}
\end{array}
\quad [5]
$$

where μ is the overall mean, a_1, a_2, and a_3 are the additive effects due to the substitution of the first, second, and third risk haplotypes by the non-risk haplotype, respectively, d_{12}, d_{13}, d_{23}, d_{10}, d_{20}, and d_{30} are the dominance effects due to the interaction between the first and second risk haplotypes, between the first and third risk haplotypes, between the second and third risk haplotypes, between the first risk and the non-risk haplotypes, between the second risk and the non-risk haplotypes, and between the third risk and the non-risk haplotypes, and i_{12}, i_{13}, i_{23}, i_{10}, i_{20}, and i_{30} are the imprinting effects due to different parental origins between the first and second risk haplotypes, between the first and third risk haplotypes, between the second and third risk haplotypes, between the first risk and the non-risk haplotypes, between the second risk and the non-risk haplotypes, and between the third risk and the non-risk haplotypes.

2.4. Model Selection

For an observed marker ($\mathbf{S}$) and phenotypic ($\mathbf{y}$) data set, we do not know real risk haplotypes and their number. These can be detected by using a combinatory approach assuming that any one or more of the four haplotypes can be a risk haplotype under the biallelic, triallelic, and quadriallelic models, respectively. The likelihoods and model selection criteria, AIC or BIC, are then calculated and compared among different models and assumptions. This can be tabulated as follows:

Model	No. Risk haplotypes	Log-likelihood	AIC or BIC
Biallelic	$B_l(l = 1, \ldots, 6)$	$\log L_{B_l}(\hat{\Theta}_q \mid y, S)$	C_{B_l}
Triallelic	$T_l(l = 1, \ldots, 7)$	$\log L_{T_l}(\hat{\Theta}_q \mid y, S)$	C_{T_l}
Auadriallelic	Q	$\log L_Q(\hat{\Theta}_q \mid y, S)$	C_Q

[6]

where $\hat{\Theta}_q$ is a vector of the estimates of unknown parameters that include genotypic values of different composite diplotypes and residual variance. The largest log-likelihood and/or the smallest AIC or BIC value calculated is thought to correspond to the most likely risk haplotypes and the optimal number of risk haplotypes.

2.5. Genetic Effects

Understanding the genetic architecture of a quantitative trait is crucial for genetic studies. The quantitative genetic architecture is specified by a set of genetic parameters that include the additive, dominance, and imprinting effects, as well as the mode of inheritance. By estimating and testing these parameters, the genetic architecture of a trait can well be studied. After an optimal model (bi-, tri-, or quadriallelic) is determined, specific genetic effects of haplotypes can be estimated from estimated genotypic values for the composite diplotypes with the formulas as follows:

Model	Additive effect	Dominance effect	Imprinting effect
Billelic	$a = \frac{1}{2}(\mu_{11} - \mu_{00})$	$d = \frac{1}{2}(\mu_{10} + \mu_{01} - \mu_{11} - \mu_{00})$	$i = \frac{1}{2}(\mu_{10} - \mu_{01})$
Triallelic	$a_1 = \frac{1}{3}[2\mu_{11} - (\mu_{22} + \mu_{00})]$	$d_{12} = \frac{1}{2}(\mu_{12} + \mu_{21} - \mu_{11} - \mu_{22})$	$i_{12} = \frac{1}{2}(\mu_{12} - \mu_{21})$
	$a_2 = \frac{1}{3}[2\mu_{22} - (\mu_{11} + \mu_{00})]$	$d_{10} = \frac{1}{2}(\mu_{10} + \mu_{01} - \mu_{11} - \mu_{00})$	$i_{10} = \frac{1}{2}(\mu_{10} - \mu_{01})$
		$d_{20} = \frac{1}{2}(\mu_{20} + \mu_{02} - \mu_{22} - \mu_{00})$	$i_{20} = \frac{1}{2}(\mu_{20} - \mu_{02})$
Quadri-allelic	$a_1 = \frac{1}{4}[3\mu_{11} - (\mu_{22} + \mu_{33} + \mu_{44})]$	$d_{12} = \frac{1}{2}(\mu_{12} + \mu_{21} - \mu_{11} - \mu_{22})$	$i_{12} = \frac{1}{2}(\mu_{12} - \mu_{21})$
	$a_2 = \frac{1}{4}[3\mu_{22} - (\mu_{11} + \mu_{33} + \mu_{44})]$	$d_{13} = \frac{1}{2}(\mu_{13} + \mu_{31} - \mu_{11} - \mu_{33})$	$i_{13} = \frac{1}{2}(\mu_{13} - \mu_{31})$
	$a_3 = \frac{1}{4}[3\mu_{33} - (\mu_{11} + \mu_{22} + \mu_{44})]$	$d_{23} = \frac{1}{2}(\mu_{23} + \mu_{32} - \mu_{22} - \mu_{33})$	$i_{23} = \frac{1}{2}(\mu_{23} - \mu_{32})$
		$d_{10} = \frac{1}{2}(\mu_{10} + \mu_{01} - \mu_{11} - \mu_{00})$	$i_{10} = \frac{1}{2}(\mu_{10} - \mu_{01})$
		$d_{20} = \frac{1}{2}(\mu_{20} + \mu_{02} - \mu_{22} - \mu_{00})$	$i_{20} = \frac{1}{2}(\mu_{20} - \mu_{02})$
		$d_{30} = \frac{1}{2}(\mu_{30} + \mu_{03} - \mu_{33} - \mu_{00})$	$i_{30} = \frac{1}{2}(\mu_{30} - \mu_{03})$

[7]

2.6. Model Extension

For simplicity, our imprinting model was based on a pair of SNPs. However, an extension to including more SNPs has been made (36). For a set of three SNPs, for example, there are eight different haplotypes, among which it is possible to have one to seven risk haplotypes. The biallelic model specifies one risk haplotype which may be composed of 1 (8 cases), 2 (24 cases), 3 (56 cases), or 4 (170 cases) haplotypes. The triallelic, quadriallelic, pentaallelic, hexaallelic, septemallelic, and octoallelic models contain 28, 56, 170, 56, 24, and 8 cases, respectively. Model selection based on a combinatorial approach is adequately powerful for determining the optimal number and combination of risk haplotypes, although this will be computationally extensive when the number of SNPs increases.

3. Methods

3.1. Model

Assume that two SNPs **A** and **B** are genotyped for a total of n subjects randomly sampled from a Hardy–Weinberg equilibrium population. Each subject is phenotyped for a quantitative trait. The two SNPs form nine genotypes with observed numbers generally expressed as $n_{r_1 r_1'/r_2 r_2'}$, where the subscripts denote the SNP alleles, $r_1 \leq r_1' = 1$ (for allele A), 2 (for allele a) and $r_2 \leq r_2' = 1$ (for allele B) and 2 (for allele b). The phenotypic value of the trait for subject i is expressed in terms of the two-SNP haplotypes as

$$y_i = \sum_{j_M=0}^{J} \sum_{j_P=0}^{J} \xi_i \mu_{j_M j_P} + e_i, \quad [8]$$

where ξ_i is the indicator variable defined as 1 if subject i has a composite diplotype constructed by maternally (j_M) and paternally inherited haplotype (j_P) and 0 otherwise, e_i is the residual error, normally distributed as $N(0, \sigma^2)$, and J is the number of risk haplotypes expressed as

$$J = \begin{cases} 4 & \text{for the biallelic model} \\ 9 & \text{for the triallelic model} \\ 16 & \text{for the quadriallelic model} \end{cases} \quad [9]$$

The genotypic values of composite diplotypes and the residual variance are arrayed by a quantitative genetic parameter vector:

$$\Theta_q = \begin{cases} (\mu, a, d, i, \sigma^2) & \text{for the biallelic model} \\ (\mu, a_1, a_2, d_{12}, d_{10}, d_{20}, i_{12}, i_{10}, i_{20}, \sigma^2) & \text{for the triallelic model} \\ (\mu, a_1, a_2, a_3, d_{12}, d_{13}, d_{23}, d_{10}, d_{20}, d_{30}, i_{12}, i_{13}, i_{23}, i_{10}, i_{20}, i_{30}, \sigma^2) & \text{for the quadriallelic model} \end{cases}$$

Below, we will present a statistical method for estimating these parameters and testing each of the genetic effects.

Table 11.1 shows how each of the nine genotypes is formed through uniting maternally and paternally inherited haplotypes with a particular mating frequency. It is possible that the haplotype frequencies are different between the maternal and paternal parents, which are denoted as p_{11}^M, p_{12}^M, p_{21}^M, and p_{22}^M in the maternal parent and as p_{11}^P, p_{12}^P, p_{21}^P, and p_{22}^P in the paternal parent for haplotypes AB, Ab, aB, and ab, respectively.

3.2. Likelihood

Considering a biallelic model under the assumption of haplotype AB as a risk haplotype, the log-likelihood of the unknown vector given the phenotypic (**y**) and SNP data (**S**) is expressed as

$$\log L_B(\Theta_q | y, S) = \sum_{i=1}^{n_{11/11}} \log f_{11}(y_i) + \sum_{i=1}^{n_{11/12}} \log[\phi_1 f_{10}(y_i) + (1 - \phi_1) f_{01}(y_i)] \quad [10]$$

$$+\sum_{i=1}^{n_{12/11}}\log[\phi_2 f_{10}(y_i)+(1-\phi_2)f_{01}(y_i)]$$
$$+\sum_{i=1}^{n_{12/12}}\log[\psi_1 f_{10}(y_i)+\psi_2 f_{01}(y_i)+(1-\psi_1-\psi_2)f_{00}(y_i)]$$
$$+\sum_{i=1}^{n_{11/22}+n_{12/22}+n_{22/11}+n_{22/12}+n_{22/22}}\log f_{00}(y_i),$$

where

$$\phi_1=\frac{p_{11}^M p_{12}^P}{p_{11}^M p_{12}^P+p_{12}^M p_{11}^P},\ \phi_2=\frac{p_{11}^M p_{21}^P}{p_{11}^M p_{21}^P+p_{21}^M p_{11}^P},$$

$$\psi_1=\frac{p_{11}^M p_{22}^P}{p_{11}^M p_{22}^P+p_{22}^M p_{11}^P+p_{12}^M p_{21}^P+p_{21}^M p_{12}^P},$$
$$\psi_2=\frac{p_{22}^M p_{11}^P}{p_{11}^M p_{22}^P+p_{22}^M p_{11}^P+p_{12}^M p_{21}^P+p_{21}^M p_{12}^P}$$

Similarly, for a biallelic model assuming that haplotypes *AB* and *Ab* are a mixed risk haplotype, the log-likelihood of the unknown vector given the phenotypic (**y**) and SNP data (**S**) is expressed as

$$\begin{aligned}\log L_B(\Theta_q|y,S)&=\sum_{i=1}^{n_{11/11}+n_{11/12}+n_{11/22}}\log f_{11}(y_i)+\sum_{i=1}^{n_{22/11}+n_{22/12}+n_{22/22}}\log f_{00}(y_i)\\&+\sum_{i=1}^{n_{12/11}}\log[\phi_1 f_{10}(y_i)+(1-\phi_1)f_{01}(y_i)]\\&+\sum_{i=1}^{n_{12/22}}\log[\phi_2 f_{10}(y_i)+(1-\phi_2)f_{01}(y_i)]\\&+\sum_{i=1}^{n_{12/12}}\log[\psi f_{10}(y_i)+(1-\psi)f_{01}(y_i)],\end{aligned}\qquad[11]$$

where

$$\phi_1=\frac{p_{11}^M p_{21}^P}{p_{11}^M p_{21}^P+p_{21}^M p_{11}^P},\phi_2=\frac{p_{12}^M p_{22}^P}{p_{12}^M p_{22}^P+p_{22}^M p_{12}^P},$$

$$\psi=\frac{p_{11}^M p_{22}^P+p_{12}^M p_{21}^P}{p_{11}^M p_{22}^P+p_{22}^M p_{11}^P+p_{12}^M p_{21}^P+p_{21}^M p_{12}^P},$$

For the triallelic model assuming that haplotypes *AB* and *Ab* are the first and second risk haplotypes, respectively, we have the likelihood,

$$\begin{aligned}\log L_T(\Theta_q|y,S)&=\sum_{i=1}^{n_{11/11}}\log f_{11}(y_i)+\sum_{i=1}^{n_{11/22}}\log f_{22}(y_i)+\sum_{i=1}^{n_{22/11}+n_{22/12}+n_{22/22}}\log f_{00}(y_i)\\&+\sum_{i=1}^{n_{11/12}}\log[\phi_1 f_{12}(y_i)+(1-\phi_1)f_{21}(y_i)]\end{aligned}\qquad[12]$$

$$+\sum_{i=1}^{n_{12/11}} \log[\phi_2 f_{10}(y_i) + (1-\phi_2) f_{01}(y_i)]$$

$$+\sum_{i=1}^{n_{12/22}} \log[\phi_3 f_{20}(y_i) + (1-\phi_3) f_{20}(y_i)]$$

$$+\sum_{i=1}^{n_{12/12}} \log[\psi_1 f_{10}(y_i) + \psi_2 f_{01}(y_i) + \psi_3 f_{20}(y_i)$$

$$+(1-\psi_1-\psi_2-\psi_3) f_{02}(y_i)],$$

where

$$\phi_1 = \frac{p_{11}^M p_{12}^P}{p_{11}^M p_{12}^P + p_{12}^M p_{11}^P}, \phi_2 = \frac{p_{11}^M p_{21}^P}{p_{11}^M p_{21}^P + p_{21}^M p_{11}^P}, \phi_3 = \frac{p_{12}^M p_{22}^P}{p_{12}^M p_{22}^P + p_{22}^M p_{12}^P}$$

$$\psi_1 = \frac{p_{11}^M p_{22}^P}{p_{11}^M p_{22}^P + p_{22}^M p_{11}^P + p_{12}^M p_{21}^P + p_{21}^M p_{12}^P},$$

$$\psi_2 = \frac{p_{22}^M p_{11}^P}{p_{11}^M p_{22}^P + p_{22}^M p_{11}^P + p_{12}^M p_{21}^P + p_{21}^M p_{12}^P},$$

$$\psi_3 = \frac{p_{12}^M p_{21}^P}{p_{11}^M p_{22}^P + p_{22}^M p_{11}^P + p_{12}^M p_{21}^P + p_{21}^M p_{12}^P}.$$

For the quadriallelic model assuming that haplotypes *AB*, *Ab*, and *aB* are the first, second and third risk haplotypes, respectively, the likelihood is written as

$$\log L_Q(\Theta_q|y,S)$$

$$= \sum_{i=1}^{n_{11/11}} \log f_{11}(y_i) + \sum_{i=1}^{n_{11/22}} \log f_{22}(y_i) + \sum_{i=1}^{n_{22/11}} \log f_{33}(y_i) + \sum_{i=1}^{n_{22/22}} \log f_{00}(y_i)$$

$$+ \sum_{i=1}^{n_{11/12}} \log[\phi_1 f_{12}(y_i) + (1-\phi_1) f_{21}(y_i)] + \sum_{i=1}^{n_{12/11}} \log[\phi_2 f_{13}(y_i) + (1-\phi_2) f_{31}(y_i)]$$

$$+ \sum_{i=1}^{n_{12/22}} \log[\phi_3 f_{20}(y_i) + (1-\phi_3) f_{02}(y_i)] + \sum_{i=1}^{n_{22/12}} \log[\phi_4 f_{30}(y_i) + (1-\phi_4) f_{03}(y_i)]$$

$$+ \sum_{i=1}^{n_{12/12}} \log[\psi_1 f_{10}(y_i) + \psi_2 f_{01}(y_i) + \psi_3 f_{23}(y_i) + (1-\psi_1-\psi_2-\psi_3) f_{32}(y_i)], \quad [13]$$

where

$$\phi_1 = \frac{p_{11}^M p_{12}^P}{p_{11}^M p_{12}^P + p_{12}^M p_{11}^P}, \phi_2 = \frac{p_{12}^M p_{11}^P}{p_{11}^M p_{12}^P + p_{12}^M p_{11}^P},$$

$$\phi_3 = \frac{p_{12}^M p_{22}^P}{p_{12}^M p_{22}^P + p_{22}^M p_{12}^P}, \phi_4 = \frac{p_{22}^M p_{12}^P}{p_{12}^M p_{22}^P + p_{22}^M p_{12}^P},$$

$$\psi_1 = \frac{pM_{11}pP_{22}}{pM_{11}pP_{22} + pM_{22}pP_{11} + pM_{12}pP_{21} + pM_{21}pP_{12}},$$

$$\psi_2 = \frac{pM_{22}pP_{11}}{pM_{11}pP_{22} + pM_{22}pP_{11} + pM_{12}pP_{21} + pM_{21}pP_{12}},$$

$$\psi_3 = \frac{p_{12}^M p_{21}^P}{p_{11}^M p_{22}^P + p_{22}^M p_{11}^P + p_{12}^M p_{21}^P + p_{21}^M p_{12}^P}.$$

In equations [9–12], and [13], $f_{jMjP}(y_i)$ is a normal distribution density function of composite diplotype $jMjP$ with mean μ_{jMjP} and variance σ^2.

3.3. EM Algorithms

3.3.1. Population Genetic Parameters

Below, the EM algorithm is described for estimating haplotype frequencies, $\Theta_p = (p_{11}, p_{12}, p_{21}, p_{22})$, constructed by a pair of SNPs (**A** with alleles A and a and **B** with alleles B and b) in a sex-specific population. Let k denote a sex, female or male. A sex-specific log-likelihood function of haplotype frequencies (Θ_p^k) given SNP observations (S_k) is constructed as

$$\begin{aligned}
&\log L_k(\Theta_p^k|S_k) = \text{constant} \\
&+2n_{11/11}^k ln p_{11}^k + n_{11/12}^k ln(2p_{11}^k p_{12}^k) + 2n_{11/22}^k ln p_{12}^k + n_{12/11}^k ln(2p_{11}^k p_{21}^k) \\
&+n_{12/12}^k ln(2p_{11}^k p_{22}^k + 2p_{12}^k p_{21}^k) + n_{12/22}^k ln(2p_{12}^k p_{22}^k) + 2n_{22/11}^k ln p_{21}^k \\
&+n_{22/12}^k ln(2p_{21}^k p_{22}^k) + 2n_{22/22}^k ln p_{22}^k.
\end{aligned}$$

The EM algorithm was derived to estimate the MLEs of Θ_p^k with the following formulas:

$$\begin{aligned}
\hat{p}_{11}^k &= \tfrac{1}{n_k}(2n_{11/11}^k + n_{11/12}^k + n_{12/11}^k + \phi_k n_{12/12}^k), \\
\hat{p}_{12}^k &= \tfrac{1}{n_k}[2n_{11/22}^k + n_{11/12}^k + n_{12/22}^k + (1-\phi_k)n_{12/12}^k], \\
\hat{p}_{21}^k &= \tfrac{1}{n_k}[2n_{22/11}^k + n_{12/11}^k + n_{22/12}^k + (1-\phi_k)n_{12/12}^k], \\
\hat{p}_{22}^k &= \tfrac{1}{n_k}(2n_{22/22}^k + n_{12/22}^k + n_{22/12}^k + \phi_k n_{12/12}^k),
\end{aligned} \quad [14]$$

where

$$\phi_k = \frac{p_{11}^k p_{22}^k}{p_{11}^k p_{22}^k + p_{12}^k p_{21}^k}. \quad [15]$$

By giving initial values for the haplotype frequencies, we calculate ϕ_k with equation [14], and then estimate the new haplotype frequencies with equation [15]. This process is iterated until the estimates of haplotype frequencies are stable.

After the haplotype frequencies are estimated, sex-specific allele frequencies at the two SNPs **A** and **B** and their sex-specific linkage disequilibrium (D) are estimated by

$$p_A^k = p_{11}^k + p_{12}^k,$$
$$p_B^k = p_{11}^k + p_{21}^k,$$
$$D_k = p_{11}^k p_{00}^k - p_{10}^k p_{01}^k.$$

It is not possible to estimate maternally and paternally inherited haplotype frequencies with a random sample from a natural population, although the estimation of Θ_q is dependent on these parent-specific haplotype frequencies, as shown in **Table 11.1**. The parent-specific haplotype frequencies can be approximated by sex-specific haplotype frequencies in the current generation. This can be proven as follows.

Let $p_{11}(t)$ and $p_{11}(t+1)$ be the frequencies of a representative haplotype AB in generations t and $t+1$, respectively. If the population undergoes random mating, the relationship of haplotype frequencies between the two generations is expressed as (40)

$$\begin{aligned} p_{11}(t+1) &= p(t+1)q(t+1) + D(t+1) && \text{for generation } t+1 \\ &= p(t)q(t) + (1-r)D(t) && \text{for generation } t \\ &\approx p(t)q(t) + D(t) && \text{for generation } t \\ &= p_{11}(t) && \text{for generation } t \end{aligned}$$

where $p(t)$ and $q(t)$ are the allele frequencies of SNPs **A** and **B** in generation t, which are constant from generation to generation, i.e., $p(t) = p(t+1)$ and $q(t) = q(t+1)$, for a Hardy–Weinberg equilibrium, D is the linkage disequilibrium between the two SNPs, which decays in a proportion of r with generation, and r is the recombination fraction between the two SNPs. Because the two SNPs are genotyped from the same region of a candidate gene, their recombination fraction should be very small and can be thought to be close to zero.

3.3.2. Quantitative Genetic Parameters

In what follows, we implement the EM algorithm to estimate the genotypic values of each composite diplotype and the residual variance under different models. We will describe the estimating procedures for the unknown parameter contained within Θ_q, separately for different types of models.

Biallelic Model: Recall the log-likelihood for a biallelic model by assuming haplotype AB as one single risk haplotype (*see* **Table 11.1**). We define the posterior probabilities with which a two-SNP genotype for subject i carries one of its underlying diplotypes by

$$\begin{aligned}
\Phi^{1}_{10|i} &= \frac{\phi_1 f_{10}(y_i)}{\phi_1 f_{10}(y_i) + (1-\phi_1) f_{01}(y_i)}, \Phi^{\bar{1}}_{01|i} = \frac{(1-\phi_1) f_{01}(y_i)}{\phi_1 f_{10}(y_i) + (1-\phi_1) f_{01}(y_i)}, \\
\Phi^{2}_{10|i} &= \frac{\phi_2 f_{10}(y_i)}{\phi_2 f_{10}(y_i) + (1-\phi_2) f_{01}(y_i)}, \Phi^{\bar{2}}_{01|i} = \frac{(1-\phi_2) f_{01}(y_i)}{\phi_2 f_{10}(y_i) + (1-\phi_2) f_{01}(y_i)} \\
\Psi_{10|i} &= \frac{\psi_1 f_{10}(y_i)}{\psi_1 f_{10}(y_i) + \psi_2 f_{01}(y_i) + (1-\psi_1-\psi_2) f_{00}(y_i)}, \\
\Psi_{01|i} &= \frac{\psi_2 f_{01}(y_i)}{\psi_1 f_{10}(y_i) + \psi_2 f_{01}(y_i) + (1-\psi_1-\psi_2) f_{00}(y_i)}, \\
\Psi_{00|i} &= \frac{(1-\psi_1-\psi_2) f_{01}(y_i)}{\psi_1 f_{10}(y_i) + \psi_2 f_{01}(y_i) + (1-\psi_1-\psi_2) f_{00}(y_i)}.
\end{aligned} \qquad [16]$$

With these posterior probabilities, we calculate the unknown parameters by

$$\begin{aligned}
2\hat{\mu}_{11} &= \frac{\sum_{i=1}^{n_{11/11}} y_i}{n_{11/11}}, \\
\hat{\mu}_{10} &= \frac{\sum_{i=1}^{n_{11/12}} \Phi^{1}_{10|i} y_i + \sum_{i=1}^{n_{12/11}} \Phi^{2}_{10|i} y_i + \sum_{i=1}^{n_{12/12}} \Psi_{10|i} y_i}{\sum_{i=1}^{n_{11/12}} \Phi^{1}_{10|i} + \sum_{i=1}^{n_{12/11}} \Phi^{2}_{10|i} + \sum_{i=1}^{n_{12/12}} \Psi_{10|i}}, \\
\hat{\mu}_{01} &= \frac{\sum_{i=1}^{n_{11/12}} \Phi^{\bar{1}}_{01|i} y_i + \sum_{i=1}^{n_{12/11}} \Phi^{\bar{2}}_{01|i} y_i + \sum_{i=1}^{n_{12/12}} \Psi_{01|i} y_i}{\sum_{i=1}^{n_{11/12}} \Phi^{\bar{1}}_{10|i} + \sum_{i=1}^{n_{12/11}} \Phi^{\bar{2}}_{10|i} + \sum_{i=1}^{n_{12/12}} \Psi_{01|i}}, \\
\hat{\mu}_{00} &= \frac{\sum_{i=1}^{n_{12/12}} \Psi_{00|i} y_i + \sum_{i=1}^{n_{11/22}+n_{12/22}+n_{22/11}+n_{22/12}+n_{22/22}} y_i}{\sum_{i=1}^{n_{12/12}} \Psi_{00|i} + n_{11/22}+n_{12/22}+n_{22/11}+n_{22/12}+n_{22/22}},
\end{aligned} \qquad [17]$$

$$\begin{aligned}
\hat{\sigma}^2 = \ & \tfrac{1}{n}\Big[\sum_{i=1}^{n_{11/11}} (y_i - \hat{\mu}_{11})^2 + \sum_{i=1}^{n_{11/12}} \Phi^{1}_{10|i} (y_i - \hat{\mu}_{10})^2 \\
& + \sum_{i=1}^{n_{12/11}} \Phi^{2}_{10|i} (y_i - \hat{\mu}_{10})^2 + \sum_{i=1}^{n_{12/12}} \Psi_{10|i} (y_i - \hat{\mu}_{10})^2 \Big] \\
& + \sum_{i=1}^{n_{11/12}} \Phi^{\bar{1}}_{01|i} (y_i - \hat{\mu}_{01})^2 + \sum_{i=1}^{n_{12/11}} \Phi^{\bar{2}}_{01|i} (y_i - \hat{\mu}_{01})^2 \\
& + \sum_{i=1}^{n_{12/12}} \Psi_{01|i} (y_i - \hat{\mu}_{01})^2 + \sum_{i=1}^{n_{12/12}} \Psi_{00|i} (y_i - \hat{\mu}_{00})^2 \\
& + \sum_{i=1}^{n_{11/22}+n_{12/22}+n_{22/11}+n_{22/12}+n_{22/22}} (y_i - \hat{\mu}_{00})^2 \Big].
\end{aligned}$$

A loop of iterations is then constructed by the E (equation [16]) and M steps (equation [17]). The MLEs of the unknown parameters are obtained when the estimates are stable in the EM iterations.

Equation [10] is the log-likelihood for a biallelic model that assumes haplotypes AB and Ab as a mixed risk haplotype. The posterior probabilities with which a two-SNP genotype for subject i carries one of its underlying diplotypes are defined as

$$
\begin{aligned}
\Phi^{1}_{10|i} &= \frac{\phi_1 f_{10}(y_i)}{\phi_1 f_{10}(y_i) + (1-\phi_1) f_{01}(y_i)}, \quad \Phi^{\bar{1}}_{10|i} = \frac{(1-\phi_1) f_{01}(y_i)}{\phi_1 f_{10}(y_i) + (1-\phi_1) f_{01}(y_i)}, \\
\Phi^{2}_{10|i} &= \frac{\phi_2 f_{10}(y_i)}{\phi_2 f_{10}(y_i) + (1-\phi_2) f_{01}(y_i)}, \quad \Phi^{\bar{2}}_{10|i} = \frac{(1-\phi_2) f_{01}(y_i)}{\phi_2 f_{10}(y_i) + (1-\phi_2) f_{01}(y_i)}, \\
\Psi_{10|i} &= \frac{\psi f_{10}(y_i)}{\psi f_{10}(y_i) + (1-\psi) f_{01}(y_i)}, \quad \Psi^{1}_{01|i} = \frac{(1-\psi) f_{01}(y_i)}{\psi f_{10}(y_i) + (1-\psi) f_{01}(y_i)}.
\end{aligned}
\qquad [18]
$$

The unknown parameters in this case are estimated by

$$
\begin{aligned}
\hat{\mu}_{11} &= \frac{\sum_{i=1}^{n_{11/11}+n_{11/12}+n_{11/22}} y_i}{n_{11/11}+n_{11/12}+n_{11/22}}, \\
\hat{\mu}_{10} &= \frac{\sum_{i=1}^{n_{12/11}} \Phi^{1}_{10|i} y_i + \sum_{i=1}^{n_{12/22}} \Phi^{2}_{10|i} y_i + \sum_{i=1}^{n_{12/12}} \Psi_{10|i} y_i}{n_{12/11}\Phi^{1}_{10|i} + \sum_{i=1}^{n_{12/22}} \Phi^{2}_{10|i} + \sum_{i=1}^{n_{12/12}} \Psi_{10|i}}, \\
\hat{\mu}_{01} &= \frac{\sum_{i=1}^{n_{12/11}} \Phi^{1}_{01|i} y_i + \sum_{i=1}^{n_{12/22}} \Phi^{2}_{01|i} y_i + \sum_{i=1}^{n_{12/12}} \Psi_{01|i} y_i}{n_{12/11}\Phi^{\bar{1}}_{01|i} + \sum_{i=1}^{n_{12/22}} \Phi^{\bar{2}}_{01|i} + \sum_{i=1}^{n_{12/12}} \Psi_{01|i}}, \\
\hat{\mu}_{00} &= \frac{\sum_{i=1}^{n_{22/11}+n_{22/12}+n_{22/22}} y_i}{n_{22/11}+n_{22/12}+n_{22/22}}, \\
\hat{\sigma}^2 &= \frac{1}{n}\Bigg[\sum_{i=1}^{n_{11/11}+n_{11/12}+n_{11/22}} (y_i-\mu_{11})^2 + \sum_{i=1}^{n_{12/11}} \Phi^{1}_{10|i}(y_i-\mu_{10})^2 \\
&+ \sum_{i=1}^{n_{12/22}} \Phi^{2}_{10|i}(y_i-\mu_{10})^2 + \sum_{i=1}^{n_{12/12}} \Psi_{10|i}(y_i-\mu_{10})^2 \\
&+ \sum_{i=1}^{n_{12/11}} \Phi^{\bar{1}}_{01|i}(y_i-\mu_{01})^2 + \sum_{i=1}^{n_{12/22}} \Phi^{\bar{2}}_{01|i}(y_i-\mu_{01})^2 + \sum_{i=1}^{n_{12/12}} \Psi_{01|i}(y_i-\mu_{01})^2 \\
&+ \sum_{i=1}^{n_{22/11}+n_{22/12}+n_{22/22}} (y_i-\mu_{00})^2 \Bigg].
\end{aligned}
\qquad [19]
$$

A loop of iterations between E (equation [18]) and M steps (equation [19]) is made to obtain the MLEs of the unknown parameters.

Triallelic Model: According to equation [11] that is the log-likelihood for a triallelic model assuming haplotypes AB and Ab as two distinct risk haplotypes. The posterior probabilities with which a two-SNP genotype for subject i carries one of its underlying diplotypes are defined as

$$\Phi^1_{12|i} = \frac{\phi_1 f_{12}(y_i)}{\phi_1 f_{12}(y_i)+(1-\phi_1)f_{21}(y_i)}, \Phi^{\bar{1}}_{21|i} = \frac{(1-\phi_1)f_{12}(y_i)}{\phi_1 f_{12}(y_i)+(1-\phi_1)f_{21}(y_i)},$$
$$\Phi^2_{10|i} = \frac{\phi_2 f_{10}(y_i)}{\phi_2 f_{10}(y_i)+(1-\phi_2)f_{01}(y_i)}, \Phi^{\bar{2}}_{01|i} = \frac{(1-\phi_2)f_{01}(y_i)}{\phi_2 f_{10}(y_i)+(1-\phi_2)f_{01}(y_i)},$$
$$\Phi^3_{20|i} = \frac{\phi_3 f_{20}(y_i)}{\phi_3 f_{20}(y_i)+(1-\phi_3)f_{02}(y_i)}, \Phi^{\bar{3}}_{02|i} = \frac{(1-\phi_3)f_{02}(y_i)}{\phi_3 f_{20}(y_i)+(1-\phi_3)f_{02}(y_i)} \quad [20]$$
$$\Psi_{10|i} = \frac{\psi_1 f_{10}(y_i)}{\psi_1 f_{10}(y_i)+\psi_2 f_{01}(y_i)+\psi_3 f_{20}(y_i)+(1-\psi_1-\psi_2-\psi_3)f_{02}(y_i)},$$
$$\Psi_{01|i} = \frac{\psi_2 f_{01}(y_i)}{\psi_1 f_{10}(y_i)+\psi_2 f_{01}(y_i)+\psi_3 f_{20}(y_i)+(1-\psi_1-\psi_2-\psi_3)f_{02}(y_i)},$$
$$\Psi_{20|i} = \frac{\psi_3 f_{20}(y_i)}{\psi_1 f_{10}(y_i)+\psi_2 f_{01}(y_i)+\psi_3 f_{20}(y_i)+(1-\psi_1-\psi_2-\psi_3)f_{02}(y_i)},$$
$$\Psi_{02|i} = \frac{(1-\psi_1-\psi_2-\psi_3)f_{02}(y_i)}{\psi_1 f_{10}(y_i)+\psi_2 f_{01}(y_i)+\psi_3 f_{20}(y_i)+(1-\psi_1-\psi_2-\psi_3)f_{02}(y_i)}.$$

The unknown parameters in this case are estimated by

$$\hat{\mu}_{11} = \frac{\sum_{i=1}^{n_{11/11}} y_i}{n_{11/11}}, \quad \hat{\mu}_{22} = \frac{\sum_{i=1}^{n_{11/22}} y_i}{n_{11/22}}, \quad \hat{\mu}_{00} = \frac{\sum_{i=1}^{n_{22/11}+n_{22/12}+n_{22/22}} y_i}{n_{22/11}+n_{22/12}+n_{22/22}},$$
$$\hat{\mu}_{12} = \frac{\sum_{i=1}^{n_{11/12}} \Phi^1_{12|i} y_i}{\sum_{i=1}^{n_{11/12}} \Phi^1_{12|i}}, \quad \hat{\mu}_{21} = \frac{\sum_{i=1}^{n_{11/12}} \Phi^{\bar{1}}_{21|i} y_i}{\sum_{i=1}^{n_{11/12}} \Phi^{\bar{1}}_{21|i}},$$
$$\hat{\mu}_{10} = \frac{\sum_{i=1}^{n_{12/11}} \Phi^2_{10|i} y_i + \sum_{i=1}^{n_{12/12}} \Psi_{10|i} y_i}{\sum_{i=1}^{n_{12/11}} \Phi^2_{10|i} + \sum_{i=1}^{n_{12/12}} \Psi_{10|i}}, \quad \hat{\mu}_{01} = \frac{\sum_{i=1}^{n_{12/11}} \Phi^{\bar{2}}_{01|i} y_i + \sum_{i=1}^{n_{12/12}} \Psi_{01|i} y_i}{\sum_{i=1}^{n_{12/11}} \Phi^{\bar{2}}_{01|i} + \sum_{i=1}^{n_{12/12}} \Psi_{01|i}},$$
$$\hat{\mu}_{20} = \frac{\sum_{i=1}^{n_{12/22}} \Phi^3_{20|i} y_i + \sum_{i=1}^{n_{12/12}} \Psi_{20|i} y_i}{\sum_{i=1}^{n_{12/22}} \Phi^3_{20|i} + \sum_{i=1}^{n_{12/12}} \Psi_{20|i}}, \quad \hat{\mu}_{02} = \frac{\sum_{i=1}^{n_{12/22}} \Phi^{\bar{3}}_{02|i} y_i + \sum_{i=1}^{n_{12/12}} \Psi_{02|i} y_i}{\sum_{i=1}^{n_{12/22}} \Phi^{\bar{3}}_{02|i} + \sum_{i=1}^{n_{12/12}} \Psi_{02|i}}, \quad [21]$$
$$\begin{aligned}\hat{\sigma}^2 = \frac{1}{n}\Bigg[&\sum_{i=1}^{n_{11/11}} (y_i-\hat{\mu}_{11})^2 + \sum_{i=1}^{n_{11/22}} (y_i-\hat{\mu}_{22})^2 + \sum_{i=1}^{n_{22/11}+n_{22/12}+n_{22/22}} (y_i-\hat{\mu}_{00})^2 \\
&+ \sum_{i=1}^{n_{11/12}} \Phi^1_{12|i}(y_i-\hat{\mu}_{12})^2 + \sum_{i=1}^{n_{11/12}} \Phi^{\bar{1}}_{21|i}(y_i-\hat{\mu}_{21})^2 + \sum_{i=1}^{n_{12/11}} \Phi^2_{10|i}(y_i-\hat{\mu}_{10})^2 \\
&+ \sum_{i=1}^{n_{12/12}} \Psi_{10|i}(y_i-\hat{\mu}_{10})^2 + \sum_{i=1}^{n_{12/11}} \Phi^{\bar{2}}_{01|i}(y_i-\hat{\mu}_{01})^2 + \sum_{i=1}^{n_{12/12}} \Psi_{01|i}(y_i-\hat{\mu}_{01})^2 \\
&+ \sum_{i=1}^{n_{12/22}} \Phi^3_{20|i}(y_i-\hat{\mu}_{20})^2 + \sum_{i=1}^{n_{12/12}} \Psi_{20|i}(y_i-\hat{\mu}_{20})^2 + \sum_{i=1}^{n_{12/22}} \Phi^{\bar{3}}_{02|i}(y_i-\hat{\mu}_{02})^2 \\
&+ \sum_{i=1}^{n_{12/12}} \Psi_{02|i}(y_i-\hat{\mu}_{02})^2\Bigg].\end{aligned}$$

A loop of iterations between E (equation [20]) and M steps (equation [21]) is made to obtain the MLEs of the unknown parameters.

Quadriallelic Model: According to equation [12] that is the log-likelihood for a quadriallelic model assuming haplotypes *AB*, *Ab*, and *aB* as three distinct risk haplotypes. The posterior probabilities with which a two-SNP genotype for subject *i* carries one of its underlying diplotypes are defined as

$$\Phi^{1}_{12|i}=\frac{\phi_1 f_{12}(y_i)}{\phi_1 f_{12}(y_i)+(1-\phi_1)f_{21}(y_i)},\ \Phi^{\bar{1}}_{21|i}=\frac{(1-\phi_1)f_{21}(y_i)}{\phi_1 f_{12}(y_i)+(1-\phi_1)f_{21}(y_i)},$$

$$\Phi^{2}_{13|i}=\frac{\phi_2 f_{13}(y_i)}{\phi_2 f_{13}(y_i)+(1-\phi_2)f_{31}(y_i)},\ \Phi^{\bar{2}}_{31|i}=\frac{(1-\phi_2)f_{31}(y_i)}{\phi_2 f_{13}(y_i)+(1-\phi_2)f_{31}(y_i)},$$

$$\Phi^{3}_{20|i}=\frac{\phi_3 f_{20}(y_i)}{\phi_3 f_{20}(y_i)+(1-\phi_3)f_{02}(y_i)},\ \Phi^{\bar{3}}_{02|i}=\frac{(1-\phi_3)f_{02}(y_i)}{\phi_3 f_{20}(y_i)+(1-\phi_3)f_{02}(y_i)},$$

$$\Phi^{4}_{30|i}=\frac{\phi_4 f_{30}(y_i)}{\phi_4 f_{30}(y_i)+(1-\phi_4)f_{03}(y_i)},\ \Phi^{\bar{4}}_{03|i}=\frac{(1-\phi_4)f_{03}(y_i)}{\phi_4 f_{30}(y_i)+(1-\phi_4)f_{03}(y_i)},\quad [22]$$

$$\Psi_{10|i}=\frac{\psi_1 f_{10}(y_i)}{\psi_1 f_{10}(y_i)+\psi_2 f_{01}(y_i)+\psi_3 f_{23}(y_i)+(1-\psi_1-\psi_2-\psi_3)f_{32}(y_i)},$$

$$\Psi_{01|i}=\frac{\psi_2 f_{01}(y_i)}{\psi_1 f_{10}(y_i)+\psi_2 f_{01}(y_i)+\psi_3 f_{23}(y_i)+(1-\psi_1-\psi_2-\psi_3)f_{32}(y_i)},$$

$$\Psi_{23|i}=\frac{\psi_3 f_{23}(y_i)}{\psi_1 f_{10}(y_i)+\psi_2 f_{01}(y_i)+\psi_3 f_{23}(y_i)+(1-\psi_1-\psi_2-\psi_3)f_{32}(y_i)},$$

$$\Psi_{32|i}=\frac{(1-\psi_1-\psi_2-\psi_3)f_{32}(y_i)}{\psi_1 f_{10}(y_i)+\psi_2 f_{01}(y_i)+\psi_3 f_{23}(y_i)+(1-\psi_1-\psi_2-\psi_3)f_{32}(y_i)}.$$

The unknown parameters in this case are estimated by

$$\hat{\mu}_{11}=\frac{\sum_{i=1}^{n_{11/11}} y_i}{n_{11/11}},\ \hat{\mu}_{22}=\frac{\sum_{i=1}^{n_{11/22}} y_i}{n_{11/22}},\ \hat{\mu}_{33}=\frac{\sum_{i=1}^{n_{22/11}} y_i}{n_{22/11}},\ \hat{\mu}_{00}=\frac{\sum_{i=1}^{n_{22/22}} y_i}{n_{22/22}},$$

$$\hat{\mu}_{12}=\frac{\sum_{i=1}^{n_{11/12}} \Phi^{1}_{12|i} y_i}{\sum_{i=1}^{n_{11/12}} \Phi^{1}_{12|i}},\ \hat{\mu}_{21}=\frac{\sum_{i=1}^{n_{11/12}} \Phi^{\bar{1}}_{21|i} y_i}{\sum_{i=1}^{n_{11/12}} \Phi^{1}\bar{1}_{12|i}},$$

$$\hat{\mu}_{13}=\frac{\sum_{i=1}^{n_{12/11}} \Phi^{2}_{13|i} y_i}{\sum_{i=1}^{n_{12/11}} \Phi^{2}_{12|i}},\ \hat{\mu}_{31}=\frac{\sum_{i=1}^{n_{12/11}} \Phi^{\bar{2}}_{31|i} y_i}{\sum_{i=1}^{n_{12/11}} \Phi^{\bar{2}}_{31|i}},\quad [23]$$

$$\hat{\mu}_{20}=\frac{\sum_{i=1}^{n_{12/22}} \Phi^{3}_{20|i} y_i}{\sum_{i=1}^{n_{12/22}} \Phi^{3}_{20|i}},\ \hat{\mu}_{02}=\frac{\sum_{i=1}^{n_{12/22}} \Phi^{\bar{3}}_{02|i} y_i}{\sum_{i=1}^{n_{12/22}} \Phi^{\bar{3}}_{02|i}},$$

$$\hat{\mu}_{30}=\frac{\sum_{i=1}^{n_{22/12}}\Phi^{4}_{30|i}y_i}{\sum_{i=1}^{n_{22/12}}\Phi^{4}_{30|i}},\quad \hat{\mu}_{03}=\frac{\sum_{i=1}^{n_{22/12}}\Phi^{\bar{4}}_{03|i}y_i}{\sum_{i=1}^{n_{22/12}}\Phi^{\bar{4}}_{03|i}},$$

$$\hat{\mu}_{10}=\frac{\sum_{i=1}^{n_{12/12}}\Psi_{10|i}y_i}{\sum_{i=1}^{n_{12/12}}\Psi_{10|i}},\quad \hat{\mu}_{01}=\frac{\sum_{i=1}^{n_{12/12}}\Phi_{01|i}y_i}{\sum_{i=1}^{n_{12/12}}\Psi_{01|i}},$$

$$\hat{\mu}_{23}=\frac{\sum_{i=1}^{n_{12/12}}\Psi_{23|i}y_i}{\sum_{i=1}^{n_{12/12}}\Psi_{23|i}},\quad \hat{\mu}_{32}=\frac{\sum_{i=1}^{n_{12/12}}\Phi_{32|i}y_i}{\sum_{i=1}^{n_{12/12}}\Psi_{32|i}}$$

$$\begin{aligned}\hat{\sigma}^2=\frac{1}{n}\Big[&\sum_{i=1}^{n_{11/11}}(y_i-\hat{\mu}_{11})^2+\sum_{i=1}^{n_{11/22}}(y_i-\hat{\mu}_{22})^2+\sum_{i=1}^{n_{22/11}}(y_i-\hat{\mu}_{33})^2+\sum_{i=1}^{n_{22/22}}(y_i-\hat{\mu}_{00})^2\\
&+\sum_{i=1}^{n_{11/12}}\Phi^{1}_{12|i}(y_i-\hat{\mu}_{12})^2+\sum_{i=1}^{n_{11/12}}\Phi^{\bar{1}}_{21|i}(y_i-\hat{\mu}_{21})^2+\sum_{i=1}^{n_{12/11}}\Phi^{2}_{13|i}(y_i-\hat{\mu}_{13})^2\\
&+\sum_{i=1}^{n_{12/11}}\Phi^{\bar{2}}_{31|i}(y_i-\hat{\mu}_{31})^2+\sum_{i=1}^{n_{12/22}}\Phi^{3}_{20|i}(y_i-\hat{\mu}_{20})^2+\sum_{i=1}^{n_{12/22}}\Phi^{\bar{3}}_{02|i}(y_i-\hat{\mu}_{02})^2\\
&+\sum_{i=1}^{n_{22/12}}\Phi^{4}_{30|i}(y_i-\hat{\mu}_{30})^2+\sum_{i=1}^{n_{22/12}}\Phi^{\bar{4}}_{03|i}(y_i-\hat{\mu}_{03})^2+\sum_{i=1}^{n_{12/12}}\Psi_{10|i}(y_i-\hat{\mu}_{10})^2\\
&+\sum_{i=1}^{n_{12/12}}\Phi_{01|i}(y_i-\hat{\mu}_{01})^2+\sum_{i=1}^{n_{12/12}}\Psi_{23|i}(y_i-\hat{\mu}_{23})^2+\sum_{i=1}^{n_{12/12}}\Phi_{32|i}(y_i-\hat{\mu}_{32})^2\Big].\end{aligned}$$

A loop of iterations between E (equation [22]) and M steps (equation [23]) is made to obtain the MLEs of the unknown parameters.

3.4. Hypothesis Tests

For a specific model, testing the existence of functional haplotypes is the first step. This can be done by formulating a null hypothesis,

$$\begin{aligned}&H_0: \quad \mu_{j_M j_P}\equiv\mu(j_M, j_P=0,1,\ldots,J)\\ &H_1: \quad \text{At least one of equality in } \mathrm{H}_0 \text{ does not hold}\end{aligned} \qquad [24]$$

The log-likelihood ratio (LR) is then calculated by plugging the estimated parameters into the likelihood under the H_0 and H_1, respectively. The LR can be viewed as being asymptotically χ^2-distributed with $(J+1)^2-1$ degrees of freedom.

After a significant haplotype effect is detected, a series of further tests are performed for the significance of additive, dominance, and imprinting effects triggered by haplotypes. The null hypotheses under each of these tests can be formulated by setting the effect being tested to be equal to zero. For example, under the triallelic model, the null hypothesis for testing the imprinting effect of the haplotypes is expressed as

$$H_0 : i_{12} = i_{10} = i_{20} = 0.$$

In practice, it is also interesting to test each of the additive genetic effects, each of the dominance effects, and each of the imprinting effects for the tri- and quadriallelic models. The estimates of the parameters under the null hypotheses can be obtained with the same EM algorithm derived for the alternative hypotheses, but with a constraint of the tested effect equal to zero. The log-likelihood ratio test statistic for each hypothesis is thought to asymptotically follow a χ^2-distribution with the degrees of freedom equal to the difference of the numbers of the parameters being tested under the null and alternative hypotheses.

4. Example of Applying the Model

Cardiovascular disease, principally heart disease and stroke, is the leading killer for both men and women among all racial and ethnic groups. As part of the ongoing cardiovascular genetics project, this study intends to investigate the genetic control of human obesity that is a causal factor for cardiovascular disease. As a major lipolytic receptor in human fat cells, the β-adrenergic receptors (β-AR) are thought to play an important role in cardiovascular function. In this study, two common polymorphisms located at codons 16 (with two alleles *A* and *G*) and 27 (with two alleles *C* and *G*) for the β-*2AR* gene were genotyped for a obesity-susceptible population mixed with 80 females and 74 males. All the subjects were measured for body masses and body heights, with which body mass indices (BMIs) were calculated. The frequencies of haplotypes *AC*, *AG*, *GC*, and *GG* were estimated separately for the female and male populations. These two SNPs were found to be significantly associated with BMI in both females ($D = 0.104$, $p<0.0001$) and males ($D = 0.151$, $p<0.0001$). In the female population, haplotypes *AG*, *GC*, and *GG* were equally predominant, whereas in the males only haplotypes *AG* and *GC* were predominant (**Table 11.2**). Overall, there was no significant difference in haplotype frequencies between the two sexes ($p = 0.145$).

There was no significant difference in body mass index (BMI) between the sexes, averaged as 29.89 (ranging from 17.96 to 52.31) in the females and 29.15 (ranging from 17.83 to 50.95) in the males. BMI approximately follows a normal distribution in both sexes (**Fig. 11.1**). By assuming one, two, or three risk haplotypes at the two SNPs typed that trigger an effect on BMI, we used the bi-, tri-, and quadriallelic models to determine the optimal risk haplotypes and their number based on the AIC and BIC

Table 11.2.
The MLEs of the haplotype frequencies at SNPs (codons 16 and 27) in 74 females and 80 males sampled from a natural population

	Frequency	
Haplotype	Female	Male
AC	0.0263	0.0240
AG	0.3379	0.4016
GC	0.3317	0.3881
GG	0.3041	0.1862
Difference	p-value = 0.1447	
$p(A)$	0.3642	0.4257
$q(G)$	0.6420	0.5878
D	0.1040	0.1514

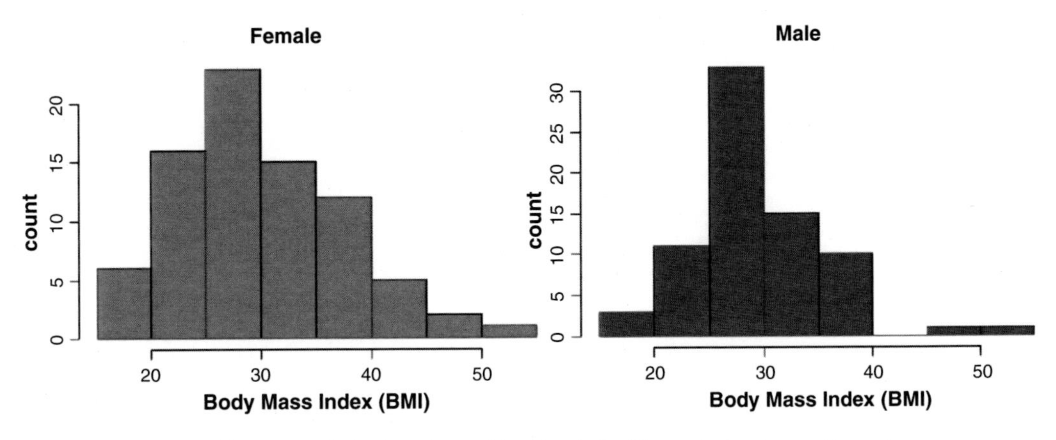

Fig. 11.1. Distribution of body mass index in males and females derived from an obesity genetics project.

criteria. Both criteria suggested that two risk haplotypes *AC* and *GG* were the optimal combination that explained the data with the triallelic model. As shown by the log-likelihood ratio test, this pair of associated SNPs had a significant effect on BMI ($p<0.001$) at the haplotype level. The estimated additive, dominance and imprinting effects of the haplotypes are tabulated in **Table 11.3**.

Additive genetic effects on BMI were found to be non-significant for risk haplotype *AC* ($p = 0.365$), but significant for risk haplotype *GG* ($p = 0.050$). These two risk haplotypes together exerted an increasing significant additive genetic effect. There was a marked interaction effect between risk haplotypes *AC* and *GG*, leading to a significant dominance effect ($p<0.0001$). The

Table 11.3.
The MLEs of the additive, dominance, and imprinting effects of the haplotypes composed of two SNPs at codons 16 and 27 within a β-adrenergic receptor, as well as the significance tests of each of these effects

Para meter	Definition	Value	*p*-value
Additive			0.0046
a_1	Substitution of haplotype *AC* by *AG* or *GC*	0.718	0.3656
a_2	Substitution of haplotype *GG* by *AG* or *GC*	–1.570	0.0506
Dominance			<0.0001
d_{12}	Interaction between haplotypes *AC* and *GG*	–2.311	<0.0001
d_{10}	Interaction between haplotypes *AC* and *AG* or *GC*	3.529	0.0003
d_{20}	Interaction between haplotypes *GG* and *AG* or *GC*	–2.929	0.0003
Imprinting			0.0806
i_{12}	Parent-dependent effect between haplotypes *AC* and *GG*	2.811	0.4134
i_{10}	Parent-dependent effect between haplotypes *AC* and *AG* or *GC*	7.301	0.0008
i_{20}	Parent-dependent effect between haplotypes *GG* and *AG* or *GC*	–1.090	0.4061

dominance effects of each of these two risk haplotypes over non-risk haplotypes *AG* or *GC* were also significant (**Table 11.3**). The most striking finding from this study is the identification of significant imprinting effects at this pair of SNPs. Risk haplotype *AC*, although its additive effect was not significant, displayed a significant imprinting effect ($p = 0.0008$) relative to the non-risk haplotypes. In other words, for any composite diplotype constituted by *AC* and *AG* or *GC*, a significant difference in BMI may be due to the parent-specific contribution of risk haplotype *AC*. The composite diplotype composed of risk haplotype *AC* derived from the paternal parent was much less obese (with a BMI smaller by 7.3) than that with risk haplotype *AC* derived from the maternal parent.

5. Concluding Remarks

Although a traditional view assumes that the maternally and paternally derived alleles of each gene are expressed simultaneously at a similar level, there are many exceptions where alleles are expressed from only one of the two parental chromosomes (1,7).

This so-called genetic imprinting or parent-of-origin effect has been thought to play a pivotal role in regulating the phenotypic variation of a complex trait (2–5). With the discovery of more imprinting genes involved in trait control through molecular and bioinformatics approaches, we will be in a position to elucidate the genetic architecture of quantitative variation for various organisms including humans.

Genetic mapping in controlled crosses has opened up a great opportunity for a genome-wide search of imprinting effects by identifying imprinted quantitative trait loci (iQTLs). This approach has successfully detected iQTLs that are responsible for body mass and diseases (14–26). Cloning of these iQTLs will require high-resolution mapping of genes, which is hardly met for traditional linkage analysis based on the production of recombinants in experimental crosses. Single-nucleotide polymorphisms (SNPs) are powerful markers that can explain interindividual differences. Multiple adjacent SNPs are especially useful to associate phenotypic variability with haplotypes (30–34). By specifying one risk haplotype, i.e., one that operates differently from the rest of haplotypes (called non-risk haplotypes), Liu et al. (35) proposed a statistical method for detecting risk haplotypes for a complex trait with a random sample drawn from a natural population. Liu et al.'s approach can be used to characterize DNA sequence variants that encode the phenotypic value of a trait. Wu et al. (39) constructed a general multiallelic model in which any number of risk haplotypes can be assumed. The best number and combination of risk haplotypes can be estimated by using the likelihoods and AIC or BIC values (40).

In this chapter, we incorporate genetic imprinting into Wu et al.'s multiallelic model to estimate the number and combination of multiple functional haplotypes that are expressed differently depending on the parental origin of these haplotypes. Because of the modeling of any possible distinct haplotypes, the multiallelic model will have more power for detecting significant haplotypes and their imprinting effects than biallelic models. The imprinting model was shown to work well in a wide range of parameter space for a modest sample size. However, a considerably large sample size is needed if there are multiple risk haplotypes that contribute to trait variation. When analyzing a real example from a cardiovascular study, the new model detected significant haplotypes composed of two SNPs within the *β-2AR* gene, a major lipolytic receptor in human fat cells, which play an important role in cardiovascular disease. Haplotype *AC* derived from the *β-2AR* gene is paternally imprinted for human obesity, leading to the reduction of body mass index (BMI) by 20%.

In practice, although the human genome contains millions of SNPs (41, 42), it is not necessary to model and analyze these SNPs simultaneously. These SNPs are often distributed in different haplotype blocks (43, 44), within each of which a particular (small) number of representative SNPs or htSNPs can uniquely explain most of the haplotype variation. A minimal subset of htSNPs, identified by several computing algorithms (28, 45, 46), can be implemented into our imprinting model to detect their imprinting effects at the haplotype level. In addition, our model can be extended to model imprinting effects in a network of interactive architecture, including haplotype–haplotype interactions from different genomic regions (38), haplotype–environment interactions, and haplotype effects regulating pharmacodynamic reactions of drugs (37). It can be expected that all extensions will require expensive computation, but this computing can be made possible if combinatorial mathematics, graphical models, and machine learning are incorporated into closed forms of parameter estimation. When these detailed extensions that take account of more realistic biological and genetic problems are made, we will be close to addressing fundamental questions about the genetic control of complex traits by collecting SNP and haplotype data.

Acknowledgments

The preparation of this manuscript is supported by Joint DMS/ NIGMS-0540745 to RW.

References

1. Wilkins, JF, Haig, D. (2003) What good is genomic imprinting: the function of parent-specific gene expression. *Nat Rev Genet* **4**, 359–368.
2. Wood, AJ, Oakey, RJ. (2006) Genomic imprinting in mammals: emerging themes and established theories. *PLoS Genet* **2**(11), e147.
3. Lewis, A, Reik, W. (2006) How imprinting centres work. Cytogenet *Genome Res* **113**, 81–89.
4. Jirtle, RL, Skinner, MK. (2007) Environmental epigenomics and disease susceptibility. *Nat Rev Genet* **8**, 253–262.
5. Feil, R, Berger, F. (2007) Convergent evolution of genomic imprinting in plants and mammals. *Trends Genet* **23**, 192–199.
6. Reik, W, Lewis, A. (2005) Co-evolution of X-chromosome inactivation and imprinting in mammals. *Nat Rev Genet* **6**, 403–410.
7. Reik, W, Walter, J. (2001) Genomic imprinting: parental influence on the genome. *Nat Rev Genet* **2**, 21–32.
8. Feinberg, AP, Tycko, B. (2004) The history of cancer epigenetics. *Nat Rev Cancer* **4**, 143–153.
9. Hanson, RL, Kobes, S, Lindsay, RS, et al. (2001) Assessment of parent-of-origin effects in linkage analysis of quantitative traits. *Am J Hum Genet* **68**, 951–962.
10. Shete, S, Amos, CI. (2002) Testing for genetic linkage in families by a variance-components approach in the presence of genomic imprinting. *Am J Hum Genet* **70**, 751–757.
11. Shete, S, Zhou, X, Amos, CI. (2003) Genomic imprinting and linkage test for quantitative trait loci in extended pedigrees. *Am J Hum Genet* **73**, 933–938.
12. Haghighi, F, Hodge, SE. (2002) Likelihood formulation of parent-of-origin effects on

segregation analysis, including ascertainment. *Am J Hum Genet* **70,** 142–156.

13. Knott, SA, Marklund, L, Haley, CS, et al. (1998) Multiple marker mapping of quantitative trait loci in a cross between outbred wild boar and large white pigs. *Genetics* **149,** 1069–1080.
14. de Koning, DJ, Rattink, AP, Harlizius, B, et al. (2000) Genome-wide scan for body composition in pigs reveals important role of imprinting. *Proc Natl Acad Sci USA* **97,** 7947–7950.
15. de Koning, DJ, Bovenhuis, H, van Arendonk, JAM. (2002) On the detection of imprinted quantitative trait loci in experimental crosses of outbred species. *Genetics* **161,** 931–938.
16. Liu, T, Todhunter, RJ, Wu, S, et al. (2007) A random model for mapping imprinted quantitative trait loci in a structured pedigree: an implication for mapping canine hip dysplasia. *Genomics* **90,** 276–284.
17. Cheverud, JM, Hager, R, Roseman, C, et al. (2008) Genomic imprinting effects on adult body composition in mice. *Proc Natl Acad Sci* **105,** 4253–4258.
18. Jeon, J-T, Carlborg, O, Tornsten, A, et al. (1999) A paternally expressed QTL affecting skeletal and cardiac muscle mass in pigs maps to the IGF2 locus. *Nat Genet* **21,** 157–158
19. Nezer, C, Moreau, L, Brouwers, B, et al. (1999) An imprinted QTL with major effect on muscle mass and fat deposition maps to the IGF2 locus in pigs. *Nat Genet* **21,** 155–156.
20. Nezer, C, Collette, CM, Brouwers, B, et al. (2003) Haplotype sharing refines the location of an imprinted quantitative trait locus with major effect on muscle mass to a 250-kb chromosome segment containing the porcine IGF2 gene. *Genetics* **165,** 277–285.
21. Van Laere, AS, Nguyen, M, Braunschweig, M, et al. (2003) A regulatory mutation in IGF2 causes a major QTL effect on muscle growth in the pig. *Nature* **425,** 832–836.
22. Tuiskula-Haavisto, M, de Koning, DJ, Honkatukia, M, et al. (2004) Quantitative trait loci with parent-of-origin effects in chicken. *Genet Res* **84,** 57–66.
23. Lewis, A, Redrup, L. (2005) Genetic imprinting: conflict at the Callipyge locus. *Curr Biol* **15,** R291–R294.
24. Cui, YH, Lu, Q, Cheverud, JM, et al. (2006) Model for mapping imprinted quantitative trait loci in an inbred F_2 design. *Genomics* **87,** 543–551.
25. Cui, YH. (2006) A statistical framework for genome-wide scanning and testing imprinted quantitative trait loci. *J Theor Biol* **244,** 115–126.
26. Cui, YH, Cheverud, JM, Wu, RL. (2007) A statistical model for dissecting genomic imprinting through genetic mapping. *Genetica* **130,** 227–239.
27. Flint, J, Valdar, W, Shifman, S, et al. (2005) Strategies for mapping and cloning quantitative trait genes in rodents. *Nat Rev Genet* **6,** 271–286.
28. Eyheramendy, S, Marchini, J, McVean, G, et al. (2007) A model-based approach to capture genetic variation for future association studies. *Genome Res* **17,** 88–95.
29. Konfortov, BA, Bankier, AT, Dear, PH. (2007) An efficient method for multi-locus molecular haplotyping. *Nucleic Acids Res* **35,** e6.
30. Judson, R, Stephens, JC, Windemuth, A. (2000) The predictive power of haplotypes in clinical response. *Pharmacogenomics* **1,** 15–26.
31. Bader, JS. (2001) The relative power of SNPs and haplotype as genetic markers for association tests. *Pharmacogenomics* **2,** 11–24.
32. Winkelmann, BR, Hoffmann, MM, Nauck, M, et al. (2003) Haplotypes of the cholesterylester transfer protein gene predict lipid-modifying response to statin therapy. *Pharmacogenomics J.* **3,** 284–296.
33. Clark, AG. (2004) The role of haplotypes in candidate gene studies. *Genet Epidemiol* **27,** 321–333.
34. Jin, GF, Miao, RF, Deng, YM, et al. (2007) Variant genotypes and haplotypes of the epidermal growth factor gene promoter are associated with a decreased risk of gastric cancer in a high-risk Chinese population. *Cancer Res* **98,** 864–868.
35. Liu, T, Johnson, JA, Casella, G, et al. (2004) Sequencing complex diseases with HapMap. *Genetics* **168,** 503–511.
36. Li, HY, Kim, BR, Wu, RL. (2006) Identification of quantitative trait nucleotides that regulate cancer growth: a simulation approach. *J Theor Biol* **242,** 426–439.
37. Lin, M, Aquilante, C, Johnson, JA, et al. (2005) Sequencing drug response with HapMap. *Pharmacogenomics J.* **5,** 149–156.
38. Lin, M, Wu, RL. (2006) Detecting sequence-sequence interactions for complex diseases. *Curr Genomics* **7,** 59–72.
39. Wu, S, Yang, J, Wang, CG, et al. (2007) A general quantitative genetic model for

haplotyping a complex trait in humans. *Curr Genomics* **8,** 343–350.

40. Lynch, M, Walsh, B. (1998) Genetics and Analysis of Quantitative Traits. Sinauer Associates, Sunderland, MA.
41. Inbar, E, Yakir, B, Darvasi, A. (2002) An efficient haplotyping method with DNA pools. *Nucleic Acids Res* **30,** e76.
42. International HapMap Consortium (2005) A haplotype map of the human genome. *Nature* **437,** 1299–1320.
43. Patil, N, Berno, AJ, Hinds, DA, et al. (2001) Blocks of limited haplotype diversity revealed by high-resolution scanning of human chromosome 21. *Science* **294,** 1719–1723.
44. Terwilliger, JD, Hiekkalinna, T. (2006) An utter refutation of the "Fundamental Theorem of the HapMap". *Eur J Hum Genet* **14,** 426–437.
45. Zhang, K, Deng, M, Chen, T, et al. (2002) A dynamic programming algorithm for haplotype block partitioning. *Proc Natl Acad Sci* **99,** 7335–7339.
46. Sebastiani, P, Lazarus, SW, Kunkel, LM, et al. (2003) Minimal haplotype tagging. *Proc Natl Acad Sci* **100,** 9900–9905.

Chapter 12

Haplotype Association Mapping in Mice

Shirng-Wern Tsaih and Ron Korstanje

Abstract

Haplotype Association Mapping (HAM) is a novel phenotype-driven approach to identify genetic loci and was originally developed for mice. This method, which is similar to Genome-Wide Association (GWA) studies in humans, looks for associations between the phenotype and the haplotypes of mouse inbred strains, treating inbred strains as individuals. Although this approach is still in development, we review the current literature, present the different methods and applications that are in use, and provide a glimpse of what is to come in the near future.

Key words: Haplotypes, association, mice.

1. Introduction

Because the experiments can be controlled, in silico association mapping studies in inbred model organisms are potentially more powerful than association mapping in outbred human populations. In Haplotype Association Mapping (HAM) in the mouse we use the large number of genetic markers and mouse inbred strains to test for association between a trait and a genotype treating different strains of mice as individuals from a population (**Fig. 12.1**).

The first paper describing the use of this concept, also known as in silico mapping, to predict linkage in inbred strains was by Grupe et al. in 2001 where they used approximately 3,350 SNPs in 15 inbred strains (1). They used the allelic distributions across the inbred strains to calculate genotypic distances between loci for a pair of mouse strains. These genetic distances were then compared with phenotypic differences between the two mouse strains and this process was repeated for all mouse strain pairs for which

K. DiPetrillo (ed.), *Cardiovascular Genomics*, Methods in Molecular Biology 573,
DOI 10.1007/978-1-60761-247-6_12, © Humana Press, a part of Springer Science+Business Media, LLC 2009

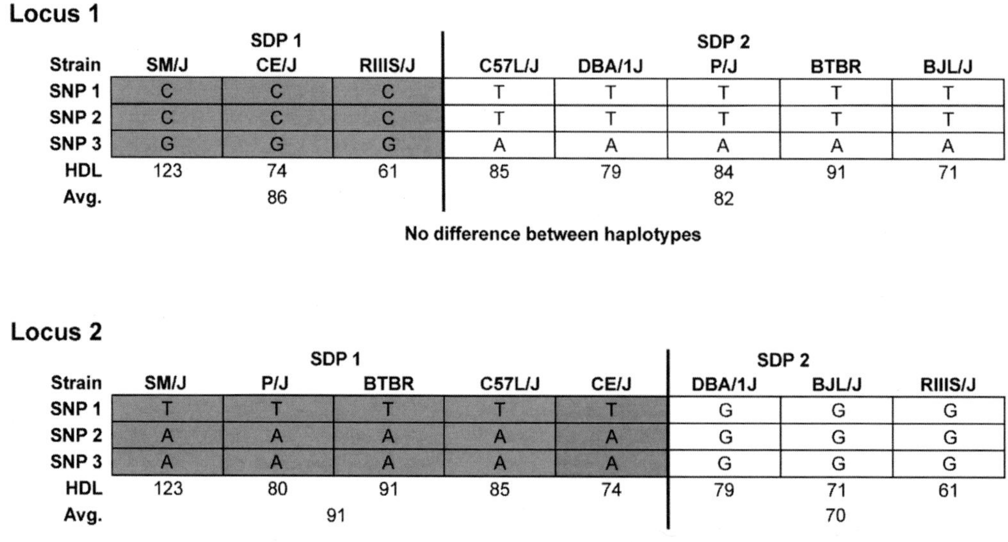

Fig. 12.1. Haplotype Association Mapping. The *top panel* represents the strain distribution pattern (SDP) across three SNPs in eight strains at locus 1. NZB, CE, and RIIIS mice share an SDP, whereas PL, DBA/2, LP, BTBR, and SJL share a different SDP. Genome-wide haplotype association first uses the SDP to infer common haplotypes among inbred strains, and then associates the haplotype with a phenotype. At locus 1, the average (Avg.) plasma HDL concentration is 85 for SDP 1 and 77 for SDP 2, so the haplotype is not linked to the phenotype. By contrast, the average phenotype for SDP 1 at locus 2 is 96 (NZB, LP, BTBR, and PL mice) versus 64 for SDP 2 (CE, DBA/2, SJL, and RIIIS mice), thus this haplotype is linked to plasma HDL concentration. Data were derived from Pletcher et al. Although this example shows two SDPs at each locus for simplicity, genome-wide haplotype association can test any number of SDPs at each locus for association with a phenotype.

phenotypic information was available. Finally, a correlation value was derived using linear regression on the phenotypic and genotypic distances for each genomic locus.

At the time, there was much debate (2, 3) about the method and the limited number of strains that Grupe et al. used, but it was an appealing idea that was further developed with the fast growing number of SNPs genotyped for an increasing number of inbred strains with phenotype data. In 2004, Pletcher et al. showed that with genotype information from 10,990 SNPs tested on 48 inbred strains and the use of alternative algorithms, they were able to reproduce previously characterized quantitative trait loci (QTL) for high-density lipoprotein and gallstone phenotypes (4) substantially narrowed the QTL interval for some of them such that strong candidate genes were identified.

In the past few years, several groups have used HAM either to identify novel QTL, identify which strains are most informative to initiate QTL studies, or to refine experimental QTL and narrow the number of candidate genes. Liu et al. mined the Mouse Phenome Database (www.jax.org/phenome) and combined 173 phenotypes with 148,062 SNPs in order to identify 937 QTL for these different phenotypes (5). They were able to refine 67% of the experimentally found QTL to a genomic region of less than 0.5 Mb with a 40-fold increase in mapping position. The same group also nicely used HAM to narrow a previously found QTL to only two genes. They used urethane-induced lung adenoma incidence in 21 strains with 123,073 SNPs and reproduced the *Pas1* locus. This locus spanned approximately 26 cM in the original cross but was narrowed to less than 0.5 Mb using HAM. This refined region covers only seven genes; among these at least two were reported as functionally important in mouse lung tumorigenesis.

Lightfoot et al. used HAM to confirm their experimentally found QTL for physical activity traits in a B6 × C3H intercross (6). Data from 27 strains genotyped with 1,272 SNPs on Chromosomes 9 and 13 not only confirmed the three significant QTL, but also identified 18 additional QTL not revealed in the F2 cohort.

At present, there is no other animal species for which the number of genetic markers and inbred strains are sufficient to perform HAM analysis. With the recent development of high-density SNP maps and genotype information for many rat inbred strains (7), HAM studies using rats are expected within the next few years. HAM has been successfully applied on plants. For example, using 8,590 SNPs in 553 lines, Belo et al. (8) identified an amino acid change in *fad2* associated with oleic acid levels in maize.

2. Data and Data Requirements

2.1. The Trait Under Study and Its Genetic Components

Most of the current HAM algorithms are based on the analysis of variance (ANOVA) or linear regression model approaches, which require the usual model assumptions: (1) the phenotype of interest should be independent (independence assumption); (2) normally distributed (normality assumption); and (3) the variance of the observations is the same in each haplotype group (assumption of variance homogeneity). Ideally, samples should have minimal population structure or familial relatedness for this type of approach. Violation of the independence assumption, in general, leads to inflated test statistics relative to expectation under the assumption of an independent sample and no genetic association with the phenotype. Assumptions of normality and homogeneity can be mildly violated without serious risk for relatively large samples (e.g., 20 or more observations from each group) when the numbers of observations selected from each population are nearly equal. When one of these two assumptions is in serious doubt, one option is to transform the data (e.g., by means of a log, square root, or other transformation) so that they more closely satisfy the assumptions. Alternatively, select a more appropriate method of analysis (e.g., nonparametric ANOVA methods or other more complicated modeling techniques).

In our experience, the success of HAM very much depends on the trait under investigation. First of all, most of the HAM algorithms take the input of a vector of strain mean phenotype values instead of per animal observation; it is important to look critically at the within strain variation to minimize the impact of outliers on strain mean estimates. Once HAM peaks are identified and the strains contributing to the peaks are determined, it is a good practice to review the number of animals used to calculate the means for these strains. Second, males and females should be analyzed separately. Not only do they often have different means within strains, but also males and females have different genetic determinants for many traits. As the differences in the mean and the sex-specific factors are different between strains, this will be completely lost when analyzing males and females combined. Third, success depends on the number of genetic factors involved in the trait and their relative contributions to it. We had good success with traits like HDL and gallstones, which are determined by many genetic factors with a relatively large contribution. In contrast, the success rate was much lower for plasma sodium levels, which are highly influenced by many environmental factors.

2.2. The Number of Genetic Markers Used

There has been an explosion in the availability of genotype data in the past few years. For mice, the sequencing of the initial four inbred strains (C57BL/6 J, A/J, DBA/2 J, and 129/SvImJ), the re-sequencing of additional strains by Perlegen, and the large SNP discovery projects by several companies and consortia resulted in over 10 million SNP genotypes in the Mouse Phenome Database for many different strains (ranging between 4 and 107 strains depending on the source). Grupe et al. worked with less than 4,000 SNPs for 15 strains in 2001, we currently perform HAM analysis with 1.5 million SNPs typed on 74 strains. A study by Cervino et al. (9) compared HAM using 20,000 and 140,000 SNPs and concluded that the increase of SNPs did not improve the QTL detection much, but that the increased quality of the newer, denser maps resulted in more accurate mapping.

2.3. The Number of Strains Used

Wang et al. looked at the effect of strain numbers on power under varying genetic models and for varying number of haplotypes (10). In their example, a sample size of 10 was powerful only when looking at very strong genetic effects, such as Mendelian traits. For a trait with a genetic effect contributing in the range of 30% to the total variance, 30 strains or more were required for acceptable power. Cervino et al. performed simulation studies using HDL cholesterol levels and simulated data as the trait, randomly selecting 30, 20, and 10 strains to investigate how often the correct

QTL region was identified as the most likely region. They showed a dramatic loss of power as the traits became more complex and the genetic effect decreased; the power decreased proportionally to the number of strains.

3. Current Methods

Several HAM algorithms are in use to measure the strength of the phenotype–genotype association. The simplest method is single-marker mapping. Each SNP is biallelic across inbred mouse strains, so a *t*-test is used to measure the strength of association. However, when using the panel of inbred mouse strains, inspection of allele patterns across multiple loci suggests that the genetic structure in many cases is more complex than two genetic groups (*see* **Note 1**). In other words, the genotype at one biallelic SNP is often insufficient to discern the genomic structure, especially for complex traits, explaining the high false-negative results using the single-marker mapping strategy. These observations led to the definition of an inferred haplotype group as a set of strains with an identical genotype pattern over a local window of multiple adjacent SNPs.

3.1. The Sliding Window Approach

The publicly available *SNPster* web tool (http://snpster.gnf.org) correlates strain-specific phenotype data with a high-density SNP map (roughly 140,000 SNPs or 1 SNP/6 kb) (4, 11). The *SNPster* algorithm infers local haplotype using a sliding 3-SNP window approach. Two strains are determined to be in the same inferred haplotype group only if their genetic pattern across three adjacent SNPs is identical. The F-statistics from ANOVA tests is used to measure the strength of association. All *p*-values are transformed using $-\log_{10}(p\text{-value})$ and reported as an association score at each locus. The F-statistic is weighted to reduce the importance of association scores that are largely driven by closely related strains, such as multiple lines from the C57-related strains (11). Genome-wide bootstrap iterations are used in the SNPster web tool to calculate a generalized family-wise error rate (gFWER) for multiple testing correction (*see* **Note 2**). Given a set of hypothesis tests, the FWER threshold is defined as the probability of observing one false positive under the null (Type I error). Hence, the gFWER is defined as the probability of *k* or more false rejections.

3.2. Hidden Markov Model Approach

The extensive linkage disequilibrium found in laboratory mice provides the basis for accurate imputation. In our practice, we use the program *hmmSNP*, a hidden Markov model (HMM)-based algorithm with a left-to-right architecture (*see* **Fig. 12.1** in Szatkiewicz et al. (12)) to impute missing genotype data (primary purpose) and to infer haplotype states (secondary purpose) (12). The number of hidden states at each SNP locus and the prior distribution of the model parameters of the HMM were optimized for genome-wide imputation accuracy. Trait data at strain mean levels are used as vectors and the HMM-smoothed haplotype states at each SNP across multiple inbred mouse strains are used as a matrix. Regression-based test statistics are computed to measure the strength of association between haplotype blocks and phenotype and significance is estimated to detect groups with different mean phenotypes. The segregation of strains into genotypic groups varies widely over haplotype blocks, therefore, *P*-values, and not the F-test statistics, are compared between haplotype blocks. All *P*-values are transformed using $-\log_{10}(P\text{-value})$ in the scan plots. Type I error rate for multiple testing due to genome-wide searching is controlled for using family-wise error rate control (13) based on permutation tests, where the strain label in the phenotype data is shuffled and the genotype data are kept intact. The minimum *P*-value is recorded on each permutation and percentiles of their distribution provide approximate multiple test-adjusted thresholds. The FWER threshold is usually estimated using 1,000 permutation tests. The analysis is carried out in the MATLAB computing environment (The Mathworks, www.mathworks.com).

3.3. Comparison of HAM Results Between SNPster and the HMM Approach

We considered HDL cholesterol levels across 41 laboratory mouse strains (32 classical inbred and 9 wild-derived inbred strains, *see* **Note 3**) fed chow diet as our test case. In our evaluation, we used male data only to illustrate the agreements and differences between the SNPster (version 3.3) and HMM approach. Male-specific HDL data (MPD 9904) at strain mean level was downloaded directly from the Mouse Phenome Database (www.jax.org/phenome). We obtained the 150 K SNPs used by SNPster and processed it with *hmmSNP* program to impute SNP genotype and to infer haplotype structure assuming six states. Based on 1,000 permutation tests, significance and suggestive association score thresholds (FWER threshold) in our HMM approach were 5.17 (alpha=0.05) and 4.36 (alpha=0.2), respectively. The HMM approach detected one suggestive HAM peak at 19 at 57.29 Mb with a score of 4.48. This haplotype region contains only two SNPs. Chromosome 2 at 5.4 Mb had the second highest peak with association score of 4.05, below our suggestive threshold level. **Table 12.1** lists the 13 genomic regions among 7 chromosomes (Chrs 1, 3, 4, 5, 7, 15, and 19) with association scores above 4.0 using SNPster and the association scores of the highest peaks from the HMM approach for comparison. Five regions found with SNPster were also found by the HMM approach, although with lower scores except on Chr 19. The major factor involved in the discrepancy between these two methods is the inferred haplotype structure. First, haplotype block size is fixed at 3-SNP windows in SNPster, but is dynamic in HMM approach. Second,

Table 12.1
Comparison of HAM results for male HDL (MPD 9904) between SNPster and HMM approach

Locus	SNPster score	HMM score (no. of SNPs)	QTL	Candidate
Chr 1:64.62–64.73	4.4		No	*Fzd5*
Chr 1:173.05–173.07	4.9	3.1 (6)	Yes	*Apoa2*
Chr 2: 5.39–5.40		4.0 (1)	No	*Camk1d*
Chr 3:63.38–63.43	4.4		Yes	*Mme*
Chr 3:120.09–120.15	4.7	3.5 (4)	Yes	
Chr 4:35.78–35.88		3.0 (9)	Yes	
Chr 4:125.01–125.10	4.4		Yes	
Chr 4:149.35–149.49	4.1		Yes	
Chr 5:87.05–87.12	4.2		Yes	*Cenpc1*
Chr 5:89.68–89.85	4.0		Yes	*Grsf1*
Chr 5:91.69–91.80	5.8	3.2 (1)	Yes	*Rassf6*
Chr 7:15.83–16.00	4.4	3.6 (1)	Yes	*Slc1a5*
Chr 9:83.58–83.58		2.9 (2)	Yes	
Chr 10:8.8		2.7 (2)		
Chr 11:89.64–89.65		3.4 (2)	Yes	
Chr 13:14.09–14.10		3.1 (2)	No	
Chr 14:8.26–8.26		3.4 (1)	Yes	
Chr 15:9.53–9.55		3.2 (3)	No	
Chr 15:58.51–58.78	4.1		Yes	*Tatdn1*
Chr 18:44.12–44.16	4.2		Yes	
Chr 18:82.37–82.43		3.7 (2)	Yes	
Chr 19:57.24–57.30	4.1	4.5 (2)	Yes	*Ablim1*

SNPster-inferred haplotype structure is subject to the missing genotype pattern in the SNP set. Strains with missing genotypes were excluded prior to the local haplotype reconstruction; therefore, the number of strains used in the analysis is not fixed at 41 in SNPster. In our HMM approach, we overcame the missing data issue by filling in missing genotypes with imputed genotypes using all strains. The size of our HMM reconstructed haplotype block is dynamic, ranging from one SNP to as many as ten SNPs.

3.4. Identify Causal SNP with the Linear Mixed Models Approach

As mentioned previously, spurious association between a genetic marker and a phenotype could arise due to the population structure and the genetic relatedness among inbred strains caused by the genealogical history of laboratory mice. As a result, the false-positive rates are inflated in conventional statistical tests. Yu et al. suggested that linear mixed models could effectively correct for population structure in genetic association studies (*14*). Applications using mixed models for association mapping have been shown to obtain fewer false positives and have higher power than methods using genomic control, structure association, or principal component analysis (14–16). Recently, Kang et al. (17) developed an efficient mixed-model association (EMMA) test that efficiently corrects for population structure in model organism association mapping studies (*17*). EMMA is orders of magnitude faster than other implementations commonly used for the linear mixed model approach. Kang et al. scanned MPD data with or without correction for population structure and posted their results at EMMA web server http://whap.cs.ucla.edu/mpad/. For the male MPD 9904 dataset, the top two peaks with correction for population structure were seen on Chrs 1 and 5. The SNP at the highest peak is located on Chr 1 at 64.80 Mb.

The linear mixed model can be written as $y = X\beta + Zu + e$, where y is an $n \times 1$ vector of observed phenotypes, and X is an $n \times q$ matrix of fixed effects including mean, SNPs, and other confounding variables. β is a $q \times 1$ vector representing coefficients of the fixed effects. Z is an $n \times t$ incidence matrix mapping each observed phenotype to one of t inbred strains. u is the random effect of the mixed model with $\mathrm{Var}(u) = \sigma_g^2 K$; where K is the $t \times t$ kinship matrix inferred from genotypes (genetic similarity matrix) as described in Kang et al. (17), and e is an $n \times n$ matrix of residual effect such that $\mathrm{Var}(e) = \sigma_e^2 I$. The overall phenotypic variance–covariance matrix can be represented as $V = \sigma_g^2 ZKZ' + \sigma_e^2 I$. A standard F-test then be performed to test H_1: $\beta \neq 0$ against H_0: $\beta = 0$.

4. Conclusion

The Haplotype Association Mapping methodology is still under considerable development and is based on statistical assumptions that are adjusted as more complete genotype data become available. Therefore, the most appropriate algorithms for use in associating phenotype with genotype in inbred mouse strains are subject to debate.

As an adjunct to QTL analysis, the association of SNPs to phenotypes within a previously defined QTL can greatly reduce the interval under consideration for follow-up studies. This type of

SNP haplotype analysis will add the advantage of increased genetic diversity among the laboratory inbred strains to the statistical power and precision of conventional QTL analysis. We can expect new methods that will take population structure and phylogeny into account and software that will make these methods available to a bigger group of scientists.

5. Notes

1. Haplotype mapping algorithms cannot fully account for the expected complexity of some regions of the genome: previous studies have shown that the inbred mouse genome is bi-modally distributed in regions of high (approximately 1 SNP/kb) and low (0.5 SNPs/10 kb) polymorphism (18–21). This complex and fluctuating pattern, combined with a small number of haplotypes at any one region, limits the ability to associate phenotype with causal genotype. For any one region of the genome, a small set of inbred strains can capture all genetic variabilities, but the composition of strains within that set will vary across the genome (20). Thus, successful Haplotype Association Mapping likely requires the combination of a large set of inbred strains with a high-density SNP map. However, even after addressing these concerns, regions of low polymorphism, sparse SNP data, high haplotype diversity, high recombination rates, or genes with very low effect sizes are all likely to inhibit accurate mapping.
2. The best method of setting significance is still unknown. Both false discovery rate (FDR) and FWER calculations are likely to be too conservative because of haplotype structure, while an adjustable gFWER, proposed by McClurg et al. (11), equaling the probability of $k + 1$ false positives must tolerate some false positives to account for small effect sizes. In our experience we have had a false discovery rate as high as 60–70%, but most of these are easy to identify by taking the SNP file and sorting the haplotypes.
3. Wild-derived mouse strains are genetic outliers compared to the classical inbred mouse strains. Imputed SNP genotypes of wild-derived mouse strains tend to have lower accuracy compared to the classical inbred mouse strains (12). However, the inclusion of the wild-derived strains improves the imputation of missing genotypes in regions where some classical inbred strains are not of *Mus*

musculus domesticus (domesticus) origin. Whether or not inclusion of the wild-derived mouse strains is appropriate in HAM is a matter of debate. If the population structure is detected, we recommend removing wild-derived inbred strains and repeating the HAM analysis or using the linear mixed models approach mentioned above.

References

1. Grupe, A, Germer, S, Usuka, J, et al. (2001) In silico mapping of complex disease-related traits in mice. Science **292**(5523), 1915–1918.
2. Chesler, EJ, Rodriguez-Zas, SL, Mogil, JS. (2001) In silico mapping of mouse quantitative trait loci. Science **294**(5551), 2423.
3. Darvasi, A. (2001) In silico mapping of mouse quantitative trait loci. Science **294**(5551), 2423.
4. Pletcher, MT, McClurg, P, Batalov, S, et al. (2004) Use of a dense single nucleotide polymorphism map for in silico mapping in the mouse. PLoS Biol **2**, e393.
5. Lui, P, Vikis, H, Lu, Y, et al. (2007) Large-Scale in silico mapping of complex quantitative traits in inbred mice. PLoS ONE **2**(7), e651.
6. Lightfoot, JT, Turner, MJ, Pomp, D, et al. (2008) Quantitative trait loci for physical activity traits in mice. Physiol Genomics **32**, 401–408.
7. Star Consortium (2008) SNP and haplotype mapping for genetic analysis in the rat. Nat Genet **40**(5), 560–566.
8. Belo, A, Zheng, P, Luck S, et al. (2008) Whole genome scan detects an allelic variant of *fad2* associated with increased oleic acid levels in maize. Mol Genet Genomics **279**, 1–10.
9. Cervino, ACL, Darvasi, A, Fallahi, M, et al. (2007) An integrated in silico gene mapping strategy in inbred mice. Genetics **175**, 321–333.
10. Wang, J, Liao, G, Usuka, J, et al. (2005) Computational genetics: from mouse to human? Trends Genet **21**(9), 526–532.
11. McClurg, P, Pletcher, MT, Wiltshire, T, et al. (2006) Comparative analysis of haplotype association mapping algorithms. BMC Bioinformatics **7**, 61.
12. Szatkiewicz, JP, Beane, GL, Ding, Y, et al. (2008) An imputed genotype resource for the laboratory mouse. Mamm Genome **19**(3), 199–208.
13. Westfall, PH, Young, SS. (1993) Re-Sampling Based Multiple Testing, Wiley, New York.
14. Yu, J, Presoir, G, Briggs, WH, et al. (2006). A unified mixed-model method for association mapping that accounts for multiple levels of relatedness. Nat Genet **38**(2), 203–208
15. Malosetti, M, van der Linden, CG, Vosman, B, et al. (2007) A mixed-model approach to association mapping using pedigree information with an illustration of resistance to Phytophthora infestans in potato. Genetics **175**, 879–899
16. Zhao, K, Aranzana, MJ, Kim, S, et al. (2007) An Arabidopsis example of association mapping in structured samples. PLoS Genet **3**(1), e4.
17. Kang, HM, Zaitlen, NA, Wade, CM, et al. (2008) Efficient control of population structure in model organism association mapping. Genetics **178**, 1709–1723
18. Wade, CM, Kulbokas, EJ, 3rd, Kirby, AW, et al. (2002) The mosaic structure of variation in the laboratory mouse genome. Nature **420**, 574–578.
19. Wiltshire, T, Pletcher, MT, Batalov, S, et al. (2003) Genome-wide single-nucleotide polymorphism analysis defines haplotype patterns in mouse. Proc Natl Acad Sci USA **100**, 3380–3385.
20. Frazer, KA, Wade, CM, Hinds, DA, et al. (2004) Segmental phylogenetic relationships of inbred mouse strains revealed by fine-scale analysis of sequence variation across 4.6 mb of mouse genome. Genome Res **14**, 1493–1500.
21. Yalcin, B, Fullerton, J, Miller S, et al. (2004). Unexpected complexity in the haplotypes of commonly used inbred strains of laboratory mice. Proc Natl Acad Sci USA **101**, 9734–9739.

Chapter 13

Candidate Gene Association Analysis

Jonathan B. Singer

Abstract

Candidate gene association study is the most common method for associating human genetic variations with the phenotypes they produce, due to the relative simplicity of acquiring patient samples and genotype data. The study design begins with identifying appropriate DNA samples and an appropriate phenotype for analysis. The candidate genes and polymorphisms must then be chosen. After genotyping the candidate genes in the DNA samples, the results are checked to ensure appropriate quality and association analysis is performed. The raw results are interpreted and placed into context and follow-up analysis is carried out to validate and refine the findings.

A wide range of software packages are available for both the association analysis and other steps in the study. This chapter describes the use of PLINK as the analysis tool in an example, as that suite has emerged as the most popular option for genetic association testing.

Key words: Association, candidate gene, genotyping, PLINK.

1. Introduction

Until recently, genetic association testing of quantitative or binary traits was synonymous with candidate gene analysis, as the number of markers required for genome-wide coverage made studies of one or a few candidate genes the only practical option. The onset of microarray genotyping has made whole-genome association analysis practical, and in many cases it is the preferable option. Hypothesis-driven candidate gene associations still remain the norm, however, due to a number of powerful attractions. Consideration of a candidate gene association strategy should take into account the various strengths and drawbacks of the method.

K. DiPetrillo (ed.), *Cardiovascular Genomics*, Methods in Molecular Biology 573,
DOI 10.1007/978-1-60761-247-6_13, © Humana Press, a part of Springer Science+Business Media, LLC 2009

The relatively sparse number of polymorphisms tested provides superior power to identify significant associations. Because the candidate polymorphisms have been specifically selected by the investigator, the credibility and interpretability of those significant associations are frequently greater than is the case for associations identified in a genome-wide study. Hypothesis-driven candidate gene studies are frequently more acceptable to investigational review boards than are whole-genome studies, and even a broad candidate gene study is typically less expensive than whole-genome analysis.

The primary disadvantage of candidate analysis, of course, is that it is limited by the accuracy of the candidate selection. By definition, it cannot identify anything unexpected, and it is constrained by the frequently poor knowledge of the genetics underlying most diseases and processes.

This chapter addresses questions on study design, quality control, and statistical analysis. It assumes a functioning genotyping facility, which can be chosen from a variety of platforms measuring different polymorphism types. Basic analyses using PLINK (1) and Haploview (2) are described in order to provide a starting point; the statistical analysis of genetic variation is a major field of study in and of itself, and is outside the scope of this discussion.

2. Materials

2.1. Patient Samples

1. Appropriately consented DNA samples from patients with the phenotype and demographic data are needed to perform the desired association analysis. The requirements for quality and quantity of DNA depend on the genotyping platform used.

2.2. Genotype Data

1. The acquisition of data for single-nucleotide polymorphisms (SNPs), restriction fragment length polymorphisms (RFLPs), microsatellites, and other polymorphisms can be performed via standard laboratory methods (3) or on a variety of commercial genotyping systems from Applied Biosystems (http://www.appliedbiosystems.com), Illumina (http://www.illumina.com), Sequenom (http://www.sequenom.com), and others.

2.3. Analysis Software

1. The choice of software for statistical analysis should be determined by personal preference and by the details of the analysis (*see* **Note 1**). In this chapter, PLINK is used for single-point analysis and quality control. PLINK has emerged as the most popular tool for whole-genome analysis of association, due to its wide range of features and superb computational

performance. This free, open-source package can be obtained from http://pngu.mgh.harvard.edu/~purcell/plink, and versions for all major operating systems are available, as is source code.

2. Haploview/Tagger is used for selecting tagging polymorphisms and haplotype analysis. This free, open-source package can be obtained from http://www.broad.mit.edu/mpg/Haploview as a cross-platform Java executable or as source code.
3. Also useful are Structure (4) for analysis of population structure and Excel (Microsoft, Redmond, WA, USA) for general quantitative analysis and statistical testing.

3. Methods

3.1. Selection of Samples and Phenotype

1. A phenotype is attractive for association of analysis if it is variable and there is evidence of its heritability. A continuous phenotype offers more statistical power in its original form than if compressed to a binary or other discontinuous form. Log-transformation or other normalization may be appropriate, however, in order for genetic models that assume linearity of effect to work properly. Samples should be chosen to maximize the degree of phenotype heterogeneity and minimize noise from population admixture. Power calculations, using simulated data, can be used to determine the minimal effect size detectable in the proposed sample set.

3.2. Selection of Genes and Polymorphisms

1. The most common source of candidate genes and candidate polymorphisms in those genes is the existing literature.
2. Additional candidate polymorphisms in identified genes can be chosen by searching the dbSNP database for missense or splice variant polymorphisms (**Fig. 13.1**). A variety of other databases exist to identify polymorphisms that are likely to cause pronounced effects (5–9) (*see* **Note 2**).
3. Tagging SNPs (**Fig. 13.2**), to test a region more comprehensively based on haplotype structure, can be selected with tools like Tagger or Stampa (10). Inclusion of the polymorphisms already chosen should be forced in order to generate a more efficient set of additional tags.

3.3. Data File Creation

1. The relatively small data files involved in candidate gene analysis can typically be generated in a spreadsheet or similar application, with no need for databases or other more elaborate informatics. Excel can usually produce the necessary files

Region	Contig position	mRNA pos	dbSNP rs# cluster id	Hetero-zygosity	Validation	3D	Clinically Associated	Function	dbSNP allele	Protein residue	Codon pos	Amino acid pos
exon_5	37643178	361	rs2069885	0.121	H			missense	T	Met [M]	2	117
								contig reference	C	Thr [T]	2	117
exon_3	37646160	176	rs2066760	0.180				synonymous	A	Leu [L]	3	55
								contig reference	C	Leu [L]	3	55
exon_1	37646518	12						start codon				1

Fig. 13.1. Coding SNPs in the IL9 gene, viewed in the dbSNP browser.

A

B

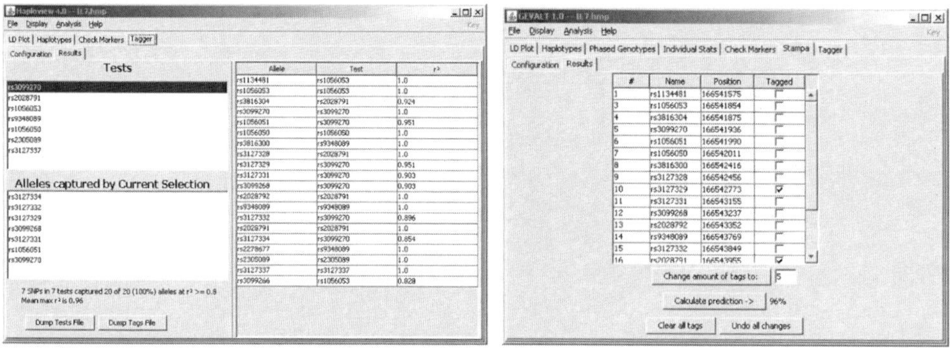

Fig. 13.2. Tagging SNP identification in the IL7 gene region, using (**A**) Tagger and (**B**) Stampa.

with some hand-editing. While PLINK supports a variety of input formats, the basic .ped and .map files are the best suited for a candidate gene study with more samples than polymorphisms. The .map format can also be readily remapped to Haploview's .info format.

3.4. Association Analysis

1. Once genotyping is complete, quality-control measures should be applied to remove unreliable data. These can be performed manually before generating a final dataset for statistical analysis or during analysis by using the QC options in analysis software (*see* **Note 3**). Polymorphisms with low call rate, samples with low call rate, and polymorphisms with minor allele rates are good candidates for exclusion.

   Example, using PLINK to perform QC removals of SNPs and patients, and then write a binary file:

   ```
   > plink --out Example --file Example --geno 0.03 --mind 0.10 --maf 0.01 --make-bed
   ```

2. Testing adherence to Hardy–Weinberg equilibrium can identify a variety of problems. Disequilibrium is likely to result from two sources, genotyping error and admixture of multiple populations in the patient set. The former is obviously

problematic and such polymorphisms should be either re-genotyped more effectively or removed from the analysis. The latter is not necessarily a problem, but can lead to erroneous associations for phenotypes that vary between the admixed populations, and results for such polymorphisms, if analyzed, should be regarded with extra scrutiny. Dividing 0.05 by the number of polymorphisms tested is a reasonable threshold for identifying significant deviations from Hardy–Weinberg equilibrium, and tools like Structure can be used to find population admixture.

3. Association analysis is then typically a straightforward function in the analysis software. The basic case–control and quantitative trait analyses in PLINK are a good starting point, and PLINK offers many options for those wishing to venture into more elaborate test designs, such as stratified analysis, regressions, and conditioned analysis.

   Example, using PLINK to perform a basic analysis on the binary file created in step 3.3.1:

   ```
   >plink --out Example --bfile Example--assoc
   ```

4. A simple way to perform haplotype association analysis on case–control data in the absence of phasing information is by using Haploview to infer haplotype blocks and perform association testing (*see* **Note 4**).

3.5. Correction of Multiple Testing

1. As long as a single phenotype is being tested, the simplest, best option to correct for the testing of multiple polymorphisms and haplotypes is to use the permutation option in PLINK (*see* **Note 5**).

   Example, using PLINK to perform a basic analysis on the binary file created in step 3.3.1, with permutation:

   ```
   > plink --out Example --bfile Example -assoc
   -mperm
   ```

2. Bonferroni correction, in which the threshold for statistical significance is divided by the number of tests, is a straightforward way to correct for more complicated project designs. (For example, a study of 12 polymorphisms and 3 phenotypes might have a threshold for significance of 0.0014 or $0.05/(12 \times 3)$.) This provides a particularly conservative correction, as it does not take into account correlation between phenotypes or genotyped loci.

3. Given the necessary programming skills, permutation can also be performed to suit any statistical design by repeatedly randomizing the genotype and phenotype data, performing the statistical analyses and determining a threshold for significance from the obtained results (*see* **Note 6**).

4. In the case of candidate loci selected from the literature, for which a specific hypothesis exists, correction for multiple testing may not be warranted if the result is to be presented as a replication of that finding (*see* **Note 7**). In such a case, fidelity to the details of the original analysis (phenotype, statistical test, covariates) is appropriate even if they differ from those used for the primary analysis.

3.6. Next Steps

1. A number of post-analysis steps can be taken to evaluate the plausibility of the top obtained results. If previous associations with a similar phenotype have been reported for a locus, consistency of direction of effect is reassuring (while inconsistency suggests the opposite). In the absence of previous publications, the direction of effect can be evaluated for consistency with the expected biological effect of the polymorphism. Apparent overdominance, where the two homozygous classes for an identified locus have similar phenotypes while the association is driven by a different phenotype for the heterozygotes, is frequently biologically implausible (*see* **Note 8**).
2. Other lines of experimental data can support the apparent associations, such as variation in gene expression or protein level consistent with the biological mechanism suggested by an association.
3. None of these, of course, serve as a substitute for replicating the association in an independent set of samples. Replication can usually use a smaller sample set, as the number of hypotheses being tested is normally smaller than in the original study.

4. Notes

1. Scores of other non-commercial and commercial software packages can be valuable for association analysis. The nature of the phenotype (a survival endpoint, for example), the desire to test multimarker associations, extensive population admixture, or other study-specific issues may demand more specialized tools. Alternatively, analyses suitable for regulatory submission may require analysis with validated statistical tools.
2. Normal guidelines for genotyping assay development apply here as well: avoid pairs of adjacent polymorphisms and use BLAST (or a similar tool) to identify repeated genomic sequences. Changing genotyping methods might be warranted depending on the perceived value of a polymorphism,

with sequencing of a difficult region considered worthwhile for a missense variant with previous reported associations, but not for a tagging SNP.

3. PLINK's defaults (minimum 90% call rate for both polymorphisms and samples, minimum 1% minor allele frequencies) are good starting points for quality control of genotype data.
4. This sort of haplotype testing has somewhat fallen out of favor as its advantages over single-point analysis are limited. It can be useful, however, and is often valuable to replicate reported associations with haplotypes.
5. Permutation can also be used to more accurately estimate *p*-values instead of relying on the asymptotic value given by a chi-square test. Given the small size of candidate gene studies, even millions of permutations can be performed in a reasonable time on a typical desktop computer.
6. For example, a study of 12 polymorphisms and 3 phenotypes, permuted 10,000 times, produces 360,000 *p*-values. The 18,000th lowest *p*-value $((12x3x10{,}000)x0.05 = 18{,}000)$ represents the threshold for study-wide significance at $p = 0.05$, and obtained *p*-values below that threshold would be deemed significant.
7. For example, after performing a complete suite of association tests, followed by correction for multiple testing, one might notice that a previously reported association fails to replicate. If one wishes to publish a negative result for that replication, repetition of the original statistical method, with no correction for multiple testing, is appropriate in order to provide the fairest test of the original report.
8. Both direction of effect and mode of inheritance can be taken into account in the association analysis software. Most of the analysis options in PLINK assume an additive model, so overdominant associations are less likely to rise to the top of results. Direction of effect could be taken into account by treating tests as one-tailed (dividing obtained *p*-values by two), but the convention in the field is to use two-tailed tests and report concordance of direction as additional evidence.

References

1. Purcell, S, Neale, B, Todd-Brown, K, et al. (2007) PLINK: a tool set for whole-genome association and population-based linkage analyses. *Am J Hum Gen* **81,** 559–575.
2. Barrett, JC, Fry, B, Maller, J, et al. (2005) Haploview: analysis and visualization of LD and haplotype maps. *Bioinformatics* **21,** 263–265.
3. Sambrook, J, Russell, DW. (2001) *Molecular Cloning: A Laboratory Manual.* Cold Spring Harbor Laboratory Press, Cold Spring Harbor, NY.

4. Pritchard, JK, Stephens, M, Donnelly, P. (2000) Inference of population structure using multilocus genotype data. *Genetics* **155,** 945–959.
5. Guryev, V, Berezikov, E, Cuppen, E. (2005) CASCAD: a database of annotated candidate single nucleotide polymorphisms associated with expressed sequences. *BMC Genomics* **6,** 10.
6. Kang, HJ, Choi, KO, Kim, BD, et al. (2005) *Nucleic Acids Res* **33,** D518–D522.
7. Karchin, R, Diekhans, M, Kelly, L, et al. (2005) LS-SNP: large-scale annotation of coding non-synonymous SNPs based on multiple information sources. *Bioinformatics* **21,** 2814–2820.
8. Yuan, HY, Chiou, JJ, Tseng, WH, et al. (2006) FASTSNP: an always up-to-date and extendable service for SNP function analysis and prioritization. *Nucleic Acids Res* **34,** W635–W641.
9. Lee, PH, Shatkay, H. (2008) F-SNP: computationally predicted functional SNPs for disease association studies. *Nucleic Acids Res* **36,** D820–D824.
10. Halperin, E, Kimmel, G, Shamir, R. (2005) Tag SNP selection in genotype data for maximizing SNP prediction accuracy. *Bioinformatics* **21**(1), i195–i203.

Chapter 14

Genome-Wide Association Study in Humans

J. Gustav Smith and Christopher Newton-Cheh

Abstract

Genome-wide association studies have opened a new era in the study of the genetic basis of common, multifactorial diseases and traits. Before the introduction of this approach only a handful of common genetic variants showed consistent association for any phenotype. Using genome-wide association, scores of novel and unsuspected loci have been discovered and later replicated for many complex traits. The principle is to genotype a dense set of common genetic variants across the genomes of individuals with phenotypic differences and examine whether genotype is associated with phenotype. Because the last common human ancestor was relatively recent and recombination events are concentrated in focal hot-spots, most common variation in the human genome can be surveyed using a few hundred thousand variants acting as proxies for ungenotyped variants. Here, we describe the different steps of genome-wide association studies and use a recent study as example.

Key words: Genome-wide association study, GWAS, whole-genome association study, WGAS, complex genetics, common variation.

1. Introduction

Genome-wide approaches offer a systematic analysis of genes with and without a priori evidence for involvement in the molecular basis of a trait or disease. The benefits of genome-wide studies, as opposed to candidate-gene-based studies, stretch beyond identifying a genetic basis for trait differences to uncovering novel molecular pathways and interactions that underlie the trait, leading to new insights into physiology and pathophysiology. Moreover, the in vivo relevance of a gene to a trait, having been established in humans at the outset, can guide development of model organisms and systems with direct applicability to human disease. Lastly, identified genes may be of special interest to the medical community as a means to predict disease risk at the population level and as novel therapeutic targets.

K. DiPetrillo (ed.), *Cardiovascular Genomics*, Methods in Molecular Biology 573,
DOI 10.1007/978-1-60761-247-6_14, © Humana Press, a part of Springer Science+Business Media, LLC 2009

Genome-wide approaches were first proposed by Botstein and colleagues (1). The principle is to genotype a number of common genetic variants spaced across the genome which can act as proxies for ungenotyped variants due to coinheritance of neighboring variants on a chromosomal segment. Botstein suggested using a panel of restriction fragment length polymorphisms (RFLPs) to examine how familial transmission of allelic variants tracks with transmission of phenotype. When a polymorphism shows significant linkage to a disease, additional markers can be genotyped in that region, termed fine-mapping, and sequencing can be performed to identify the responsible gene and variant. This genome-wide approach examining familial transmission, termed linkage analysis, has been successfully utilized to identify the rare variants underlying more than 2,000 monogenic disorders to date (http://www.ncbi.nlm.nih.gov/Omim/mimstats.html). Monogenic (i.e., Mendelian) disorders, such as familial hypercholesterolemia, are relatively less common in the population and are typically caused by rare variants, often leading to altered amino acid sequence, which have strong (nearly deterministic) effects producing overt disease. Complex traits, by contrast, result from multiple genetic and environmental contributors and are typically much more common, such as hypercholesterolemia. As predicted by Risch and Merikangas, linkage has largely failed to identify common variants in common, multifactorial traits due to inherent methodological limitations (2).

In the last year we have witnessed the success of genome-wide association (GWA) studies, proposed by Risch and Merikangas in the same seminal perspective (2), in identifying dozens of common genetic variants associated with common diseases and traits. These discoveries have provided new insights into the genetic architecture of common traits by providing evidence that common variation explains at least a portion of the variation in risk of common disease, known as the common disease–common variant hypothesis (3).

Genome-wide association studies make use of abundant and easily genotyped molecular markers, single-nucleotide polymorphisms (SNPs), which are single-base substitutions with minor allele frequencies above 0.01. Fundamentally, such association studies are quite simple and test the hypothesis that allele frequency differs between individuals with differences in phenotype. The facts that the out-of-Africa migration approximately 50,000 years ago, from which all modern non-African human populations were founded, was relatively recent on an evolutionary scale (4) and that recombination across the human genome is highly concentrated in hotspots (5) have resulted in a high degree of correlation between adjacent variants (termed linkage disequilibrium). This linkage disequilibrium allows the majority of common variation to be assayed with 500,000–1,000,000 correctly chosen proxy SNPs (6). Although proposed more than a decade ago, these studies have only recently been made possible by the completion of the human genome reference sequence (7, 8), the deposit of millions of SNPs into public databases (9), developments in high-throughput genotyping technology (10), and the completion of the Human Haplotype Map (HapMap) (11).

Before the era of GWA studies, only a handful of the thousands of genetic polymorphisms examined in candidate genes showed consistent replication (12, 13). This has been attributed in part to poor study design, genotyping error, and population stratification, but most importantly to overly permissive *p*-value thresholds in the original samples (resulting in false-positive results) and inadequately powered sample sizes in replication cohorts (resulting in false-negative results). To address these problems, a working group assembled by the US National Cancer Institute and the National Human Genome Research Institute suggested a set of standards for performing genetic association studies (14). Here, we follow the framework outlined by the working group, highlight steps where there is emerging consensus, and suggest how to proceed where there are no standards. It should also be stressed that since GWA studies were made possible only recently and have gone through rapid developments, there is little consensus for many steps of the process. We cite the central publications that have shaped current GWA methodology for the interested reader, but cite only a few of the large number of GWA studies published. For a collection of all GWA studies published we refer to the homepage of the NHGRI Office of Population Genomics (http://www.genome.gov/26525384).

A GWA study raises many issues in computational management and statistical analysis of the large amount of genotype data generated, which is reflected in the contents of this chapter. We describe the procedures for (1) careful study design before data collection where the critical issue is to clearly define the trait and reduce the impact of potential biases (2); data acquisition, including DNA genotyping, genotype calling, and the quality control of results (3); statistical analysis; and (4) follow-up of results to validate and replicate interesting loci, identify causal genes and variants, and improve power in meta-analysis. To provide an example, we use the Diabetes Genetics Initiative (15), one of the early GWA studies which helped establish standards and identified problems and advantages of the method.

2. Materials

2.1. A Statistics Software Package

R, SAS, SPSS, Stata, StatView.

2.2. Specific Applications for Statistical Genetics (Suggestions, See Note 1)

1. PLINK
   [http://pngu.mgh.harvard.edu/~purcell/plink/]
   Software for management and analysis of GWA datasets (42).
2. snpMatrix
   [http://www.bioconductor.org/]
   Application for management and analysis of GWA datasets implemented in Bioconductor (16).
3. SAS/Genetics (with SAS by SAS Institute, Inc., Cary, NC, USA)
   Statistical genetics analysis package for SAS.
4. R Genetics
   [http://rgenetics.org/]
   Statistical genetics analysis package for Bioconductor.
5. QTDT family-based association test
   [http://www.sph.umich.edu/csg/abecasis/QTDT]
   Application for family-based association analysis (17).
6. FBAT and PBAT
   [http://www.biostat.harvard.edu/~fbat/default.html]
   [http://www.biostat.harvard.edu/~clange/default.htm]
   Applications for family-based association analysis (18, 19).
7. HaploView
   [http://www.broad.mit.edu/mpg/haploview/]
   Software for analysis of correlation patterns in SNP data and visualization of haplotypes and GWA results (20).
8. EIGENSOFT
   [http://genepath.med.harvard.edu/~reich/Software.htm]
   Software for Principal Components Analysis of population substructures (21).
9. STRUCTURE
   [http://pritch.bsd.uchicago.edu/structure.html]
   Software for structured analysis of population substructures (41).
10. MACH
    [http://www.sph.umich.edu/csg/abecasis/MACH/]
    Software for imputation of SNPs not genotyped from genotyped SNPs using LD data (22).
11. IMPUTE (Marchini 2007)
    [https://mathgen.stats.ox.ac.uk/impute/impute.html]
    Software for imputation of SNPs not genotyped from genotyped SNPs using LD data used by the WTCCC (23).

12. WGAViewer (Ge, 2008)
[http://www.genome.duke.edu/centers/pg2/downloads/wgaviewer.php]
Software for visualization and functional annotation of WGA data (24).
13. ANCESTRYMAP
[http://genepath.med.harvard.edu/~reich/Software.htm]
Software for ancestry mapping (25).

2.3. Webpages

1. UCSC Genome Browser
[http://genome.ucsc.edu/]
Genome browser maintained by the University of California Santa Cruz.
2. Ensembl
[http://www.ensembl.org/]
Genome browser maintained by EMBL-EBI and the Sanger Institute.
3. dbSNP and dbGene
[http://www.ncbi.nlm.nih.gov/sites/entrez]
Databases maintained by the National Center for Biotechnology Information containing data on genes and SNPs deposited by researchers.
4. The Human Haplotype Map
[http://www.hapmap.org/]
Genome browser annotated with results from the HapMap project (6, 11).
5. Genetic Power Calculator
[http://pngu.mgh.harvard.edu/~purcell/gpc/]
Online calculator of power in genetic association studies (26).
6. CaTS
[http://www.sph.umich.edu/csg/abecasis/CaTS/]
Online calculator of power in genetic association studies with multistage designs (27).
7. SNP Annotation and Proxy Search (SNAP)
[http://www.broad.mit.edu/mpg/snap/]
Online application for SNP annotation and proxy searching.

3. Methods

3.1. Study Design

3.1.1. Define the Phenotype: Quantitative and Qualitative Traits

As in all epidemiological studies, the first step is careful trait definition and a decision whether to use a qualitative or quantitative trait (*see* **Note 2**). Quantitative traits generally increase power through larger information content, but also may decrease power if imprecise measurements are made, which can be detectable by low inter-reader or intra-reader reliability. Certain traits, such as diseases, are by definition qualitative but can be measured quantitatively (i.e., when analyzing survival time to an endpoint). Similarly, quantitative traits can be analyzed qualitatively by applying a threshold value; for example, when comparing individuals in the extremes of a distribution to reduce trait heterogeneity.

When using a qualitative trait with cases and controls (i.e., those with and without type 2 diabetes), trait ascertainment should be performed in such a way as to reduce misclassification bias (incorrect case–control assignment) and selection bias (sample not representative of population). Finally, as in all studies of etiology, it may be important to define the phenotype as precisely as possible regarding fundamental mechanism (if well understood) to increase power by reducing heterogeneity, while retaining simplicity to facilitate replication studies and not reducing sample size excessively. The appropriate balance between very selective entry criteria and maximal inclusiveness to increase sample size needs to be considered on a trait-by-trait basis.

Example Several traits were studied in the Diabetes Genetics Initiative (DGI). Here, we focus on one quantitative trait, low-density lipoprotein cholesterol (LDL), and one qualitative trait, the disease type 2 diabetes (T2D), which are both well-studied phenotypes. T2D was defined according to internationally standardized criteria established by WHO. To reduce etiologic heterogeneity, steps were taken to exclude individuals with type 1 diabetes, which is thought to have a different pathophysiology, and individuals with monogenic diabetes (e.g. maturity onset diabetes of the young). Control subjects were defined as having normal glucose tolerance in testing to exclude diabetes or pre-diabetes. Also, they did not have any first-degree relatives with known diabetes. LDL was estimated using the clinical standard of the Friedewald formula, in which measurements of total blood lipid content are performed and non-LDL lipid types (high-density lipoprotein cholesterol, triglycerides) are subtracted. Measurements were performed on standard clinical assays. Lipid lowering therapy, present in a minority of subjects, was an exclusion criterion.

3.1.2. Estimate the Contribution of Genetic Variation to Phenotypic Variation

It is likely that most human traits that exhibit variability have a genetic component. However, it is well known that the degree of heritability differs greatly between traits and heritability reflects the total effect of genetic variation as well as environmental and behavioral factors segregating in families. Hence, it can be useful to assess the amount of trait variability explained by heritability before embarking on a costly genome-wide association study. That is, if the heritability of a trait is low, problems with trait definition, measurement error, or other systematic problems may bias genetic association studies to the null. In such cases, attempts could instead be made to enrich for genetic effects by examining precisely defined subphenotypes (e.g., measures of insulin resistance or beta cell function as opposed to the clinical definition of type 2 diabetes).

To examine whether evidence of heritability exists, perform a critical literature review and perform heritability analyses in the sample used if family data exist. Studies that examine heritability typically examine familial aggregation, concordance between monozygotic as compared to dizygotic twins (MZ >> DZ concordance suggests a genetic basis), and sibling recurrence risk. Heritability analysis is well-discussed elsewhere (28).

Example Both T2D and LDL are well known to have substantial heritable components. A literature review reveals that T2D heritability estimates vary widely among studies, but most estimates range between 30 and 70%. Sibling recurrence risk ratio is estimated to be relatively high at $\lambda_s \sim 3.0$ and monozygotic twins have higher concordance than dizygotic twins, consistent with a strong genetic basis. A literature review of LDL reveals heritability estimates in most studies around 50% and that monozygotic twins have higher concordance than dizygotic twins.

Estimation of trait heritability was not performed in the DGI population given the absence of adequate family data.

3.1.3. Determine Sample Structure: Case–Control, Cohort, Family-Based, Isolates

The four most common sample structures used are case–control studies, cohort studies, family-based studies, and studies in population isolates. Generally, the choice of sample structure should be based on the prevalence and familial segregation of the phenotype of interest (*see* **Note 2**).

Case–control collections are the most widely used samples since they are comparatively easy and inexpensive to collect, but this design is also the most sensitive to different forms of bias, including selection bias and confounding by ancestry (population stratification). To avoid problems of population stratification, controls should always be selected from the same population as cases. Population stratification can be adjusted for in the data analysis, but steps should be taken to minimize the potential problem up front in the study design. Further matching for sex and other dichotomous covariates with a strong impact on phenotype can also be performed to increase power, but by reducing non-genetic contributions to trait variation. Further gain in power may result from a focus on cases with a family history of a condition, called enrichment sampling, but this remains to be proven. The ratio of controls to cases is a matter of availability of resources. If one has a choice between increasing the ratio of controls to cases versus keeping the ratio at 1:1 but increasing the number of case–control pairs, it is more powerful to increase the number of case–control pairs in 1:1 matching. However, for a trait such as sudden cardiac death, in which cases are not necessarily available to increase the number of case–control sets, increasing the number of controls to two to three times the number of cases provides some gain in power. Beyond a ratio of 3:1, there is minimal incremental power and it is typically not advisable given the fixed cost of genotyping platforms. If existing and appropriate control samples are available (without additional genotyping cost) there is no downside to increasing the number of controls.

Cohort studies are more difficult and costly to collect since they involve phenotyping a large number of individuals, often with lengthy follow-up to detect incident events. They have the advantage of being more representative of the population and hence less prone to selection bias, but the trait under study must be relatively common in the population for a well-powered GWA study, rendering this study design impossible for rare diseases. While the lack of selection on trait status may reduce potential biases, it may also reduce the precision of a trait measurement compared to a study focused explicitly on a single trait. One advantage is that individuals can be examined and followed for large numbers of traits as exemplified by the multiple GWA studies performed in the Framingham Heart Study (29).

If the phenotype is qualitative, relatively rare, and shows marked segregation in families, a family-based approach is likely to be of great value. However, familial aggregation may be due to shared environmental factors and rare genetic variants. Family-based approaches focus on transmission of alleles from heterozygous parents to affected individuals more often than expected by chance. Methods exist to study only this within-family component, a design with the benefits of not requiring phenotypic information in the parents and resistance to population stratification, but with power reduced by the requirement for informative parent–offspring trios. In addition, methods have been developed to incorporate both the within-family and between-family (as in studies of unrelated individuals) components with a gain in power; the cost is greater susceptibility to population stratification, but current methods to adjust for population structure render this less of an issue. Family-based studies are particularly advantageous in pediatric diseases, with the major disadvantage being the difficulty of ascertaining familial genotypes. In addition, all studies of familial transmission are more sensitive to false-positive association due to genotyping errors than studies of unrelated individuals (30).

Population isolates may have experienced strong founding effects, in which genetic variation carried into an isolate from the source population is reduced and linkage disequilibrium increased. Such samples have proven to be valuable tools in identifying genes for Mendelian traits using linkage analysis, due to the low genetic and environmental heterogeneity. Their value in genome-wide association studies is less clear since the sample sizes are typically small, comparability with other populations is uncertain, and excess relatedness introduces analytical difficulties.

Example In DGI, a case–control approach was used with both population-based and family-based subsamples that were first analyzed separately and then pooled for joint analysis. In the unrelated subsample, cases and controls were matched on gender, age of diabetes onset, collection locale, and body mass index. Up to two cases were matched to up to two controls.

In the family-based subsample one or two discordant sib pairs were included per family. The samples were collected in a number of towns in Finland and Sweden.

3.1.4. Calculate Power and Determine Sample Size

One of the lessons learned from the first wave of GWA studies is the need for large sample sizes. The power to detect an association with sufficient statistical support to distinguish a result from a chance finding depends on sample size, allele frequency, effect size, and coverage of genetic variation on the genotyping array. For common variants, the theoretically expected effect sizes (2) are genotypic relative risks below 2.0, which is consistent with effect sizes seen in most GWA studies (typically 1.1–1.5). Minor allele frequencies examined in most GWA studies range between 0.01 and <0.50, but power is much lower to detect effects at the low end of the frequency spectrum. The genomic coverage reflects the proportion of all genetic variants for which the genotyped variants alone or in aggregate (as haplotypes) act as proxies at a certain level of linkage disequilibrium measured as the coefficient of determination, r^2, the proportion of variation of SNP explained by another SNP.

Since the effect estimates seen for SNPs in GWA studies are typically modest, it follows that sample sizes have to be very large in order to achieve *p*-values below the significance thresholds currently recommended (5×10^{-8} to 1×10^{-7}; *see* **Section 3.3.2**). As a guideline, for SNPs with minor allele frequency >0.2 sample sizes of 1,000 cases and 1,000 controls are required for odds ratios of 1.5 with 80% power. Estimates of the sample size needed for certain levels of power can be obtained using the online applications Genetic Power Calculator or CaTS (*see* **Section 2**).

Genome-wide association analysis can be performed in two stages to reduce genotyping costs while retaining power (27). A large number of markers are first genotyped in a random subset of a sample and the strongest associations are then genotyped in the full sample, reducing costs compared to genotyping all markers in the entire sample. However, arrays with fixed marker sets are becoming less costly and may soon be more cost-effective than genotyping large numbers of individual markers. Power in one- and two-stage designs can be estimated using the software CaTS (*see* **Section 2**), but in general favor somewhat larger second-stage samples.

Example Using CaTS, Zondervan and Cardon estimated that a sample size of about 6,500 cases and 6,500 controls was needed to detect association with a SNP conferring a relative risk of 1.2 and an allele frequency of 0.2, assuming 300,000 independent tests at 80% power at a significance threshold of 1×10^{-7} (31).

3.2. Data Acquisition and Quality Control

3.2.1. Genotyping and Genotype Calling

Several high-throughput genotyping platforms assaying marker sets of different density and selection method are currently available, mainly from three companies: Affymetrix, Illumina, and Perlegen. Marker densities and selection methods correspond to different coverage of genome-wide variation, which is a determinant of the power to detect a true association if sample size is adequate (32). Marker density currently ranges from 100,000 to 1,000,000 variants per array selected using one of four methods: (a) randomly; (b) through a tagging approach to maximize coverage based on linkage disequilibrium patterns; (c) a combination of the first two; or (d) focusing only on SNPs known or likely to be functional. The SNPs on most arrays are a mix of variants that are polymorphic in populations of differing ancestry such that a number of SNPs will always be uninformative because they are monomorphic or rare in the population genotyped. Generally, the choice of array should focus on maximizing power taking into account parameters of sample size, ancestral origin, and marker selection methods. With adequate sample size, increased marker density increases power up to a threshold of about 500,000 markers. One million SNPs are likely to be of only modest value in populations other than those of African ancestry, in which linkage disequilibrium is slightly lower and genetic diversity slightly higher. Power calculations for different arrays have been published (33–35) and currently available arrays have recently been reviewed (*see* **Notes 3 and 4**) (10).

Sample collection and handling should be as uniform as possible to minimize bias from batch effects from different sources of DNA, extraction protocols, genotyping procedures, and plate effects. To avoid plate effects (e.g., poor genotyping specific to an individual plate), assignment of individuals to genotyping plates should be randomized blinded to case–control status, while preserving case–control clusters within plates to avoid differential effects within case–control clusters across plates.

A number of different algorithms for allelic discrimination from the intensity data generated in the genotyping procedure, termed genotype calling, have been proposed. The basic principle is clustering of results with similar signal intensities into discrete genotypes. Most genotyping assays come with a recommended calling algorithm. For example, Affymetrix currently recommends the Robust Linear Modeling using Mahalanobis distance (RLMM) procedure described by Rabbee and Speed (36) and developed further by inclusion of a Bayesian estimation component, called BRLMM. Other well-known calling algorithms include CHIAMO, developed by the WTCCC, and Birdseed, developed at the Broad Institute.

Example In DGI, genotyping was performed on the Affymetrix Human Mapping 500 K GeneChip, which contains oligonucleotides to assay 500,568 SNPs. A total amount of 1 μg of genomic DNA from cases and controls were interspersed equally across 96-well plates. Genotyping was performed according to recommendations from the manufacturer. Using BRLMM, a call rate of 99.2% was achieved for 386,731 SNPs passing quality control.

3.2.2. Genotype Quality Control

Genotyping errors are a potential cause of spurious associations and must be carefully sought and addressed, both on the DNA and SNP levels.

On the sample side, potential errors to assess include inadvertent sample duplication, contamination, and low DNA quality. An efficient way to address duplications and sample contaminations is to compare genome-wide SNP data between each pair of individuals in the sample using the population genetic tools of identity-by-descent (IBD) probabilities and coefficient of inbreeding estimates (F). IBD probabilities are calculated from pairwise identity-by-state (IBS) distances, which can also be used to assess population substructure and cryptic relatedness (discussed later in **Section 3.2.4**). IBD probabilities for each pair of individuals can be assessed using the Pihat method, which estimates the proportion of the genomic variation shared IBD. Identical twins, and duplicates, are 100% identical by descent (Pihat=1.0), first-degree relatives are 50% IBD (Pihat=0.5), second-degree relatives are 25% IBD (Pihat=0.25), and third-degree relatives are 12.5% equal IBD (Pihat=0.125). F can be calculated from homozygous genotypes using PLINK. Whether duplications or twins, most studies will randomly include only one individual from pairs of Pihat 1.0. Contamination can be detected by a pattern of low-level relatedness to many people, a strong negative F, and an elevated genome-wide heterozygosity estimate. Individuals with a high rate of missing genotypes should be excluded, as this is an indication of low DNA quality. Commonly used approaches involve exclusion of all individuals with <2–5% missing genotypes, depending on the platform and sample quality. As an additional sample quality control check, most arrays offer a sex confirmation assay that should be compared with self-reported gender.

On the SNP level, errors in genotyping need to be assessed. A good indicator of genotype quality is the SNP call rate, so SNPs with a high proportion of missing genotypes should be excluded. Commonly used thresholds range from 2 to 5% missing genotypes. SNPs with a low population frequency (minor allele frequency, MAF) may also be excluded, as power to detect association with these is low and genotype calling is more prone to errors, especially for case–control analysis. A commonly used threshold is to exclude SNPs with minor allele frequency below 1–5% depending on the sample size (larger sample sizes have better power at lower frequencies). Another quality control test for genotyping errors is deviation from Hardy–Weinberg equilibrium (HWE). The benefits of excluding SNPs that fail an HWE test should be carefully weighed as disequilibrium typically is underpowered for genotyping error detection, but may be caused by population stratification, excess relatedness, positive or negative selection, or true association signals. If HWE is used, it should only be calculated in controls if possible by applying stringent thresholds, in the range of 10^{-6}, as true association signals in merged case–control data can generate only modest deviations from HWE (37, 38). HWE can be calculated using Pearson's Chi-square test or Fisher's exact test, as implemented in most statistical analysis applications. A quantile–quantile plot can be useful to determine whether general HWE disequilibrium appears likely to be caused by population stratification, excess relatedness, selection, or association signals in the same manner as described in **Section 3.2.4** for association test statistics. An additional genotyping quality control sometimes used is deliberate duplication of a number of individuals for comparison of concordance rates.

In family-based studies, an additional quality-control filter is to exclude SNPs which fail to show transmission according to Mendelian inheritance laws. The threshold for Mendelian error filtering of DNA samples or SNPs is based on the number of possible errors given the number of related individuals and number of SNPs examined.

After genotype quality control, we recommend examining whether genotyping failure rates differ between cases and controls using a Pearson's Chi-square test with one degree of freedom to exclude false associations due to systematic differences in quality of genotyping or DNA that can lead to spurious association (39). As a final quality control of SNP genotype calling, visual inspection of cluster plots should be performed after analysis for associations that are considered significant (14).

Example In the DGI analysis, individuals with genotyping call rates below 0.95 were excluded. IBD probability estimates from PLINK were used to verify reported familial relationships. Individuals in the whole sample with cryptic (unrecognized) first-degree relatedness were excluded. Individuals whose sex based on genotypes was discrepant from that self-reported were also excluded. After these exclusions, genotypes for 2,931 individuals remained.

Exclusion criteria for SNPs in DGI included: mapping to multiple locations in the genome (3,605 SNPs), call rate < 0.95 (34,532 SNPs) or < 0.90 in either the family-based or the population-based sample (229 SNPs), MAF < 0.01 (66,787 SNPs) or MAF < 0.01 in either subsample (2,909 SNPs), and Hardy–Weinberg disequilibrium in controls with $p < 1 \times 10^{-6}$ (5,775 SNPs). After quality-control filters were applied, genotypes for 386,731 SNPs were obtained. Testing for differences in missing data between cases and controls was not significant.

3.2.3. Phenotype Quality Control

As with all datasets, quality control must be performed on the phenotype dataset. All individuals should match predefined criteria for inclusion and exclusion. All categorical variables should be examined for improperly coded results, and continuous data should be checked for outliers with consideration to exclude individuals with extreme values. Continuous variables should also be assessed for relative normality in distribution. Helpful tools to this end are visual inspection of a histogram and measures of skewness and kurtosis. If a right-skewed distribution is observed, a logarithm-transformation is often performed to normalize the dataset.

Example The absence of individuals with exclusion criteria and presence of individuals with inclusion criteria was confirmed. No aberrant categories in T2D were detected. LDL values exceeding the 99.5 percentile (>7.423 mmol/L) were excluded as outliers (n=15 individuals) and 197 individuals receiving lipid-lowering therapy were excluded. The LDL distribution was considered near-normal upon inspection of the histogram, examination of skewness and kurtosis, and comparison of mean and median.

3.2.4. Handling Population Stratification and Relatedness

Population stratification, also called confounding by ancestry, arises when samples are composed of subgroups with differences in allele frequencies, due to either different ancestry or excess relatedness, leading to phenotypic differences. This can confound a study, resulting in both false-positive and false-negative findings (39, 40). For example, in Campbell et al. (40), the coincidence of gradients of high to low frequency in the lactase persistence allele and of high to low height from northern to southern Europe was shown to result in spurious association of height with the lactase persistence allele. Hence, it is recommended to test for population structure in samples using several tools (*see* **Note 5**). A high degree of shared ancestry and relatedness in the study sample makes a family-based study design more tractable (discussed in **Section 3.3.2**).

Of the several tests developed to examine population stratification using genome-wide SNP data, we recommend first estimating the extent of association test statistic inflation using a quantile–quantile (QQ) plot for both HWE and association tests. In these plots, observed p-values for each SNP are ranked in order from smallest to largest and plotted against expected values from a theoretical distribution, such as the χ^2 distribution. Population stratification can be seen as a global excess of higher observed *p*-values than expected throughout the span of the plot. This stems from a simplifying hypothesis that only a modest number of variants are associated with a given trait and association of thousands of SNPs indicates population stratification. In addition, a quantitative measure of this deviation can be derived from the genomic inflation factor, λ_{GC}, which is easily calculated as the ratio of the median association test statistic over the theoretical median test statistic of a null distribution (for χ^2 this is 0.675^2) for one degree of freedom tests (41). If using a genotype test – in which each genotype class *AA*, *Aa*, *aa* is tested for a difference in phenotype – the procedure accommodates the two degree of freedom test (42). Under a completely null distribution, observed *p*-values correspond exactly to the expected λ_{GC} of 1. A λ_{GC} that exceeds 1 represents global inflation of the test statistic and can arise from biased genotyping failures between cases and controls or, more commonly, population stratification not accounted for in the genotype–phenotype procedure. Such population stratification can be corrected for by dividing the χ^2 statistic of each SNP by the λ_{GC} value, which makes less extreme all observed *p*-values and ensures that the mean is one, a procedure termed genomic control (41). Genomic control may increase the risk of a false-negative association (type 2 error) somewhat because of its global approach to adjusting *p*-values, but can increase power overall by enabling the use of different, stratified samples. An alternate, less commonly used approach is to calculate λ_{GC} from a number of "null" SNPs not expected to show an association with the trait under study.

One can also detect differences in ancestry among individuals in a study and adjust the analysis using any of the following approaches: a principal components analysis approach called the EIGENSTRAT method implemented in EIGENSOFT (21); examination of the coefficient of relationship (*F*); a structured approach (43); and an approach based on identity-by-state distance and multidimensional scaling implemented in PLINK (44). If a subset of individuals with ancestry that is distinguishable from the overall is identified, they can be excluded or, if clustering into several subgroups, can be analyzed together in a stratified analysis such as the Cochran–Mantel–Haenszel test (discussed in **Sections 3.3.2** and **3.4.3**).

Example To account for the excess relatedness in the family-based subsample of the Diabetes Genetics Initiative, association testing for T2D was performed separately in the family-based sample and the unrelated sample after which the results were pooled in a meta-analysis as described in **Sections 3.3.2** and **3.4.3**.

To examine population stratification and excess relatedness, *p*-value distributions were first examined in QQ-plots and genomic inflation factors were calculated. The genomic inflation factors were λ_{GC}=1.05 for T2D and λ_{GC}=1.07 for LDL, corresponding to very mild overdispersions of test statistics as compared to the null distribution. Next, population stratification was assessed using the EIGENSTRAT method implemented in EIGENSOFT. After clustering outlier samples identified by EIGENSTRAT and meta-analysis using a Cochran–Mantel–Haenszel stratified test, *p*-values were very similar to *p*-values attained by genomic control as evidenced by r^2=0.95. Hence, genomic control was used to adjust for population stratification in both traits because of its simplicity.

3.2.5. Imputation of Ungenotyped SNPs

Genotypes of untested SNPs can be imputed from genotyped SNPs using LD patterns from the International Human Haplotype Map (HapMap) to expand coverage over the 3 million HapMap SNPs and allow meta-analyses across platforms using different marker sets. To date, the value of imputation in increasing power has not been clearly demonstrated, but imputation has successfully facilitated several meta-analyses using fixed genotyping arrays with different coverage. Several applications for imputation, using different methodologies, have been developed, including MACH, IMPUTE, and a method implemented in PLINK (*see* **Section 2**).

Quality-control metrics for imputation provided by the imputation applications are generally a function of the ratio of the observed to the expected variance on the imputed allele count. The ability to impute a genotype for a given SNP is a function of the accuracy of the directly genotyped SNPs in the region and their correlation to the untyped SNP. Note that the increased number of tests with imputation does not demand additional correction for multiple testing if using a significance threshold based on the total number of independent tests in the human genome (*see* **Section 3.3.2**).

Example Untyped HapMap phase II SNPs were imputed in the DGI population to increase power and allow pooling of data from different marker sets in a meta-analysis with other genome-wide datasets of T2D (45). A total of 2,202,892 SNPs were imputed using MACH and passed quality control across all three component studies. DGI and WTCCC both used the Affymetrix Human Mapping 500 K (22), and the FUSION study used the Illumina HumanHap300 BeadChip (46). SNPs with an estimated MAF<0.01 in the pooled or any individual sample were excluded. The imputed SNPs showing the strongest association were directly genotyped in DGI using a different method for validation and showed good agreement with imputed genotypes (r^2=0.84).

3.3. Statistical Analysis

3.3.1. Phenotype Modeling

The goal of phenotype modeling is to increase power through reduction of non-genetic contributions to trait variation by adjusting the trait of interest for variation attributable to known covariates. Generally, it is difficult to establish whether covariates are etiologically independent from the genetic pathways sought, but most continuous traits vary with age and sex and are often adjusted for these variables. Age, in particular, was recently found to modify the effect of association findings (47). A thorough literature review should be conducted to identify potential covariates for adjustment. In general, it is desirable to include covariates that are reproducibly measured, available in other potential replication cohorts, and are unlikely to be in the causal pathway between genotype and disease outcome.

For continuous traits, adjusted models can be created through regression modeling with univariate and stepwise regression. The proportion of variability explained by a model can be estimated as the coefficient of determination, r^2. Residuals (the observed trait value minus that predicted from a regression model) can be used as the trait for genotype–phenotype analysis in the GWA study.

Example LDL cholesterol was adjusted for covariates with previous evidence, high reproducibility, and significant association using multivariable linear regression modeling. Included variables were age, age^2, sex, enrollment center, and diabetes status. Residuals were standardized to a mean of 0 and a standard deviation of 1.

T2D was not adjusted for any covariates since individuals with other recognized etiologies were excluded in the design stage and cases and controls were matched on age, sex, and body mass index.

3.3.2. Association and Statistical Inference

The association between phenotype and genotype is examined by comparing allelic patterns between individuals with different phenotypes. All inheritance models can be examined (additive, dominant, recessive) in separate allelic tests with one degree of freedom, which improves statistical power to detect effects that follow one model more than another, but increases the number of hypotheses tested. Hence, most studies of qualitative traits with unknown allelic inheritance patterns have used either a general genotype test or an additive (allelic trend) test. The general genotype test compares distribution among frequencies of the three genotypes (minor homozygous, major homozygous, heterozygous) between cases and controls and most often uses a Pearson's Chi-square test with two degrees of freedom or Fisher's exact test. The allelic trend test assumes an additive genetic model and tests whether the slope of a fitted regression line, with the three genotypes as independent variables, differs from zero and can be adjusted for covariates. The trend test most often used is the Cochran–Armitage test, whereas Pearson's Chi-square test with one degree of freedom for minor allele frequency is not recommended (48).

For quantitative traits, statistical inference is performed using either linear regression, which assumes an additive model, or ANOVA, which is based on a general model. Both tests require the trait to be approximately normal distributed, which may be achieved by transformation (e.g., $\log_{10}$) if the trait is positively skewed and by adjustment for relevant covariates.

A number of tests for hypothesis testing in family-based samples have been developed. The transmission disequilibrium test (TDT) has one of the simplest designs and examines transmission of qualitative traits within trios (parents and one offspring), with the advantage that only offspring need to be phenotyped. FBAT (family-based association test) implements the TDT for both quantitative and qualitative traits (18). These methods offer the advantage of being resistant to population stratification, but suffer from reduced power because only within-family components are examined and only some parental genotype patterns are informative. Family-based methods with increased power test both within-family and between-family variations for association with genotype, but with a consequence of some sensitivity to population stratification. Such methods are used in DFAM (implemented in PLINK) for qualitative traits, while QTDT (17), QFAM (implemented in PLINK), and GEE (49) model quantitative traits, and PBAT handles both quantitative and qualitative traits. Note that large pedigrees must be broken down to computationally tractable units for these tests. These or related methods have been utilized in recent GWA studies (29, 50–52*)*. Most of the tests described above have been implemented in the statistical analysis applications named in **Section 2**.

The mass significance problem caused by the multiple tests of a GWA study is currently the subject of some debate (*see* **Note 6**). Although Bayesian approaches incorporating information on power and expected number of true positives received much interest early on (53), the frequentist approach of adjusting for a number of independent tests, as originally proposed by Risch and Merikangas (2), is most often used. The Wellcome Trust Case Control Consortium settled on a genome-wide significance threshold of 5×10^{-7}, based on power calculations showing that their study of 2,000 cases and 3,000 controls reached 80% power to detect an association at this significance level for SNPs with MAF>0.05 and relative risk of 1.5, with a drop in power to 43% for relative risks of 1.3 (23). Others have suggested a significance level of 5×10^{-8} in populations of European ancestry based on a overall genome-wide significance threshold of 0.05 adjusted for an estimated 1 million independent SNPs in the genome (11, 54, 55), by the Bonferroni method, paralleling the linkage thresholds proposed by Lander and Kruglyak for genome-wide linkage analysis (56). Independent SNPs for Asian populations sampled in HapMap represent a similar number of independent tests, whereas African-derived samples represent more tests, based on the extent of variation and LD (6, 11).

As discussed in **Section 3.1.4**, the typically small effect sizes seen for common variants necessitate large sample sizes to reach significance thresholds such as 5×10^{-8}. To reduce genotyping costs, a two-stage design is often utilized where genome-wide marker sets are analyzed in a smaller number of individuals in stage 1, SNPs below a less stringent threshold are genotyped in additional individuals in stage 2, and then SNPs are analyzed jointly in all individuals (27).

The effect size of a SNP is calculated as the odds ratio or relative risk per genotype for qualitative traits and as the beta coefficient from a regression model, which can be interpreted as the effect of each additional allele, for quantitative traits.

Example For T2D the population-based sample was analyzed with a Cochran–Mantel–Haenszel stratified test, as implemented in PLINK, based on the subsample matching criteria (BMI, gender, geographic region) as described in **Section 3.1.3**. The family-based sample was analyzed using the DFAM procedure in PLINK conditioning on sibship as strata. p-values from the two subsamples were then converted to Z-scores and combined with a weighted Z-meta-analysis. A significance threshold of 5×10^{-8} was used.

For LDL, residuals from the phenotype modeling described in **Section 3.3.1** were examined for association in PLINK with genotypes using linear regression modeling with genotypes as the independent variable.

The most significant SNP from the T2D analysis was located in *TCF7L2*, which has previously been associated with T2D, and reached a p-value of 5.4×10^{-6}. The most significant SNP for LDL was located in the *APOE* gene cluster that is known to influence LDL metabolism (p=3.4×10^{-13} after performing genomic control). The second and third strongest SNPs for LDL reached p-values of 2.3×10^{-8} and 8.9×10^{-8}, respectively. Thus, for these traits two SNPs at two loci reached genome-wide statistical significance in the primary analysis of DGI.

3.4. Interpretation and Follow-Up of Results

3.4.1. Validation, Replication, and Extension to Other Populations

Because of the difficulties of specifying a significance threshold in genome-wide association studies and the multitude of potential methodological errors, ultimate proof of association must come from replication of significantly associated SNPs in independent samples (14). Even for SNPs near genes that constitute strong biological candidates, from previous experimental studies or similar Mendelian traits, replication is still necessary for convincing association.

Technical validation refers to reanalysis of associated SNPs on a different genotyping platform and provides evidence that an observed association signal is not due to systematic genotyping errors. To show this, concordance of genotypes between the assays is calculated. Technical validation may be considered before replication studies are undertaken, but multiple correlated SNPs at a locus with comparable association argue against genotyping artifact as the source of apparent association. A replication study of a putative association tests the same direction of effect for the same polymorphism in the same phenotype and in a sample from the same ancestral origin and similar ascertainment, often termed exact replication. Studies examining the same locus in populations of different ancestral origin are necessary to demonstrate the relevance of a specific allele to individuals of other ancestries. Care should be taken to examine the linkage disequilibrium patterns before specifying SNPs for follow-up in samples of other ancestries. For example, LD breaks down over shorter distances in individuals of African ancestry relative to those of European ancestry based on the HapMap samples. Following up findings from European-derived samples in African-derived samples may require genotyping of additional SNPs that might be highly correlated on European chromosomes but poorly correlated on African chromosomes. This can be determined using any tagging program, such as the Tagger function in HaploView (*see* **Section 2**).

The power of the replication study to confirm the findings in light of the allele frequency and the effect size observed in the original study should be considered. True positive results that just achieve significance (by whatever appropriate threshold) tend to overestimate the true effect size due to random chance. This will hinder replication of a finding in a second sample if the sample is inadequately powered to find a weaker effect. Power and sample size for replication samples can be calculated in the software packages Genetic Power Calculator or CaTS (*see* **Section 2**). Failure to confirm a finding in an adequately powered second sample could also suggest a false-positive result in the original sample. Iterative meta-analysis of results, as replication samples are genotyped, can help clarify whether a signal is getting stronger or weaker with analysis in independent samples. As the meta-analyzed samples get larger, true-positive results will get more significant, while false-positive results will get less significant.

Example One hundred and fourteen SNPs from the extreme tail of *p*-values for T2D in DGI were genotyped with an independent technology for validation. Concordance rates were 99.5%, indicating that genotyping artifacts did not explain the lowest *p*-values observed.

For T2D, the 59 SNPs with the strongest associations were selected for replication and were genotyped in 10,850 individuals from three case–control samples of European ancestry from Sweden, the United States, and Poland. T2D definitions were the same as those used in the original GWA studies. Combined analysis of all four samples was performed using the Mantel–Haenszel odds ratio meta-analysis. This analysis identified three genome-wide significant SNPs at three loci.

For LDL, the 196 SNPs with the strongest associations were selected for replication in 18,554 individuals of European ancestry from Sweden, Finland, and Norway. LDL was measured using standard methods and the same exclusions and adjustments for covariates as in the GWA study were applied. Association was examined using linear regression modeling as in the GWA study and results were summarized using inverse-variance-weighted meta-analysis with fixed effects. Two loci were replicated. The findings were also examined in a population of Asian ancestry by genotyping in 4,259 individuals from Singapore. Neither SNP was significant in this population, but it was clearly underpowered to detect association.

3.4.2. Identifying the Causal Variant

Since only a fraction of common genetic variation is genotyped and SNPs act as proxies for untyped SNPs and other genetic variants, it is unlikely that a significantly associated SNP is directly causal. Consequently, GWA studies cannot provide unambiguous identification of causal genes, even if an associated SNP is situated close to or in a gene. Indeed, several GWA studies have found strong association signals from genomic regions containing multiple genes of strong biological candidacy, while others have found regions of association without any known genes, possibly reflecting remote regulatory elements, long-range LD, or incomplete gene annotation.

The first step toward identifying the causal variant is to examine whether the SNP showing the strongest association at a locus or any SNP to which it is highly correlated may be a functional SNP (e.g., missense or nonsense). Information on SNP function relative to nearby genes is available in public databases, such as dbSNP, and can be efficiently retrieved using bioinformatic tools such as the WGAViewer (24), SNAP, or a function implemented in PLINK. Typically multiple SNPs at a locus show association and examining the correlation among associated SNPs in HapMap can show whether correlation between SNPs explains the *p*-values seen across the locus. Some loci have been found to contain several causal variants.

Second, fine-mapping of additional SNPs known to be correlated in HapMap above a specified threshold with the most significant SNP at a locus can be performed both to narrow down the locus for resequencing and to examine correlated functional SNPs. Correlation among SNPs is measured by the coefficient of determination, r^2. Another option is to perform fine-mapping in silico without further genotyping by imputing untyped variants (discussed in **Section 3.2.5**).

Ultimately, resequencing of the associated locus is necessary to identify the set of all potential variants which could explain a SNP's association. Molecular biologic approaches are then required to determine which among them is most likely to be causal (*see* **Note 7**).

Example None of the SNPs for LDL or T2D with significant association were likely to be functional based on encoding an amino acid change when searching GeneCruiser using SNAP (*see* **Section 2**) and are presumed to be in LD with the causal variants. Fine-mapping and resequencing of significantly associated loci are in progress.

3.4.3. Meta-analysis

Meta-analysis is the statistical procedure of jointly analyzing multiple studies to increase power. For GWA studies, this approach has identified additional common variants of smaller effect for several traits, which may still provide new insights into pathophysiology and physiology although the individual predictive values are small.

Before meta-analysis, all studies included should be examined for possible sources of heterogeneity, both in design and results of included studies. Most studies include only individuals of the same continental ancestry (*see* **Note 5**). Summary association statistics from both population-based and family-based subsamples can be pooled (57). Heterogeneity of results between studies can be examined by plots (Forest plot, Galbraith plot, or funnel plot) or with formal tests such as the Cochran's Q and the I^2 statistics (58).

Phenotypes and genotypes should be quality controlled prior to analysis as described in **Sections 3.3.2–3.2.4** before pooling. Meta-analyses are then performed either under the assumption that variation between studies is due to random variation in which large studies are expected to give identical results (termed fixed-effects models) or under the assumption that variation between studies is due to different underlying effects for each study (termed random-effects). When statistical heterogeneity is detected, random-effects models should be used. Under either model, meta-analysis is performed by combining the data after weighting the results from each study. Combined analysis can be performed with effect estimates (odds ratios or beta estimates) weighted by the inverse variance, with p-values weighted using Fisher's-combined method, or by converting p-values to Z-scores based on the p-value and effect direction. For qualitative traits, meta-analysis can also be performed by pooling genotype counts using the Cochran–Mantel–Haenszel stratified method. Pooling of results at the individual level is problematic in primary screens as they increase the probability of false-positive association due to population stratification. However, examination of epistatic or gene–environment interactions can be facilitated by pooling individual-level data and by performing association analysis across all individuals, but this is conditional on achieving a significant result upon meta-analysis of summary results from separate analyses.

Example Meta-analysis for T2D was performed with two other studies from the United Kingdom (n=1,924 cases and 2,938 controls) and Finland (n=1,161 cases and 1,174 controls), resulting in a pooled sample size of 10,128 individuals (45). Using data from the HapMap CEU sample of European ancestry, 2,202,892 SNPs were imputed using the MACH software package, enabling comparison across samples with greater coverage. Each study was corrected for population stratification, cryptic relatedness, and technical artifacts before meta-analysis. Heterogeneity among studies was examined for SNP odds ratios using Cochran's Q and the I^2 statistics. Meta-analysis was then performed using Z-statistics derived from p-values and weighted so that squared weights sum to 1; each sample-specific weight was proportional to the square root of the effective number of individuals in the sample. p-values for genotyped SNPs were used where available and results for imputed SNP were used for the remainder. Weighted Z-statistics were summed across studies and converted to p-values. Odds ratios and confidence intervals for the pooled sample were obtained from a fixed-effects inverse variance model. Genomic control was performed on the combined results, although the meta-analysis had a λ_{GC} of only 1.04. Population stratification was also examined using principal components analysis, as described in **Section 3.2.4**.

Imputed variants showing association were directly genotyped for validation before replication of all SNPs in two-replication stages with a total sample size of 53,975 individuals. Based on these analyses, six novel loci were identified that met a genome-wide significance of $p<5 \times 10^{-8}$, most of them with very small effect sizes (odds ratios between 1.1 and 1.2), although effect sizes for the actual causal variants are likely to be larger.

Meta-analysis for LDL was performed with two studies from Finland (n=1,874) and Sardinia (n=4,184), resulting in a pooled sample size of 8,816 individuals (59, 60). Imputation was performed as described for T2D. Analyses were performed using linear regression modeling in each sample and then analyzed jointly using inverse variance-weighted meta-analysis with fixed effects. Heterogeneity was examined in the same way as for T2D; 30 SNPs, representing novel loci with $p<10^{-4}$ from the meta-analysis, were selected for replication in 18,554 individuals. This pooled analysis identified five additional loci associated with LDL.

4. Notes

1. As can be seen in **Section 2**, several applications of high quality for management and analysis of data have been developed by different groups. For newcomers to the field, we recommend the software PLINK, which can be downloaded freely, is regularly updated by leaders in the field, has a logical structure, and has thorough documentation provided on its webpage. PLINK output files can also be visualized in HaploView.
2. Ideally, the study design should be carefully planned before sample collection and genotyping as discussed in **Section 1**. In reality, samples collected previously for other purposes have been used in most genome-wide association studies. We find that the issues discussed in the sections on quality control of genotype and phenotype data as well as population stratification may increase power in these samples. Heritability estimates and coefficients of determination for trait models can be useful indicators of data quality. Most often, the sample available is also the major determinant of whether a case–control, cohort, or family-based design is to be implemented.
3. GWA studies remain an expensive undertaking, but prices are dropping, especially as whole-genome resequencing studies draw near. To reduce genotyping costs, we recommend considering three strategies. First, instead of selecting an expensive array with high coverage, invest in increasing sample size using more arrays with moderate coverage. Coverage can then be increased using imputation procedures (61). Calculations of cost efficiency can be performed using online tools (http://www.well.ox.ac.uk/~carl/gwa/cost-efficiency). Second, the two-stage approach discussed in **Section 3.1.4** can be used. Third, DNA pooling can be performed when genotyping to reduce costs. However, this approach suffers from reduced power and genotype quality-control difficulties (62).
4. Genome-wide SNP data can also be used for other analyses, bringing additional value to the high costs of these analyses. One such analysis is homozygosity mapping, in which low-frequency recessive variants can be examined for trait association. The method involves identification of segments of homozygosity in genome-wide SNP data using hidden Markov models or a method implemented in PLINK (63). These segments can then be examined for trait association. Similarly, haplotypes can be inferred from SNP data and haplotype association tests performed.

Another analysis possible for traits caused by genetic differences between ethnicities is admixture mapping, which can be used to identify the genetic determinants for these traits in admixed populations (25, 64).

Finally, it has been suggested that pathway-based association analyses, similar to the GSEA method used in gene expression array analyses, may be able to identify additional variants (24).

5. We recommend that individual genome-wide association studies be performed in populations of similar ancestry. It is known, through the HapMap project and studies of population genetics, that allele frequencies, and to a lesser extent LD patterns, differ between ancestries. Also, founder populations are known to differ from other populations. Even after conditioning on a different continental ancestry, assessment of population structure using the approaches mentioned in **Section 3.2.4** is highly recommended to avoid false-positive signals (39, 40). The appropriateness of meta-analyzing results across samples of differing ancestry remains unclear, but it may be reasonable to do so if similar correlation patterns exist.
6. GWA studies have been finding reproducible loci associated with complex traits and have largely avoided the flood of non-replicable findings that were seen in candidate gene-based association studies (12) and when genome-wide linkage analysis was first applied to complex diseases (65). This is likely due to the large sample sizes examined, careful study design and quality control, stringent thresholds for statistical significance, and use of independent samples for validation and replication (66).

There is no doubt that large samples are necessary, that study design and quality control is paramount to reduce bias, and that replication and validation are key to identify significant associations, but the problem of correcting for multiple hypotheses remains a subject of debate. Several approaches for multiple hypothesis correction other than the frequentist Bonferroni and Sidak corrections have been proposed, including the False Discovery Rate (67), which controls the false-positive rate, and Bayesian approaches such as the False Positive Report Probability (53), which determines the probability of an association being falsely positive based on *p*-value, power, and estimates of prior probability. Other approaches include permutation of phenotypes or genotypes to empirically determine the nominal *p*-value that corresponds to a study-wide *p*-value of 0.05. Until these approaches have become established and validated we recommend using the stringent threshold of 5×10^{-8} based on a an overall genome-wide significance threshold of 0.05 corrected by the Bonferroni method for estimates from HapMap of 1 million independent tests (11, 54, 55), in populations of European and Asian ancestries. The number of independent tests in African populations is likely to be higher.

7. For the hundreds of loci that have been discovered to date, the road from SNP identification to biology is long. As discussed above, identifying the causal variant requires resequencing of the locus in most cases. Furthermore, definitive identification of the gene affected by the causal variant requires functional studies, as genetic variants can be located in regulatory sites and affect distantly located genes as shown in the ENCODE project (68). Functional analyses may prove difficult because variants can affect gene products in multiple ways, including changes in protein sequence or changes in expression due to different gene copy number, regulation of gene transcription, and regulation of RNA post-processing (capping, splicing, polyadenylation). To date, databases of association of SNPs with transcriptome-wide expression patterns (69, 70) and alternative splicing (71) in specific cell lines are available.

   Databases of copy number polymorphisms (CNPs), another form of genetic variation in LD with SNPs that has recently been shown to have a large impact on interindividual genetic variation, are also available and the most recent microarrays include assays for large numbers of CNPs (72). The relative importance of different genetic variants such as SNPs and CNPs, or for that matter common and rare variations, is currently unknown, but will be elucidated once resequencing of genomes becomes more common. This will enable characterization of the full spectrum of genetic variation that contributes to interindividual differences for any given trait.

   Genetic variants may also act in concert with each other, termed epistasis, or with environmental factors to cause trait changes. Studies of such interactions require even larger sample sizes than GWA studies and are likely to follow from the current collaborative projects of pooling datasets.

8. For more information on general methodology see excellent recently published reviews (48, 62, 73–75). For specific points on study design see a recently published protocol (31) and for more information on the history of the population genetic theory leading to GWA study approaches see a recent discussion (76). For examples of genome-wide association scans using different methodologies we refer to the collection of all published studies maintained by the NHGRI office of population genomics at their homepage (www.genome.gov).

References

1. Botstein, D, White, RL, Skolnick, M, et al. (1980) Construction of a genetic linkage map in man using restriction fragment length polymorphisms. *Am J Hum Genet* **32:** 314–331.
2. Risch, N, Merikangas, K. (1996) The future of genetic studies of complex human diseases. *Science* **273:** 1516–1517.
3. Lander, ES. (1996) The new genomics: global views of biology. *Science* **274**: 536–539.
4. Reich, DE, Cargill, M, Bolk, S, et al. (2001) Linkage disequilibrium in the human genome. *Nature* **411:** 199–204.
5. McVean, GA, Myers, SR, Hunt, S, et al. (2004) The fine-scale structure of recombination rate variation in the human genome. *Science* **304:** 581–584.
6. Frazer, KA, Ballinger, DG, Cox, DR, et al. (2007) A second generation human haplotype map of over 3.1 million SNPs. *Nature* **449:** 851–861.
7. Lander, ES, Linton, LM, Birren, B, et al. (2001) Initial sequencing and analysis of the human genome. Nature **409**: 860–921.
8. Venter, JC, Adams, MD, Myers, EW, et al. (2001) The sequence of the human genome. *Science* **291:** 1304–1351.
9. Sachidanandam, R, Weissman, D, Schmidt, SC, et al. (2001) A map of human genome sequence variation containing 1.42 million single nucleotide polymorphisms. *Nature* **409:** 928–933.
10. Perkel, J. (2008) SNP genotyping: six technologies that keyed a revolution. *Nat Methods* **5:** 447–453.
11. Altshuler, D, Brooks, LD, Chakravarti, A, et al. (2005) A haplotype map of the human genome. *Nature* **437:** 1299–1320.
12. Hirschhorn, JN, Lohmueller, K, Byrne, E, et al. (2002) A comprehensive review of genetic association studies. *Genet Med* **4:** 45–61.
13. Lohmueller, KE, Pearce, CL, Pike, M, et al. (2003) Meta-analysis of genetic association studies supports a contribution of common variants to susceptibility to common disease. *Nat Genet* **33:** 177–182.
14. Chanock, SJ, Manolio, T, Boehnke, M, et al. (2007) Replicating genotype-phenotype associations. *Nature* **447:** 655–660.
15. Saxena, R, Voight, BF, Lyssenko, V, et al. (2007) Genome-wide association analysis identifies loci for type 2 diabetes and triglyceride levels. *Science* **316:** 1331–1336.
16. Clayton, D, Leung, HT. (2007) An R package for analysis of whole-genome association studies. *Hum Hered* **64:** 45–51.
17. Abecasis, GR, Cardon, LR, Cookson, WO. (2000) A general test of association for quantitative traits in nuclear families. *Am J Hum Genet* **66:** 279–292.
18. Laird, NM, Horvath, S, Xu, X. (2000) Implementing a unified approach to family-based tests of association. *Genet Epidemiol* **19** Suppl 1: S36–S42.
19. Lange, C, DeMeo, D, Silverman, EK, et al. (2004) PBAT: tools for family-based association studies. *Am J Hum Genet* **74:** 367–369.
20. Barrett, JC, Fry, B, Maller, J, et al. (2005) Haploview: analysis and visualization of LD and haplotype maps. *Bioinformatics* **21:** 263–265.
21. Price, AL, Patterson, NJ, Plenge, RM, et al. (2006) Principal components analysis corrects for stratification in genome-wide association studies. *Nat Genet* **38:** 904–909.
22. Li, Y, Abecasis, GR. (2006) Mach 1.0: rapid haplotype reconstruction and missing genotype inference. *Am J Hum Genet* S79: 2290.
23. The Wellcome Trust Case Control Consortium (2007). Genome-wide association study of 14,000 cases of seven common diseases and 3,000 shared controls *Nature* **447:** 661–678.
24. Ge, D, Zhang, K, Need, AC, et al. (2008) WGAViewer: software for genomic annotation of whole genome association studies. *Genome Res* **18:** 640–643.
25. Patterson, N, Hattangadi, N, Lane, B, et al. (2004) Methods for high-density admixture mapping of disease genes. *Am J Hum Genet* **74:** 979–1000.
26. Purcell, S, Cherny, SS, Sham, PC. (2003) Genetic power calculator: design of linkage and association genetic mapping studies of complex traits. *Bioinformatics* **19:** 149–150.
27. Skol, AD, Scott, LJ, Abecasis, GR, et al. (2006) Joint analysis is more efficient than replication-based analysis for two-stage genome-wide association studies. *Nat Genet* **38:** 209–213.
28. Visscher, PM, Hill, WG, Wray, NR. (2008) Heritability in the genomics era–concepts and misconceptions. *Nat Rev Genet* **9:** 255–266.
29. Cupples, LA, Arruda, HT, Benjamin, EJ, et al. (2007) The Framingham Heart Study 100 K SNP genome-wide association study

resource: overview of 17 phenotype working group reports BMC. *Med Genet* 8 Suppl **1:** S1–S4.

30. Mitchell, AA, Cutler, DJ, Chakravarti, A. (2003) Undetected genotyping errors cause apparent overtransmission of common alleles in the transmission/disequilibrium test. *Am J Hum Genet* **72:** 598–610.
31. Zondervan, KT, Cardon, LR. (2007) Designing candidate gene and genome-wide case-control association studies. *Nat Protoc* **2:** 2492–2501.
32. de Bakker, PI, Yelensky, R, Pe'er, I, et al. (2005) Efficiency and power in genetic association studies. *Nat Genet* **37:** 1217–1223.
33. Pe'er, I, de Bakker, PI, Maller, J, et al. (2006) Evaluating and improving power in whole-genome association studies using fixed marker sets. *Nat Genet* **38:** 663–667.
34. Barrett, JC, Cardon, LR. (2006) Evaluating coverage of genome-wide association studies. *Nat Genet* **38:** 659–662.
35. Bhangale, TR, Rieder, MJ, Nickerson, DA. (2008) Estimating coverage and power for genetic association studies using near-complete variation data. *Nat Genet* **40:** 841–843.
36. Rabbee, N, Speed, TP. (2006) A genotype calling algorithm for Affymetrix SNP arrays. *Bioinformatics* **22:** 7–12.
37. Wittke-Thompson, JK, Pluzhnikov, A, Cox, NJ. (2005) Rational inferences about departures from Hardy-Weinberg equilibrium. *Am J Hum Genet* **76:** 967–986.
38. Cox, DG, Kraft, P. (2006) Quantification of the power of Hardy-Weinberg equilibrium testing to detect genotyping error. *Hum Hered* **61:** 10–14.
39. Clayton, DG, Walker, NM, Smyth, DJ, et al. (2005) Population structure, differential bias and genomic control in a large-scale, case-control association study. *Nat Genet* **37:** 1243–1246.
40. Campbell, CD, Ogburn, EL, Lunetta, KL, et al. (2005) Demonstrating stratification in a European American population. *Nat Genet* **37:** 868–872.
41. Devlin, B, Roeder, K. (1999) Genomic control for association studies. *Biometrics* **55:** 997–1004.
42. Zheng, G, Freidlin, B, Gastwirth, JL. (2006) Robust genomic control for association studies. *Am J Hum Genet* **78:** 350–6.
43. Pritchard, JK, Stephens, M, Rosenberg, NA, et al. (2000) Association mapping in structured populations. *Am J Hum Genet* **67:** 170–181.
44. Purcell, S, Neale, B, Todd-Brown, K, et al. (2007) PLINK: a tool set for whole-genome association and population-based linkage analyses. *Am J Hum Genet* **81:** 559–575.
45. Zeggini, E, Weedon, MN, Lindgren, CM, et al. (2007) Replication of genome-wide association signals in UK samples reveals risk loci for type 2 diabetes. *Science* **316:** 1336–1341.
46. Scott, LJ, Mohlke, KL, Bonnycastle, LL, et al. (2007) A genome-wide association study of type 2 diabetes in Finns detects multiple susceptibility variants. *Science* **316:** 1341–1345.
47. Lasky-Su, J, Lyon, HN, Emilsson, V, et al. (2008) On the replication of genetic associations: timing can be everything! *Am J Hum Genet* **82:** 849–858.
48. Balding, DJ. (2006) A tutorial on statistical methods for population association studies. *Nat Rev Genet* 7: 781–791.
49. Liang, KY, Zeger, SL. (1986) Longitudinal data analysis using generalized estimating linear models. *Biometrika* **73:** 12–22.
50. Pilia, G, Chen, WM, Scuteri, A, et al. (2006) Heritability of cardiovascular and personality traits in 6,148 Sardinians. *PLoS Genet* **2:** e132.
51. Lowe, JK, Maller, JB, Pe'er l, et al. (2009) *PLoS Genet* **5:** e1000365.
52. Smith, JG, Lowe, JK, Kovvali, S, et al. (2009) *Heart Rhythm* **6:** 634–641.
53. Wacholder, S, Chanock, S, Garcia-Closas, M, et al. (2004) Assessing the probability that a positive report is false: an approach for molecular epidemiology studies. *J Natl Cancer Inst* **96:** 434–442.
54. Dudbridge, F, Gusnanto, A. (2008) Estimation of significance thresholds for genome-wide association scans. *Genet Epidemiol* **32:** 227–234.
55. Pe'er, I, Yelensky, R, Altshuler, D, et al. (2008) Estimation of the multiple testing burden for genomewide association studies of nearly all common variants. *Genet Epidemiol* **32:** 381–385.
56. Lander, E, Kruglyak, L. (1995) Genetic dissection of complex traits: guidelines for interpreting and reporting linkage results. *Nat Genet* **11:** 241–247.
57. Kazeem, GR, Farrall, M. (2005) Integrating case-control and TDT studies. *Ann Hum Genet* **69:** 329–335.

58. Higgins, JP, Thompson, SG, Deeks, JJ, et al. (2003) Measuring inconsistency in meta-analyses. *BMJ* **327:** 557–560.
59. Willer, CJ, Sanna, S, Jackson, AU, et al. (2008) Newly identified loci that influence lipid concentrations and risk of coronary artery disease. *Nat Genet* **40:** 161–169.
60. Kathiresan, S, Melander, O, Guiducci, C, et al. (2008) Six new loci associated with blood low-density lipoprotein cholesterol, high-density lipoprotein cholesterol or triglycerides in humans. *Nat Genet* **40:** 189–197.
61. Anderson, CA, Pettersson, FH, Barrett, JC, et al. (2008) Evaluating the effects of imputation on the power, coverage, and cost efficiency of genome-wide SNP platforms. *Am J Hum Genet* **83:** 112–119.
62. McCarthy, MI, Abecasis, GR, Cardon, LR, et al. (2008) Genome-wide association studies for complex traits: consensus, uncertainty and challenges. *Nat Rev Genet* **9:** 356–369.
63. Lander, ES, Green, P, Abrahamson, J, et al. (1987) MAPMAKER: an interactive computer package for constructing primary genetic linkage maps of experimental and natural populations. *Genomics* **1:** 174–181.
64. Reich, D, Patterson, N, De Jager, PL, et al. (2005) A whole-genome admixture scan finds a candidate locus for multiple sclerosis susceptibility. *Nat Genet* **37:** 1113–1118.
65. Altmuller, J, Palmer, LJ, Fischer, G, et al. (2001) Genomewide scans of complex human diseases: true linkage is hard to find. *Am J Hum Genet* **69:** 936–950.
66. Altshuler, D, Daly, M. (2007) Guilt beyond a reasonable doubt. *Nat Genet* **39:** 813–815.
67. Benjamini, Y, Hochberg, Y. (1995) Controlling the false discovery rate: a practical and powerful approach to multiple testing. *J R Stat Soc* **57:** 289–300.
68. Birney, E, Stamatoyannopoulos, JA, Dutta, A, et al. (2007) Identification and analysis of functional elements in 1% of the human genome by the ENCODE pilot project. *Nature* **447:** 799–816.
69. Dixon, AL, Liang, L, Moffatt, MF, et al. (2007) A genome-wide association study of global gene expression. *Nat Genet* **39:** 1202–1207.
70. Stranger, BE, Nica, AC, Forrest, MS, et al. (2007) Population genomics of human gene expression. *Nat Genet* **39:** 1217–1224.
71. Kwan, T, Benovoy, D, Dias, C, et al. (2008) Genome-wide analysis of transcript isoform variation in humans. *Nat Genet* **40:** 225–231.
72. McCarroll, SA, Altshuler, DM. (2007) Copy-number variation and association studies of human disease. *Nat Genet* **39:** S37–S42.
73. Hirschhorn, JN, Daly, MJ. (2005) Genome-wide association studies for common diseases and complex traits. *Nat Rev Genet* **6:** 95–108.
74. Wang, WY, Barratt, BJ, Clayton, DG, et al. (2005) Genome-wide association studies: theoretical and practical concerns. *Nat Rev Genet* **6:** 109–118.
75. Pearson, TA, Manolio, TA. (2008) How to interpret a genome-wide association study. *JAMA* **299:** 1335–1344.
76. Kruglyak, L. (2008) The road to genome-wide association studies. *Nat Rev Genet* **9:** 314–318.

Chapter 15

Bioinformatics Analysis of Microarray Data

Yunyu Zhang, Joseph Szustakowski, and Martina Schinke

Abstract

Gene expression profiling provides unprecedented opportunities to study patterns of gene expression regulation, for example, in diseases or developmental processes. Bioinformatics analysis plays an important part of processing the information embedded in large-scale expression profiling studies and for laying the foundation for biological interpretation.

Over the past years, numerous tools have emerged for microarray data analysis. One of the most popular platforms is Bioconductor, an open source and open development software project for the analysis and comprehension of genomic data, based on the R programming language.

In this chapter, we use Bioconductor analysis packages on a heart development dataset to demonstrate the workflow of microarray data analysis from annotation, normalization, expression index calculation, and diagnostic plots to pathway analysis, leading to a meaningful visualization and interpretation of the data.

Key words: Annotation, normalization, gene filtering, moderated F-test, GSEA, pathway analysis, affymetrix GeneChip^{TM}, sigPathway.

1. Introduction

The purpose of this chapter is to provide an understanding of the routine steps for microarray data analysis using Bioconductor (1) packages written in R (2), a widely used open source programming language and environment for statistical computing and graphics. Both R and Bioconductor are under active development by a dedicated team of researchers with a commitment to good documentation and software design. We assume that the reader has a basic understanding about data structures and functions in R programming. However, all of the analysis steps and tools described in this chapter have also been implemented in other software packages (summarized in **Section 4**). The workflow

K. DiPetrillo (ed.), *Cardiovascular Genomics,* Methods in Molecular Biology 573,
DOI 10.1007/978-1-60761-247-6_15, © Humana Press, a part of Springer Science+Business Media, LLC 2009

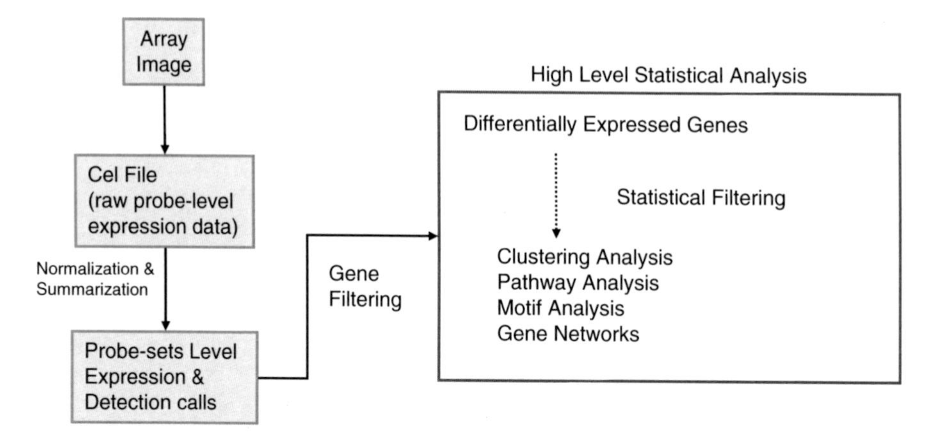

Fig. 15.1. Microarray data analysis work flow for Affymetrix GeneChip™ arrays.

shown in **Fig. 15.1** facilitates the understanding of the basic procedures in microarray data analysis and serves as an outline of this chapter.

2. Materials

2.1. Software

R can be downloaded from http://www.r-project.org and be installed on all three mainstream operating systems (Windows, Mac, Unix/Linux). The general installation manual and introductory tutorials can be obtained from the same website. Similar to other statistical software packages, R provides a statistical framework and terminal-based interface for users to input commands for data manipulation. Additional packages (**Table 15.1**) from

Table 15.1
List of add-on R packages required for analysis

Package	Description
Affy (31)	Basic functions for low-level analysis of Affymetrix GeneChip™ oligonucleotide arrays
PLIER (5)	Normalize and summarize the Affymetrix probe-level expression data using the PLIER method
LIMMA (32)	Linear model for microarray analysis
sigPathway (12)	Pathway (Gene-Set) analysis for high-throughput data
mm74av1mmentregcdf	Entrez Gene-based chip definition file (CDF) for Affymetrix MG-74AV1 platform
org.Mm.eg.db	Annotation mapping based on mouse Entrez Gene identifiers

Bioconductor (http://www.bioconductor.org) are required prior to starting the analysis. Details about the package installation can be found in **Section 3.1**. The R terminal output is highlighted throughout the chapter in `courier` font.

2.2. Dataset

A gene expression profiling experiment of heart ventricles at various stages of cardiac development generated by the CardioGenomics Program for Genomic Applications (PGA) was used as a test dataset. This dataset can be downloaded from NCBI Gene Expression Omnibus (GEO; accession number GSE75). It includes seven time-points covering gene expression in the heart from embryonic stages through adolescence into adulthood (**Table 15.2**). Though this study was performed with an earlier Affymetrix platform (MGU-74Av1), the design of this study and quality of the data make this a valuable test dataset to this date.

Table 15.2
Experimental design of the heart development dataset

Time point	Abbreviation	Number of GeneChip™ arrays
Embryonic day 12.5 d.p.c.	E12.5	3
Neonatal (Day 1 post-birth)	NN	3
1 week of age	A1w	3
4 weeks of age	A4w	3
3 months of age	A3m	3
5 months of age	A5m	3
1 year of age	A1y	6

3. Methods

3.1. R Package Installation

After downloading and installing R software (*see* **Note 1**), an R terminal can be started to install the required Bioconductor core and additional packages (*see* **Note 2**).

```
>source("http://www.bioconductor.org/biocLite.R")
>biocLite()
Running biocinstall version 2.1.11 with R version 2.6.1
Your version of R requires version 2.1 of Bioconductor.
Will install the following packages:
 [1] "affy"      "affydata"    "affyPLM"   "annaffy"    "annotate"
 [6] "Biobase"   "Biostrings"  "DynDoc"  "gcrma"     "genefilter"
[11] "geneplotter"   "hgu95av2"   "limma"    "marray"    "matchprobes"
[16] "multtest"    "ROC"     "vsn"     "xtable"     "affyQCReport"
```

```
Please wait...
also installing the dependencies 'DBI', 'RSQLite', 'affyio',
'preprocessCore', 'GO', 'KEGG', 'AnnotationDbi', 'simpleaffy'
```

"affy" and "limma" are already included in the above core packages. We can install the rest of the packages in **Table 15.1** by specifying the names as the argument using the "biocLite" function.

```
>pkgs<-c("plier", "sigPathway", "mm74av1mmentrezgcdf",
"mm74av1mmentrezgprobe", "org.Mm.eg.db")
>biocLite(pkgs)
```

3.2. Preparation for Data and Result File Storage

Organizing data and results is very helpful for flexible use of the scripts. For this project, we created a directory "cardiac_dev" and the following subdirectories to store the raw and intermediate data files and the analysis results.

1. "cel": To store the cel files
2. "obj": To store R-object
3. "gp.cmp": For group comparison results and outputs
4. "limma": To store the group comparison results
5. "img": To store the images
6. "pathway": To store the pathway analysis results

The raw data, packed in a compressed file named "GSE75_RAW.tar," can be downloaded from the GEO ftp site. The individual cel files are extracted from this file and decompressed using the WinZip program on the Windows platform. On the Linux/Unix platform, the "tar -vxf" followed by "gzip" command is used to extract and decompress the cel files.

3.3. Annotations for Entrez Gene Probe-Sets

Since we used an Entrez Gene-based chip definition file (CDF) to generate the probe-set level gene expression values, only a minimal set of annotations (including gene name and gene symbol mapped from Entrez Gene IDs) need to be readily available to obtain an initial biological impression of the results. Here, we built a data frame that contains the gene symbol and name based on the Entrez Gene IDs.

First, all probe-sets (or Entrez Gene identifiers (IDs) included in this CDF file were retrieved. Their corresponding gene IDs can be retrieved by removing the ending "_at" according to custom CDFs naming convention.

```
> library(mm74av1mmentrezgprobe)
> probe.set<-
unique(as.data.frame(mm74av1mmentrezgprobe)$Probe.Set.Name)
> length(probe.set)
[1] 7070
> probe.set [grep("_st$", probe.set)]<-paste(probe.set[grep("_st$",
probe.set)],
+                                   "at", sep="_")
> head(probe.set)
```

```
[1] "AFFX-18SRNAMur/X00686_3_at" "AFFX-18SRNAMur/X00686_5_at"
[3] "AFFX-18SRNAMur/X00686_M_at" "AFFX-BioB-3_at"
[5] "AFFX-BioB-3_st" "AFFX-BioB-5_at"
> gene.id<-sub("_at", "", probe.set)
> length(grep("AFFX", gene.id))
[1] 66
```

A total of 7,070 probe-sets are defined in this CDF, including 66 Affymetrix control probe-sets and 7,004 Entrez Gene IDs. The annotations can be retrieved using package "org.Mm.eg.db." This package is maintained by the Bioconductor core team and routinely updated. The local version can be synchronized to the updated one by the function "`update.packages.`" To view the available annotations based on the Entrez Gene IDs:

```
>library(org.Mm.eg.db)
>ls("package:org.Mm.eg.db")
 [1] "org.Mm.eg_dbconn"     "org.Mm.eg_dbfile"     "org.Mm.eg_dbInfo"
 [4] "org.Mm.eg_dbschema"   "org.Mm.egACCNUM"      "org.Mm.egACCNUM2EG"
 [7] "org.Mm.egALIAS2EG"    "org.Mm.egCHR"         "org.Mm.egCHRLENGTHS"
[10] "org.Mm.egCHRLOC"      "org.Mm.egENZYME"      "org.Mm.egENZYME2EG"
[13] "org.Mm.egGENENAME"    "org.Mm.egGO"          "org.Mm.egGO2ALLEGS"
[16] "org.Mm.egGO2EG"       "org.Mm.egMAP"         "org.Mm.egMAP2EG"
[19] "org.Mm.egMAPCOUNTS"   "org.Mm.egORGANISM"    "org.Mm.egPATH"
[22] "org.Mm.egPATH2EG"     "org.Mm.egPFAM"        "org.Mm.egPMID"
[25] "org.Mm.egPMID2EG"     "org.Mm.egPROSITE"     "org.Mm.egREFSEQ"
[28] "org.Mm.egREFSEQ2EG"   "org.Mm.egSYMBOL"      "org.Mm.egSYMBOL2EG"
[31] "org.Mm.egUNIGENE"     "org.Mm.egUNIGENE2EG"
```

A data frame named "`ann`" with probe-set ID as row names is created to store the annotations.

```
> ann<-as.data.frame(matrix(nrow=length(gene.id), ncol=4))
> dimnames(ann)<-list(probe.set, c("ProbeSet", "GeneID", "Symbol",
"GeneName"))
> ann$ProbeSet<-probe.set
> ann$GeneID<-gene.id
```

To integrate the gene symbol and names into the data frame:

```
> ann$GeneName<-unlist(unlist(as.list(org.Mm.egGENENAME)))[gene.id]
> ann$Symbol<-unlist(unlist(as.list(org.Mm.egSYMBOL)))[gene.id]
> save(ann, file="obj/ann.RData")
> tail(ann)
         ProbeSet GeneID         Symbol
99377_at 99377_at  99377          Sall4
99571_at 99571_at  99571            Fgg
99650_at 99650_at  99650  4933434E20Rik
99683_at 99683_at  99683         Sec24b
99887_at 99887_at  99887         Tmem56
99929_at 99929_at  99929         Tiparp
                                                        GeneName
99377_at                                  sal-like 4 (Drosophila)
99571_at                             fibrinogen, gamma polypeptide
99650_at                                 RIKEN cDNA 4933434E20 gene
99683_at  SEC24 related gene family, member B (S. cerevisiae)
99887_at                                  transmembrane protein 56
99929_at              TCDD-inducible poly(ADP-ribose) polymerase
```

3.4. Preparing Sample Information

Sample information is needed for high-level statistical analysis. As a simple approach, we created an R data frame object to store this information, which can be started with a tab-delimited file prepared in Excel. For this dataset, the phenotype information was copied from the GEO website with some simple text manipulation (copy/paste, replace with * as wild card, concatenate) to generate a tab-delimited file as shown in **Table 15.3**. The first column has to contain the exact cel file names, while the remaining columns can contain any additional information.

Table 15.3
Example sample information file in tab-delimited format

File name	Sample name	Group	BS
GSM2189.CEL	FVB_E12.5_1-2-3-m5	E12.5	1-2-3-m5
GSM2190.CEL	FVB_E12.5_4-5-6-m5	E12.5	4-5-6-m5
GSM2191.CEL	FVB_E12.5_7-8-9-m5	E12.5	7-8-9-m5
GSM2192.CEL	FVB_NN_1-2-m5	NN	1-2-m5
GSM2193.CEL	FVB_NN_7-8-m5	NN	7-8-m5
GSM2194.CEL	FVB_NN_9-10-m5	NN	9-10-m5
GSM2088.CEL	FVB_1w_801-m5	A1w	801-m5
GSM2178.CEL	FVB_1w_804-m5	A1w	804-m5
GSM2179.CEL	FVB_1w_805-m5	A1w	805-m5
GSM2183.CEL	FVB_4w_11293-m5	A4w	11293-m5
GSM2184.CEL	FVB_4w_11294-m5	A4w	11294-m5
GSM2185.CEL	FVB_4w_11295-m5	A4w	11295-m5
GSM2334.CEL	FVB_3m_1f-m5	A3m	1f-m5
GSM2335.CEL	FVB_3m_2f-m5	A3m	2f-m5
GSM2336.CEL	FVB_3m_3f-m5	A3m	3f-m5
GSM2186.CEL	FVB_5m_731m-m5	A5m	731m-m5
GSM2187.CEL	FVB_5m_732m-m5	A5m	732m-m5
GSM2188.CEL	FVB_5m_733m-m5	A5m	733m-m5
GSM2180.CEL	FVB_1y_511m-m5	A1y	511m-m5
GSM2181.CEL	FVB_1y_5m-m5	A1y	5m-m5
GSM2182.CEL	FVB_1y_6m-m5	A1y	6m-m5
GSM2337.CEL	FVB_1y_529f-m5	A1y	529f-m5
GSM2338.CEL	FVB_1y_530f-m5	A1y	530f-m5
GSM2339.CEL	FVB_1y_544f-m5	A1y	544f-m5

To read this file into a data frame object in R:

```
> info<-read.delim("sampleInfo.txt", as.is=T, row.names=1, quote="\"",
fill=F)
> head(info)
                       SampleName Group        BS
GSM2189.CEL FVB_E12.5_1-2-3-m5 E12.5  1-2-3-m5
GSM2190.CEL FVB_E12.5_4-5-6-m5 E12.5  4-5-6-m5
GSM2191.CEL FVB_E12.5_7-8-9-m5 E12.5  7-8-9-m5
GSM2192.CEL       FVB_NN_1-2-m5    NN    1-2-m5
GSM2193.CEL       FVB_NN_7-8-m5    NN    7-8-m5
GSM2194.CEL      FVB_NN_9-10-m5    NN   9-10-m5
```

We used the cel file names as row names of the data frame for easy manipulation in conjunction with the expression matrix later on. For easy understanding and model fitting, groups were transformed into factors from characters and arranged in a time-ordered fashion, which more appropriately describes the data.

```
>info$Group<-factor(info$Group,
                     levels=c("E12.5", "NN", "A1w", "A4w", "A3m",
"A5m", "A1y")
>summary(info)

SampleName            Group          BS
 Length:24            E12.5:3     Length :24
 Class :character     NN   :3     Class  :character
 Mode  :character     A1w  :3     Mode   :character
                      A4w  :3
                      A3m  :3
                      A5m  :3
                      A1y  :6
```

3.5. Low-Level Data Processing

3.5.1. Normalization and Summarization with Entrez Gene CDF

There have been a number of efforts to provide accurate, up-to-date annotations for microarray platforms to supplement those provided by the microarray manufacturers. Each effort aims to address specific challenges, including volatile gene predictions, changes in genomic assemblies, and probe-set redundancies (3, 4). In this example, we used a custom CDF (3) for the MG-U74av1 chip. The custom CDF attempts to address these limitations by re-defining the probe-sets using a public identifier like Entrez Gene or Refseq and by re-aligning the individual probe sequence to the latest genome annotations of the corresponding organism. Additionally, the Affymetrix platform always contains multiple probe-sets mapping to the same gene. This redundancy creates noise and errors in the pathway analysis. Using Entrez Gene-based custom CDF will generate only one expression value per gene, which improves the accuracy of the pathway analysis.

Here, we show how to use an Entrez Gene ID-based custom CDF to generate the probe-set level expression values (*see* **Note 3**). Start an R terminal in the project directory containing the Affymetrix cel files. First, the cel files are read into an AffyBatch object:

```
>library(affy)
>batch<-ReadAffy(celfile.path="cel")
>cdfName(batch)
[1] "MG_U74A"
```

Changing the CDF name of the AffyBatch object to "mm74av1mmentrezg" will enable the data processing using the custom CDF file "mm74av1mmentrezg."

```
>library(mm74av1mmentrezg)
>cdfName(batch)<-"mm74av1mmentrezg"
>save(batch, file="obj/batch.RData")
```

The probe logarithmic intensity error (PLIER) method with quantile normalization and mismatch correction was used to generate more accurate results (5–7), especially for probe-sets with low expression. PLIER produces an improved signal (a summary value for a probe set) by accounting for experimentally observed patterns for feature behavior and handling error at low and high abundances across multiple arrays. For more information, please see the Affymetrix PLIER technical note (5).

```
>library(plier)
>eset<-justPlier(batch, normalize=T)
>exp<-exprs(eset)
>pairs(exp[,1:3])
```

The last command generates a pair–pair scatter plot of the first three arrays. As shown in **Fig. 15.2A**, there are some expression values ranging from 0 to 1 with exaggerated variance in log 2 scale. However, it's common practice to perform statistical analysis on a log-transformed scale. One simple solution is to add a small constant

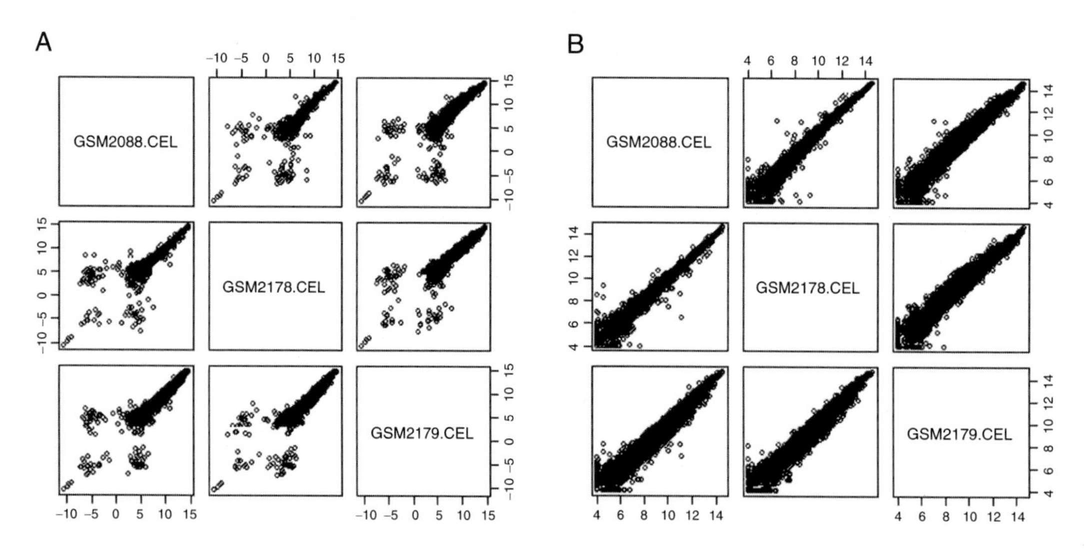

Fig. 15.2. Pair-wise scatter plot of expression values from microarrays 1 to 3 after low-level data processing. **(A)** Plot before flooring with a constant value. Large variance is observed for values between 0 and 1. **(B)** Data were plotted after flooring with a constant value.

number to floor the data. This can effectively reduce nuisance variation after transformation (**Fig. 15.2B**) with little impact on highly expressed genes. This method is also recommended in the PLIER technical note (*5*).

```
>exp<-log2(2^exp+16)
>pairs(exp[ ,1:3])
>save(exp, file="obj/exp.RData")
```

3.5.2. Gene Filtering with MAS5 Calls

Several studies have shown that using a threshold fraction of present detection calls generated from the Affymetrix MAS5 algorithm can effectively eliminate unreliable probe-sets and improve the ratio of true positives to false positives (*8, 9*). To generate the MAS5 detection calls for all probe-set:

```
>calls<-exprs(mas5calls(batch))
```

The relationship between calls and expression level and the distribution of the presence calls within a gene can be viewed using boxplots.

```
>boxplot(exp~calls)
```

As shown in **Fig. 15.3**, the detection calls are correlated with the expression level, but there is no clear cut difference among the three groups. Since we have a minimum of three samples for each group in this dataset, we applied a filtering step to keep only those probe-sets that have a present call on at least three arrays.

```
>row.calls<-rowSums(calls=="P")
>barplot(row.calls(table(row.calls)))
>exp<-exp[ row.calls>=3, rownames(info)]
>dim(exp)
[1] 3832 24
```

The 3,832 probe-sets that passed the filtering criteria are used for the high-level statistical analysis.

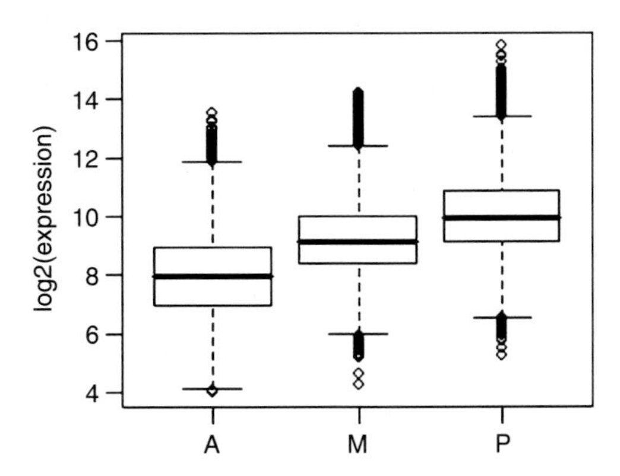

Fig. 15.3. The boxplot of log 2 (expression) vs. MAS5 detection calls. "A" – absent; "M" – marginal; "P" – present.

3.6. Principal Component Analysis (PCA)

Principal component analysis is usually performed as the first step after low-level data processing to obtain a "big picture" of the data. It is designed to capture the variance in a dataset in terms of principal components (PCs). PCA helps to dissect the source of the variance and identify the sample outliers in the dataset by reducing the dimensionality of the data. Since the number of genes (rows) is much larger than the number of samples (columns) in microarray data, function "`prcomp`" (instead of "`princomp`") is called and the expression matrix is transposed before being fed into the function. The argument "`scale.`" is explicitly turned on so that all the genes contribute equally to the analysis regardless of the magnitude of the change.

```
> dim(t(exp))
[1] 24 3832
> pca.res<-prcomp(t(exp), scale.=T, retx=T)
> names(pca.res)
[1] "sdev"  "rotation"  "center"  "scale"  "x"
>dim(pca.res$x)
[1] 24 24
```

The "`sdev`" in the result object is a list containing the standard deviations from all principal components (PCs). The variance of the first ten principal components can be plotted as (**Fig.15.4**):

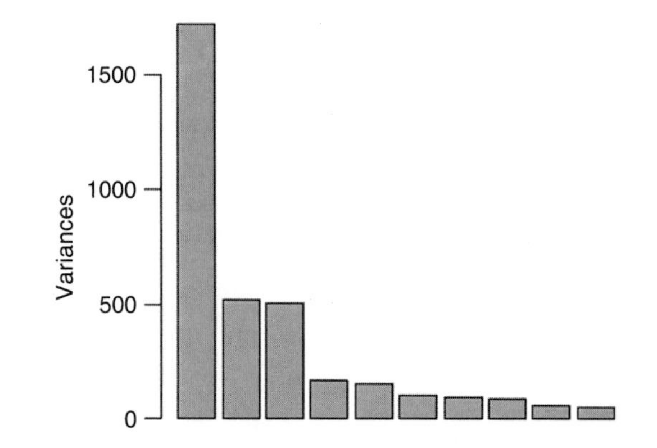

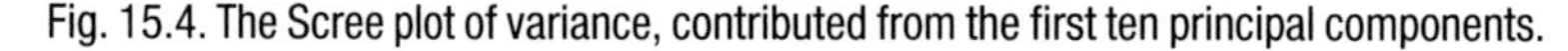

Fig. 15.4. The Scree plot of variance, contributed from the first ten principal components.

```
> plot(pca.res, las=1)
```

The percentage of the variation "`pc.per`" contributed from each PC can be calculated as

```
> pc.var<-pca.res$sdev^2
> pc.per<-round(pc.var/sum(pc.var)*100, 1)
> pc.per
 [1] 45.0 13.4 13.1 4.3 4.0 2.7 2.5 2.3 1.6 1.3 1.2
[12]  1.0  1.0  0.8 0.8 0.7 0.7 0.7 0.6 0.6 0.6 0.5
[23]  0.5  0.0
```

"x" in the results is a matrix that contains the coordinates of all samples projected onto the PCs. We can then plot the samples on to the first two PCs that carry the most variance, and label them by time-point.

```
> plot.pch<-(1:length(levels(info$Group)))[ as.numeric(info$Group)]
> plot(pca.res$x, col=1, pch=plot.pch, las=1, cex=2,xlab=paste("PC1 (",
pc.per[ 1] , "%)", sep=""),ylab=paste("PC2 (", pc.per[ 2] , "%)", sep=""))
> grid()
> legend(-90, 0, levels(info$Group), pch=1:length(levels(info$Group)),
pt.cex=1.5)
```

Visual inspection of the PCA plot yielded a straightforward diagnosis of the sources of variance in this dataset. As shown in **Fig. 15.5**, samples from the same group clustered together. The main variation in the dataset (45%) correlates with the time point of cardiac development. The second PC (13.4%) does not sort the samples by developmental stage, but seems to distinguish the neonatal (NN) and 1 week of age (A1w) groups from the other time points. One sample from group "A1w" was separated in space from the other two samples in this group, but still allowed separation from the other groups. This sample was therefore not considered an outlier and was included in the analysis.

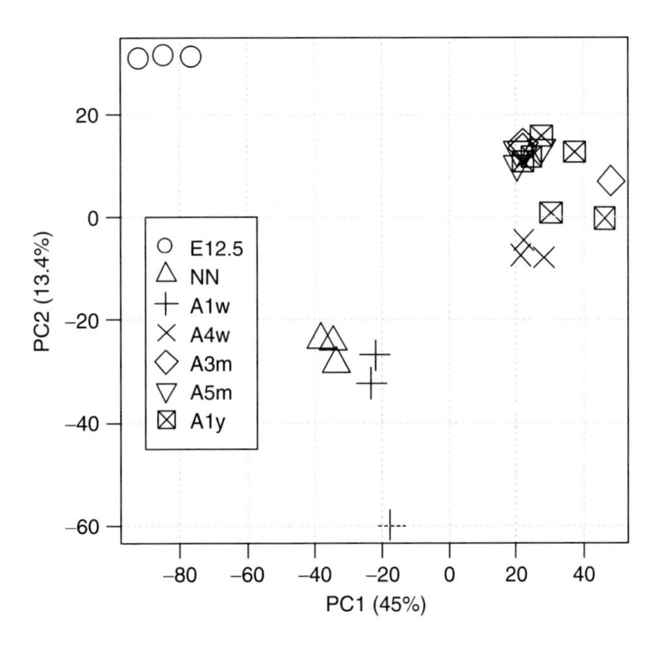

Fig. 15.5. Sample projection onto the first two PCs. The percent variance described by the corresponding PC is marked along the axes.

3.7. Identify Differentially Expressed Genes Using LIMMA

Linear Models for Microarray Data (LIMMA) is an R package that uses linear models to analyze microarray experiments (5; *see* **Note 4**). Microarray experiments frequently employ a small number of replicates per condition ($n \leq 6$), which makes estimating the variance of a gene's expression level difficult. Consequently, traditional statistical methods such as the *t*-test can be unreliable. LIMMA leverages the large number of observations in a microarray experiment to moderate the variance estimates in a data dependent fashion. The output of LIMMA is therefore similar to the output of a *t*-test but stabilized against the effects of small sample sizes. Our purpose was to use this package to identify significantly differentially expressed genes across different time points. To fit the data with the linear model, we constructed a design matrix from a "target" vector which contains the grouping information (i.e., the "Group" column in the info data frame in this example).

```
>library(limma)
>levels(info$Group)
[1] "E12.5" "NN"   "1w"   "4w"   "3m"   "5m"   "1y"
> lev<-levels(info$Group)
> design<-model.matrix(~0+info$Group)
> colnames(design)<-lev
> dim(design)
[1] 24 7
> head(design)
  E12.5 NN A1w A4w A3m A5m A1y
1     1  0   0   0   0   0   0
2     1  0   0   0   0   0   0
3     1  0   0   0   0   0   0
4     0  1   0   0   0   0   0
5     0  1   0   0   0   0   0
6     0  1   0   0   0   0   0
```

To fit the linear model with the design matrix:

```
> fit<-lmFit(exp, design)
> names(fit)
 [1] "coefficients"  "rank"   "assign"            "qr"
 [5] "df.residual"   "sigma"  "cov.coefficients"  "stdev.unscaled"
 [9] "pivot"         "genes"  "method"            "design"
```

Here, the design matrix is in a group means parameterization, where the coefficients are the mean expression of each group. To find the differences among these coefficients, an explicitly defined contrast matrix is required. For this dataset, we generated all the pair-wise comparisons.

```
> contr.str<-c()
> len<-length(lev)
> for(i in 1:(len-1))
+ contr.str<-c(contr.str, paste(lev[ (i+1):len], lev[ i], sep="-"))
> contr.str
[1] "NN-E12.5" "A1w-E12.5" "A4w-E12.5" "A3m-E12.5" "A5m-E12.5" "A1y-
E12.5"
 [7] "A1w-NN"  "A4w-NN"  "A3m-NN"  "A5m-NN"  "A1y-NN"  "A4w-A1w"
[13] "A3m-A1w" "A5m-A1w" "A1y-A1w" "A3m-A4w" "A5m-A4w" "A1y-A4w"
[19] "A5m-A3m" "A1y-A3m" "A1y-A5m"
> contr.mat<-makeContrasts(contrasts=contr.str, levels=lev)
> fit2<-contrasts.fit(fit, contr.mat)
> fit2<-eBayes(fit2)
> names(fit2)
 [1] "coefficients" "rank"    "assign"            "qr"
 [5] "df.residual"  "sigma"   "cov.coefficients"  "stdev.unscaled"
```

```
[ 9]  "genes"        "method"    "design"         "contrasts"
[13]  "df.prior"     "s2.prior"  "var.prior"      "proportion"
[17]  "s2.post"      "t"         "p.value"        "lods"
[21]  "F"            "F.p.value"
```

Now "coefficients" in the "fit2" contains difference, or log 2 fold change, and "p.value" is the moderated *t*-test *p*-value associated with all the pair-wise comparisons. "F" and "F.p.value" is the moderated *F*-test given for all those comparisons. The statistics for the changes of all genes across all groups can be retrieved, sorted by *F*-test *p*-values, integrated with gene annotation and output into a tab-delimited file:

```
> f.top<-topTableF(fit2, number=nrow(exp))
> f.top<-cbind(ann [f.top[[1]] 1:4], f.top[,c(2:7, 23:25)])
> write.table(f.top, file="f.top.txt", sep="\t", row.names=F, quote=F)
> head(f.top)
         ProbeSet GeneID Symbol
14955_at 14955_at  14955    H19
16002_at 16002_at  16002   Igf2
12797_at 12797_at  12797   Cnn1
15126_at 15126_at  15126  Hba-x
98932_at 98932_at  98932   Myl9
20250_at 20250_at  20250   Scd2
                                                            GeneName
14955_at                                           H19 fetal liver mRNA
16002_at                                    insulin-like growth factor 2
12797_at                                                   calponin 1
15126_at hemoglobin X, alpha-like embryonic chain in Hba complex
98932_at                    myosin, light polypeptide 9, regulatory
20250_at                          stearoyl-Coenzyme A desaturase 2
           NN.E12.5 A1w.E12.5 A4w.E12.5 A3m.E12.5
14955_at -0.1262318 -1.368427 -4.639086 -5.983330
16002_at -0.2315497 -1.185102 -3.823067 -4.825221
12797_at -5.6486004 -5.710477 -6.103997 -6.375261
15126_at -4.8768229 -5.507456 -5.037622 -5.267489
98932_at -0.4566940 -1.719379 -3.931495 -4.218323
20250_at -0.7785607 -1.832313 -3.600636 -3.994683
         A5m.E12.5 A1y.E12.5         F      P.Value
14955_at -5.572897 -5.783935 1087.2210 2.099165e-25
16002_at -4.705602 -4.761241  819.2029 4.251770e-24
12797_at -6.160193 -6.254116  654.1071 4.635634e-23
15126_at -5.100509 -5.210385  567.7905 2.078625e-22
98932_at -4.957873 -4.828438  430.2656 3.916275e-21
20250_at -3.860944 -4.052382  429.1944 4.020864e-21
            adj.P.Val
14955_at 8.043999e-22
16002_at 8.146390e-21
12797_at 5.921249e-20
15126_at 1.991323e-19
98932_at 2.243523e-18
20250_at 2.243523e-18
```

The middle columns (nos. 5–10) of the table contain the log 2 fold changes of all other time points vs. embryonic 12.5 d.p.c. We can loop through all two-group comparisons and output the results:

```
> for(i in 1:ncol(contr.mat)){
+ t.top<-topTable(fit2, coef=i, number=nrow(exp))
+ t.top<-cbind(ann[t.top[[1]],], t.top[, 2:ncol(t.top)])
+ write.table(t.top, sep="\t", row.names=F, quote=F,
```

```
+            file=file.path("limma", paste(contr.str[ i], "top.txt",
sep=".")))
+}
```

3.8. Clustering Analysis

Clustering analysis has been widely applied to gene expression data for pattern discovery. Hierarchical clustering is a frequently used method that does not require the user to specify the number of clusters a priori (*see* **Note 5**). Since all genes contribute equally, genes with no changes between groups only add noise to the clustering. Thus, statistical filters are often applied to eliminate such genes prior to the clustering procedure. In our dataset, there were a large number of filtered genes (2,663, 69.6%) with a statistically significant change above a Benjamini–Hochberg (BH) (10) adjusted p-value cutoff of <0.01. Consequently, we further limited the clustering to those genes that showed at least a twofold difference between any two of the seven time-points.

```
> g.2f<-(rowSums(abs(fit2[[ 1]][ rownames(exp),])>1)>0) &
+         (f.top[ rownames(exp),] $adj.P.Val<0.01)
> sum(g.2f)
[1] 1680
```

A total of 1,680 genes passed this criteria. The gene expression matrix was standardized before calculating the Euclidean distances between the genes.

```
> exp.std<-sweep(exp, 1, rowMeans(exp))
> row.sd<-apply(exp, 1, sd)
> exp.std<-sweep(exp.std, 1, row.sd, "/")
> save(exp.std, file="obj/exp.std.RData")
> exp.cl<-exp.std[ g.2f,]
```

Several different clustering methods are provided in the "`hclust`" function. Here, we chose *Ward's* minimum variance method which aims to find compact, spherical clusters (**Fig. 15.6**).

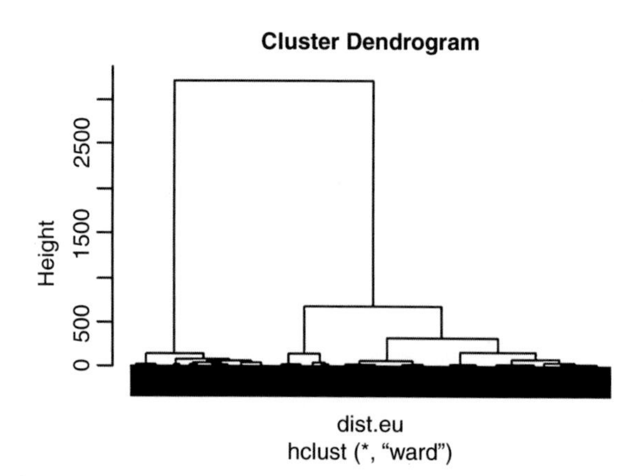

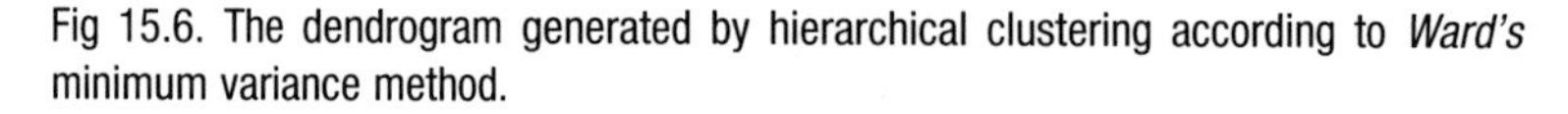
Fig 15.6. The dendrogram generated by hierarchical clustering according to *Ward's* minimum variance method.

```
> hc.res<-hclust(dist.eu, method="ward")
> plot(hc.res, labels=F)
```

The tree can be cut into branches (clusters) by specifying the height or number of branches desired. Usually, we cut the tree right above the height where the branches become dense. In this example, the dendrogram was cut into seven final clusters. The gene expression data can be displayed in a heatmap in the order of the dendrogram (**Fig. 15.7**).

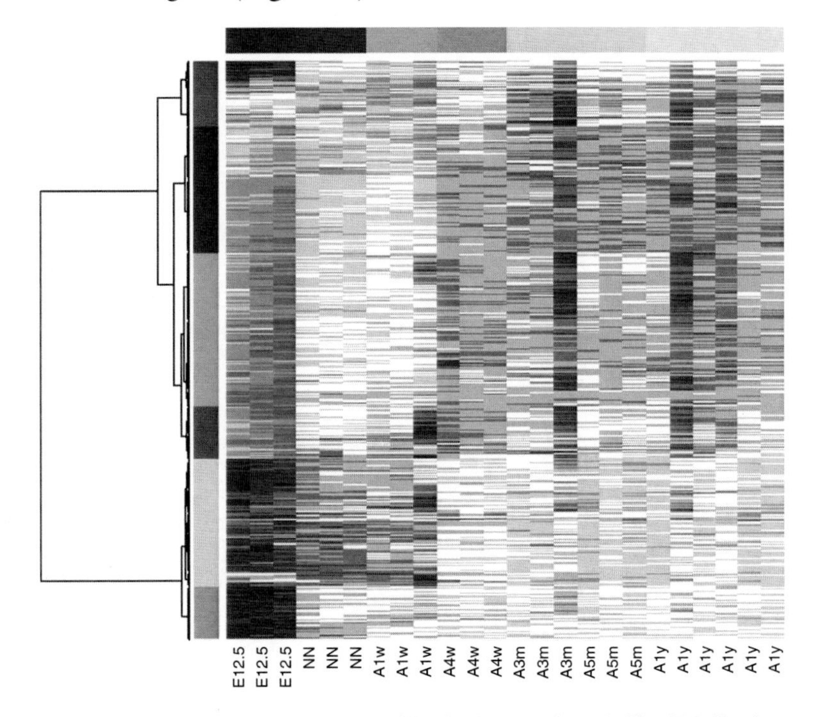

Fig. 15.7. Heatmap of genes in the order of the dendrogram shown in **Fig. 15.6**. The time points and the clusters are indicated by the *row* and *column side shadings*, respectively. Time points are sorted by increasing age (E12.5 d.p.c. to 1 year of age) from *left* to *right*.

```
> clus.res<-cutree(hc.res, k=7)
> hclust.ward<-function(d,...){hclust(d, method="ward",...)}
> heat.res<-heatmap(exp.cl, scale="none", labRow="",
+                   labCol=as.character(info$Group),
+                   Colv=NA, hclustfun = hclust.ward,
+ ColSideColors=topo.colors(7)[as.numeric(info$Group)],
+                    RowSideColors=rainbow(7)[clus.res],
+                    col=dChip.colors(10))
> names(heat.res)
[1] "rowInd" "colInd" "Rowv" "Colv"
> head(clus.res[heat.res$rowInd], n=4)
    17069_at      18830_at 100040340_at    78330_at
           3             3            3           3
> tail(clus.res[heat.res$rowInd], n=4)
 20200_at  104130_at  69094_at  56213_at
        7          7         7         7
```

The cluster number associated with each gene can be extracted and output into a text file. For easy interpretation of the clusters, genes were ordered exactly the same as they appeared in the heatmap with the clustered numbers 1–7 from top to bottom.

```
>  clus.order<-unique(clus.res[ heat.res$rowInd] )
> clus.order ##this is from bottom to top in the heatmap
[1] 3 2 1 4 6 5 7
> gene.clus.order<-match(clus.res[ heat.res$rowInd] , clus.order)
> names(gene.clus.order)<-names(clus.res[ heat.res$rowInd] )
> head(gene.clus.order)
17069_at  18830_at  100040340_at  78330_at  101540_at  18032_at
   1     1        1      1      1     1
> gene.clus.order<-rev(gene.clus.order)
> gene.clus.order<-max(gene.clus.order)-gene.clus.order+1
> clus.ann<-ann[ names(gene.clus.order),]
> clus.ann$cluster<-gene.clus.order
> head(clus.ann[ ,c(2:3,5)] )
        GeneID  Symbol cluster
56213_at  56213  Htra1    1
69094_at  69094 Tmem160    1
104130_at 104130 Ndufb11    1
20200_at  20200  S100a6    1
26968_at  26968  Islr    1
81877_at  81877   Tnxb    1
> write.table(clus.ann, file="clus.ann.txt", sep="\t", row.names=F,
na="")
```

The functional enrichment for each cluster can be calculated using the "GOstats" package from Bioconductor, or using the web-tool DAVID (*D*atabase for *A*nnotation, *V*isualization, and *I*ntegrated *D*iscovery, http://david.abcc.ncifcrf.gov (11).

3.9. Pathway (Gene-Set) Analysis

Gene-set analysis is especially helpful for identifying the biological themes related to changes between two conditions or for correlation with a specific numeric phenotypical measurement. Following Mootha's Gene-Set Enrichment Analysis (GSEA), Tian et al. (12) proposed to rank gene-sets based on two statistics, *NTk* and *NEk*, and estimate *q*-values for each pathway or gene-set to address two different aspects in pathway analysis. Given a gene-set *g*, *NTk* computes whether *g* is significantly changed compared to all other gene-sets. *NEk* serves as an indicator of whether the genes within *g* as a whole group are significantly correlated with the phenotype. "sigPathway" is an R package implementation of the method (*see* **Note 6**). Here, we show how to use sigPathway in order to identify pathways that are statistically significantly different between two developmental stages, NN and E12.5 d.p.c.

3.9.1. Construct Gene-Sets Object for sigPathway

First, we constructed an R list object that contains the gene-set annotation list we would like to use for the calculations. sigPathway will calculate the composite statistics *NTk* and *NEk* for each gene-set within this list. If the annotation list is named *G*, each element of *G* is an R list object representing one gene list and should include three essential elements:

1. source: the source of gene-set, e.g., GO, BP, or KEGG;
2. title: the title for the gene-set, e.g., "ABC transporters";
3. probes: a unique set of probe(-set)s that belong to the gene-set.

The annotation list can be from any source, including user-defined lists. Gene Ontology (GO) annotation is usually considered the most inclusive and fastest-growing public source for grouping functionally relevant genes. The following procedure shows how to build an up-to-date gene-set annotation list from scratch, starting with the "org.Mm.egGO2ALLEGS" and "GO" packages from Bioconductor.

First, a list of GO terms to Entrez Gene ID mapping is created.

```
> library(sigPathway)
> library(GO)
> x<-as.list(org.Mm.egGO2ALLEGS)
> length(x)
[1] 7938
> head(names(x))
[1] "GO:0008150" "GO:0008152" "GO:0006464" "GO:0006468" "GO:0006793"
[6] "GO:0006796"
> len<-unlist(lapply(x, length))
> head(len)
GO:0008150 GO:0008152 GO:0006464 GO:0006468 GO:0006793 GO:0006796
     21965       8832       1505        650        870        870
```

To exclude genes that are not included in this CDF (mm74av1mmentrezg):

```
> x<-lapply(x, function(y, all.gene.id){
+    y<-y[y %in% all.gene.id]
+    unique(y)
+ },    unique(ann$GeneID))
> len<-unlist(lapply(x, length))
> summary(len)
   Min.  1st Qu.   Median     Mean  3rd Qu.     Max.
   0.00     1.00     2.00    32.82     8.00  5853.00
```

The number of genes in a gene-set ranges from 0 to 5,853 genes, but we usually limit our analysis to include gene-sets with about 5–500 genes. If there are too few genes in the gene-sets, the results could be driven by only one or two genes with large expression changes and not fairly reflect the whole pathway. On the other hand, it can be difficult to interpret the biological meaning when the number of genes in a gene-set is too large.

```
> lo<-5
> hi<-500
> x<-x[len>=lo & len<=hi]
> length(x)
[1] 3452
```

The list "x" contains the probe-sets for 3,452 GO IDs. The titles and definitions for the GO terms can be obtained using the GO package:

```
> library(GO)
> gt<-as.list(GOTERM)
> length(gt)
[1] 23679
> gt[1:2]
$'GO:0019980'
GOID: GO:0019980
Term: interleukin-5 binding
```

```
Ontology: MF
Definition: Interacting selectively with interleukin-5.
Synonym: IL-5 binding

$'GO:0004213'
GOID: GO:0004213
Term: cathepsin B activity
Ontology: MF

Definition:   Catalysis of the hydrolysis of peptide bonds with a broad
              specificity. Preferentially cleaves the terminal bond of -
              Arg-Arg-Xaa motifs in small molecule substrates (thus
              differing from cathepsin L). In addition to being an
              endopeptidase shows peptidyl-dipeptidase activity
              liberating C-terminal dipeptides.
```

To generate the gene-set annotation list *G*:

```
> G<-list()
> for(i in names(x)){
+    G[[i]]<-list()
+    G[[i]]$src<-gt[[i]]@Ontology
+    G[[i]]$name<-paste(i, gt[[i]]@Term)
+    G[[i]]$probes<-paste(x[[i]], "at", sep="_")
+ }
> save(G, file="G.RData")
> length(G)
[1] 2708
> head(names(G))
[1] "GO:0006468" "GO:0006793" "GO:0006796" "GO:0006915" "GO:0008219"
[6] "GO:0012501"
> class(G)
[1] "list"
> class(G[[1]])
[1] "list"
> names(G[[1]])
[1] "src" "title" "probes"
> G[[1]][1:2]
$src
[1] "BP"

$title
[1] "GO:0006468 protein amino acid phosphorylation"
```

This list "G" is saved and can be used later for any dataset pre-processed with the CDF mm74mmav1entrezg. For this analysis, we restricted the gene-sets to those with 5–200 probe-sets that are present in the filtered expression data by using the "selectGeneSets" function. The list that was used is recorded in list "g" without altering the annotation list object "G."

```
> exp<-exp[-c(grep("AFFX", rownames(exp))),]
> g<-selectGeneSets(G, rownames(exp), minNPS=5, maxNPS=200)
> names(g)
[1] "nprobesV" "indexV"     "indGused"
> length(g[[1]])
[1] 1652
```

Gene-sets (1,652) were used in the analysis after this filter. The next step was to construct the expression data matrix for comparing the neonatal stage "NN" vs. embryonic stage "E12.5" and calculate the *NTk* and *NEk* statistics.

```
> samples.ref<-rownames(info)[ info$Group=="E12.5"]
> exp.ref<-exp[ , samples.ref]
> samples.test<-rownames(info)[ info$Group=="NN"]
> exp.test<-exp[ , samples.test]
> phenotype<-rep(c(0, 1), c(length(samples.ref), length(samples.test)))
> tab<-cbind(exp.ref, exp.test)
> NTk<-calculate.NTk(tab, phenotype, g)
> NEk<-calculate.NEk(tab, phenotype, g)
'nsim' is greater than the number of unique permutations
 Changing 'nsim' to 19, excluding the unpermuted case
```

To view the *NEk*/*NTk* distributions and their relationship:

```
> par(mfrow=c(2,2), mex=0.7, ps=9)
> qqnorm(NTk$t.set.new, main="NTk Q-Q Normal")
> qqline(NTk$t.set.new, col=2)
> qqnorm(NEk$t.set.new, main="NEk Q-Q Normal")
> qqline(NEk$t.set.new, col=2)
> plot(NTk$t.set.new, NEk$t.set.new, cex=0.7)
> abline(h=0)
> abline(v=0)
> grid()
```

As shown in **Fig. 15.8A and B**, both the *NEk* and *NTk* statistics are symmetrically distributed, but *NTk* has a longer "tail" and *NEk* a shorter "tail" than a normal distribution. The two statistics are positively correlated. By default, the top 25 enriched gene-sets ranked by averaging the individual ranks of both, *NTk* and *NEk* rankings, can be retrieved as

```
> path.res<-rankPathways(NTk, NEk, G, tab, phenotype, g, ngroups=2,
+                        methodNames=c("NTk", "NEk"), allpathways=T)
> names(path.res)
 [1] "IndexG"          "Gene Set Category"      "Pathway"
 [4] "Set Size"        "Percent Up"             "NTk Stat"
 [7] "NTk q-value"     "NTk Rank"               "NEk Stat"
[10] "NEk q-value"     "NEk Rank"
> path.res$Pathway
 [1] "GO:0006817 phosphate transport"
 [2] "GO:0005581 collagen"
 [3] "GO:0030020 extracellular matrix structural constituent conferring
tensile strength"
 [4] "GO:0005201 extracellular matrix structural constituent"
 [5] "GO:0044420 extracellular matrix part"
 [6] "GO:0005605 basal lamina"
 [7] "GO:0005578 proteinaceous extracellular matrix"
 [8] "GO:0031012 extracellular matrix"
 [9] "GO:0004364 glutathione transferase activity"
[10] "GO:0015698 inorganic anion transport"
[11] "GO:0006820 anion transport"
[12] "GO:0006084 acetyl-CoA metabolic process"
[13] "GO:0006270 DNA replication initiation"
[14] "GO:0005604 basement membrane"
[15] "GO:0008094 DNA-dependent ATPase activity"
[16] "GO:0006631 fatty acid metabolic process"
[17] "GO:0006638 neutral lipid metabolic process"
[18] "GO:0006639 acylglycerol metabolic process"
[19] "GO:0006099 tricarboxylic acid cycle"
[20] "GO:0009060 aerobic respiration"
[21] "GO:0009109 coenzyme catabolic process"
[22] "GO:0046356 acetyl-CoA catabolic process"
[23] "GO:0051187 cofactor catabolic process"
[24] "GO:0044445 cytosolic part"
[25] "GO:0032787 monocarboxylic acid metabolic process"
```

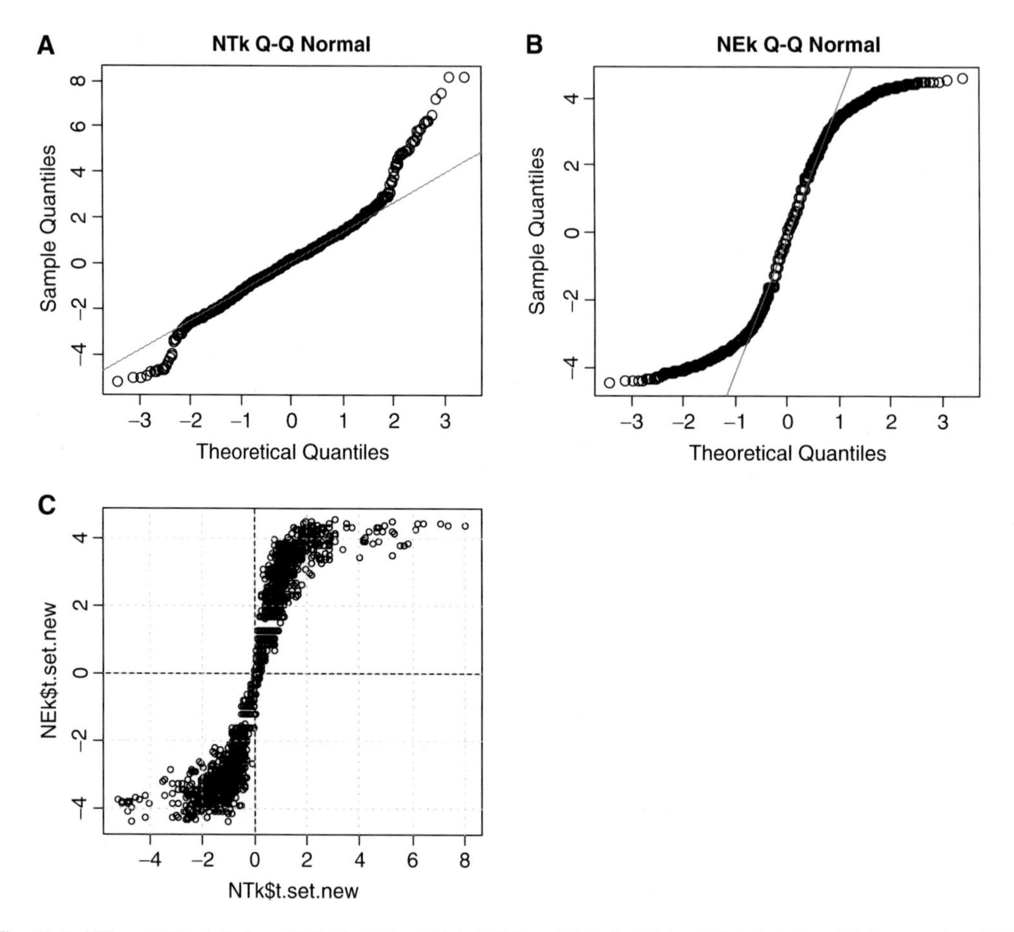

Fig. 15.8. *NTk* and *NEk* statistics. **(A)** Q–Q plot for *NTk* *t*-statistics. **(B)** Q–Q plot for *NEk* *t*-statistics. **(C)** Scatter plot of *NEk* *t*-statistics vs. *NTk* *t*-statistics.

The related pathway statistics can be viewed as

```
> path.res[ ,6:11]
   NTk Stat  NTk q-value  NTk Rank  NEk Stat  NEk q-value  NEk Rank
1      6.33   0.00000000       4.0      4.43            0         6
2      6.14   0.00000000       6.0      4.44            0         5
3      6.14   0.00000000       6.0      4.44            0         5
4      6.89   0.00000000       3.0      4.40            0        12
5      7.19   0.00000000       2.0      4.37            0        14
6      5.22   0.00000000      13.0      4.46            0         3
7      7.92   0.00000000       1.0      4.36            0        19
8      7.92   0.00000000       1.0      4.36            0        19
9      4.41   0.00000000      31.0      4.53            0         1
10     5.02   0.00000000      15.0      4.33            0        21
11     4.80   0.00000000      22.0      4.32            0        23
12     6.17   0.00000000       5.0      4.26            0        43
13    -3.09   0.03690476      51.5     -4.42            0         7
14     4.74   0.00000000      24.0      4.29            0        37
15    -3.69   0.00000000      41.0     -4.31            0        30
16     5.13   0.00000000      14.0      4.16            0        65
17     2.88   0.05470588      72.0      4.41            0        10
18     2.88   0.05470588      72.0      4.41            0        10
```

```
19      4.92   0.00000000    20.0     4.16        0      64
20      4.92   0.00000000    20.0     4.16        0      64
21      4.92   0.00000000    20.0     4.16        0      64
22      4.92   0.00000000    20.0     4.16        0      64
23      4.85   0.00000000    21.0     4.17        0      63
24     -3.09   0.03690476    51.5    -4.29        0      33
25      5.27   0.00000000    12.0     4.14        0      73
```

The extracellular matrix (ECM) and fatty acid metabolism gene-sets were the most up-regulated and DNA replication/cell cycle gene-sets were most down-regulated when comparing NN vs. E12.5 d.p.c. developmental stages. The gene-sets with the highest *NTk* rank is "`GO:0031012 extracellular matrix.`" The *NTk* *t*-statistic of 7.92 was much higher than the second ranked gene-set "`GO:0006817 phosphate transport`" (*NTk* 6.33), however, the *NEk* *t*-statistics of the two gene-sets were about the same. This scenario is clearly displayed in **Fig. 15.8C**, in which the tails of the *NTk* *t*-statistics spread wider than the tails of the *NEk* *t*-statistics Thus, it is more meaningful to rank the significance of the gene-sets based on the mean of *NEk* and *NTk* *t*-statistics rather than based on the average ranking of the two, which is the default. This can be done by specifying the argument "`npath`" in "`rankPathways`" function to the number of total gene-sets, and then reordering the data frame:

```
> path.res<-rankPathways(NTk, NEk, G, tab, phenotype, g, ngroups=2,
+                        methodNames=c("NTk", "NEk"), npath=NTk$ngs)
> ave.stat<-rowMeans(path.res[ ,c(6,9)] )
> path.res<-path.res[ order(abs(ave.stat), decreasing=T),]
> head(path.res[ [ 3]] )
[ 1] "GO:0005578 proteinaceous extracellular matrix"
[ 2] "GO:0031012 extracellular matrix"
[ 3] "GO:0044420 extracellular matrix part"
[ 4] "GO:0005201 extracellular matrix structural constituent"
[ 5] "GO:0006817 phosphate transport"
[ 6] "GO:0005581 collagen"
```

Finally, it is of interest to view the genes that contributed to the changes, especially for those top-ranked gene-sets. For example, we retrieved the statistics for all the genes in gene-set “GO:0006817 phosphate transport,” which is ranked on fifth place on the "`path.res`" `list`.

```
> t.stat<-calcTStatFast(tab, phenotype, ngroups=2)
> path.stat<-getPathwayStatistics(tab, phenotype, G, path.res$IndexG,
+                                 statList=t.stat)
> stl<-path.stat[[ 5]]
> head(stl)
     Probes     Mean_0     Mean_1     StDev_0     StDev_1
1  11487_at   9.560018   8.922373  0.18504668  0.15379436
2  11490_at   9.925778  10.564280  0.12777095  0.18693417
3  11492_at   9.407763  10.108199  0.09667248  0.24242213
4  11603_at  11.535603  11.362591  0.17646941  0.06322808
5  12111_at  10.942530  12.093049  0.23738825  0.13703260
6  12159_at  11.454837   9.619270  0.27848105  0.08053406
```

```
  T-Statistic      p-value
1    -4.590067 0.010918577
2     4.884183 0.011090014
3     4.648474 0.025147668
4    -1.598605 0.225637401
5     7.270171 0.004308710
6   -10.967170 0.004659806
> stl$logFC<-stl [,3]-stl [,2]
```

Finally, we added the gene annotations from "ann" to the output "stl". Be aware that the first column in "stl" becomes a vector of factors instead of characters.

```
> stl<-cbind(ann[as.character(stl[[1]]),], stl[,c("logFC", "p-value")])
> stl<-stl[order(stl$logFC, decreasing=T),]
> names(stl)
[1] "ProbeSet" "GeneID" "Symbol" "GeneName" "logFC" "p-value"
> stl[,c(3, 5, 6)
           Symbol       logFC      p-value
12819_at  Col15a1  5.96300996 3.109213e-03
12834_at   Col6a2  2.54466441 1.403811e-03
12842_at   Col1a1  2.42742733 7.991941e-04
12843_at   Col1a2  2.39452560 9.757436e-04
12262_at     C1qc  1.96001216 2.240684e-03
12833_at   Col6a1  1.94317393 9.974864e-05
12827_at   Col4a2  1.58503721 1.619273e-04
12259_at     C1qa  1.57040004 1.658915e-02
12831_at   Col5a1  1.42895064 2.076524e-03
12260_at     C1qb  1.34830464 1.363401e-02
12826_at   Col4a1  1.23069086 3.217859e-02
11732_at      Ank  0.93207373 2.774580e-04
12825_at   Col3a1  0.91370870 1.071566e-02
11450_at   Adipoq  0.83960637 5.935469e-03
12837_at   Col8a1  0.26383598 1.718501e-01
20505_at  Slc34a1  0.21389501 1.774224e-01
12813_at  Col10a1  0.12015915 4.409476e-01
12840_at   Col9a2 -0.07155123 5.553861e-01
140709_at   Emid2 -0.25438285 1.177430e-01
12832_at   Col5a2 -0.32661506 1.530350e-01
20515_at  Slc20a1 -2.28412220 4.464655e-04
```

Inspecting the gene expression changes that contribute to significantly changed pathways or gene-sets of interest can help to group the relevant genes together, and to prioritize the gene list based on the pathway changes.

3.10. Summary

In this chapter, we demonstrated a general workflow of bioinformatics analysis of Affymetrix GeneChip™ data using Bioconductor software packages on a public test dataset. Bioconductor is a widely used open source and open development software project for the analysis and comprehension of high-throughput data from different platforms. Bioconductor is rooted in the open source statistical computing environment R. The main advantage of Bioconductor is access to a wide range of powerful statistical and graphical methods for the analysis of genomic data, and the rapid development of extensible software based on the most advanced and updated analysis algorithms. For users not familiar with R, other software tools are available to execute the diverse analysis steps, as summarized in **Note 7**.

We focused our analysis on commonly used strategies for normalization, probe-set summarization, gene filtering, statistical analysis, and pathway prediction. However, many other analysis options can be used for each step and have been extensively discussed (6, 13, 14). A list of the available options can also be found under the Bioconductor Task View.

Additional high-level data analyses that allow a more in-depth understanding of the biology underlying gene expression changes include motif analysis for co-expressed genes (15) and gene network/topology analysis (16, 17). However, a detailed description of these analyses is beyond the scope of this introductory chapter.

4. Notes

1. R and Bioconductor package installation

   R installs and updates its packages using an HTTP protocol. When installed behind a firewall, an http proxy environment variable must be first set to enable access to the Internet from within R.

   ```
   > Sys.setenv(''http_proxy'' = ''http://my.proxy.net:9999'';).
   ```

2. Additional packages can also be installed from the R terminal menu "Packages" → "select the CRAN mirror" → "select repositories."

3. Low-level Affymetrix data processing

   Numerous methods have been published for normalization and summarization of the probe-level data. The Robust Multi-chip Average (RMA) (18), GC-RMA (19) and MBEI *(24)* are popular methods. "LIMMA" package also provides a GUI which requires minimal R programming.

4. Statistical Analysis of differentially expressed genes

   Other methods for finding differentially expressed genes can be found on the Bioconductor website using "Task View" of "DifferentialExpression" under the download section for the specific release.

5. Clustering using the HOPACH method

   Beyond classical hierarchical clustering, the "Hierarchical Ordered Partitioning and Collapsing Hybrid" (HOPACH) package uses the Mean/Median Split Silhouette (MSS) criteria to identify the level of the tree with maximally homogeneous clusters (20). In this case users do not have to pre-specify the number of clusters. This method usually identifies a larger number of homogeneous cluster with smaller size using the default setting.

6. Gene-Set Enrichment Analysis (GSEA) using sigPathway package
   a. The significance of the results depends on the collection of gene-sets available. It is important that the pathways examined are relevant to the study.
   b. Other pathways like KEGG can be constructed in the similar manner as the GO packages. The package also provides functions to import gene-sets in other formats. It is recommended to calculate the pathway statistics separately for each source due to the redundancy among gene-sets from different sources.
   c. The function "`writeSigPathway`" in the sigPathway package outputs all gene-sets in html format. This function requires the chip annotation package with accession numbers, which "`org.Mm.eg.db`" does not provide. However, for datasets summarized with Affymetrix CDF, this is a nice utility to view the results in a more user-friendly format.
7. Other statistical software for microarray data analysis

   There are many other commercial data analysis packages and open software available for microarray data analysis for users not familiar with R programming. Some of the popular packages and tools include the following:
   - Complete Analysis (normalization, group comparison, clustering, etc.)
     - GenePattern (21) (Broad Institute)
     - geWorkbench (NCI)
       http://wiki.c2b2.columbia.edu/workbench/index.php/Home
     - GeneSpring from Agilent Technologies
     - TM4 (22) – http://www.tm4.org/
     - dChip (23, 24)
       http://biosun1.harvard.edu/complab/dchip/
   - Differentially expressed genes
     - SAM (25) – http://www-stat.stanford.edu/%7Etibs/SAM/index.html
     - PaGE (26) – http://www.cbil.upenn.edu/PaGE/
   - Pathway/gene-set analysis
     - GSEA-P (27, 28):
       http://www.broad.mit.edu/gsea/
     - GeneTrail (29)
       http://genetrail.bioinf.uni-sb.de/

- Ingenuity Pathway Analysis Tool – http://www.ingenuity.com/
- MetaCore – http://www.genego.com
- GenMaPP (30) – http://www.genmapp.org/
- Collection of GO Analysis Tools: http://www.geneontology.org/GO.tools.microarray.shtml

References

1. Reimers, M, Carey, VJ. (2006). Bioconductor: an open source framework for bioinformatics and computational biology. *Method Enzymol* **411**, 119–134.
2. Team, RDC. (2007). R: A language and environment for statistical computing. R Foundation for Statistical Computing, Vienna, Austria.
3. Dai, M, Wang, P, Boyd, AD, et al. (2005). Evolving gene/transcript definitions significantly alter the interpretation of GeneChip data. *Nucleic Acids Res* **33**, e175.
4. Liu, H, Zeeberg, BR, Qu, G, et al. (2007). AffyProbeMiner: a web resource for computing or retrieving accurately redefined Affymetrix probe sets. *Bioinformatics* **23**, 2385–2390.
5. Hubbell, E, Liu, WM, Mei, R. Guide to Probe Logarithmic Intensity Error (PLIER) Estimation. http://www.affymetrix.com/support/technical/technotes/plier_techno te.pdf
6. Choe, SE, Boutros, M, Michelson, AM. (2005). Preferred analysis methods for Affymetrix GeneChips revealed by a wholly defined control dataset. *Genome Biol* **6,** R16.
7. Seo, J, Hoffman, EP. (2006). Probe set algorithms: is there a rational best bet? *BMC Bioinformatics* 7, 395.
8. McClintick, JN, Edenberg, HJ. (2006). Effects of filtering by Present call on analysis of microarray experiments. *BMC Bioinformatics* 7, 49.
9. Pepper, SD, Saunders, EK, Edwards, LE, et al. (2007). The utility of MAS5 expression summary and detection call algorithms. *BMC Bioinformatics* **8**, 273.
10. Benjamini, Y, Hochberg, Y. (1995). Controlling the false discovery rate: a practical and powerful approach to multiple testing. *J R Stat Soc Series* **57**, 289–300.
11. Dennis, G, Jr., Sherman, BT, Hosack, J, et al. (2003). DAVID: database for Annotation, Visualization, and Integrated Discovery. *Genome Biol* **4,** P3.
12. Tian, L, Greenberg, SA, Kong, SW, et al. (2005). Discovering statistically significant pathways in expression profiling studies. *Proc Natl Acad Sci USA* **102**, 13544–13549.
13. Nam, D, Kim, SY. (2008). Gene-set approach for expression pattern analysis. *Brief Bioinform* **9**, 189–197.
14. Raghavan, N, De Bondt, AM, Talloen, W, et al. (2007). The high-level similarity of some disparate gene expression measures. *Bioinformatics* **23**, 3032–3038.
15. Mootha, VK, Handschin, C, Arlow, D, et al. (2004). Erralpha and Gabpa/b specify PGC-1alpha-dependent oxidative phosphorylation gene expression that is altered in diabetic muscle. *Proc Natl Acad Sci USA* **101**, 6570–6575.
16. Baitaluk, M, Qian, X, Godbole, S, et al. (2006). PathSys: integrating molecular interaction graphs for systems biology. *BMC Bioinformatics* 7, 55.
17. Draghici, S, Khatri, P, Tarca, AL, et al. (2007). A systems biology approach for pathway level analysis. *Genome Res* **17**, 1537–1545.
18. Irizarry, RA, Bolstad, BM, Collin, F, et al. (2003). Summaries of Affymetrix GeneChip probe level data. *Nucleic Acids Res* **31**, e15.
19. Wu, Z, Irizarry, RA. (2005). Stochastic models inspired by hybridization theory for short oligonucleotide arrays. *J Comput Biol* **12**, 882–893.
20. van der Laan, M, Dudoit, S, Pollard, K. (2003). Hybrid clustering of gene expression data with visualization and bootstrap. *J Stat Plan Inference* **117**,275–303.
21. Reich, M, Liefeld, T, Gould, J, et al. (2006). GenePattern 2.0. *Nat Genet* **38**, 500–501.
22. Saeed, AI, Sharov, V, White, J, et al. (2003). TM4: a free, open-source system for

microarray data management and analysis. *Biotechniques* **34**,374–378.

23. Li, C, Wong, WH. (2001). Model-based analysis of oligonucleotide arrays: model validation, design issues and standard error application. *Genome Biol* 2(8), 0032.1–0032.11.
24. Li, C, Wong, WH. (2001). Model-based analysis of oligonucleotide arrays: expression index computation and outlier detection. *Proc Natl Acad Sci USA* **98**, 31–36.
25. Tusher, VG, Tibshirani, R, Chu, G. (2001). Significance analysis of microarrays applied to the ionizing radiation response. Proc Natl *Acad Sci USA* **98**, 5116–5121.
26. Manduchi, E, Grant, GR, McKenzie, SE, et al. (2000). Generation of patterns from gene expression data by assigning confidence to differentially expressed genes. *Bioinformatics* **16**, 685–698.
27. Mootha, VK, Lindgren, CM, Eriksson, KF, et al. (2003). PGC-1alpha-responsive genes involved in oxidative phosphorylation are coordinately downregulated in human diabetes. *Nat Genet* **34**, 267–273.
28. Subramanian, A, Kuehn, H, Gould, J, et al. (2007). GSEA-P: a desktop application for Gene Set Enrichment Analysis. *Bioinformatics* **23**, 3251–3253.
29. Backes, C, Keller, A, Kuentzer, J, et al. (2007). GeneTrail–advanced gene set enrichment analysis. *Nucleic Acids Res* **35**, W186–192.
30. Dahlquist, KD, Salomonis, N, Vranizan, K, et al. (2002). GenMAPP, a new tool for viewing and analyzing microarray data on biological pathways. *Nat Genet* **31**, 19–20.
31. Gautier, L, Cope, L, Bolstad, BM, et al. (2004). Affy–analysis of Affymetrix GeneChip data at the probe level. *Bioinformatics* **20**, 307–315.
32. Smyth, GK (2004). Linear models and empirical bayes methods for assessing differential expression in microarray experiments. *Stat Appl Genet Mol Biol* **3(1)**, Article 3.

Chapter 16

eQTL Analysis in Mice and Rats

Bruno M. Tesson and Ritsert C. Jansen

Abstract

Since the introduction of genetical genomics in 2001, many studies have been published on various organisms, including mouse and rat. Genetical genomics makes use of the latest microarray profiling technologies and combines vast amounts of genotype and gene expression information, a strategy that has proven very successful in inbred line crosses. The data are analyzed using standard tools for linkage analysis to map the genetic determinants of gene expression variation. Typically, studies have singled out hundreds of genomic loci regulating the expression of nearby and distant genes (called local and distant expression quantitative trait loci, respectively; eQTLs). In this chapter, we provide a step-by-step guide to performing genome-wide linkage analysis in an eQTL mapping experiment by using the R statistical software framework.

Key words: eQTL, genetical genomics, mice, rats, microarray, ANOVA.

1. Introduction

A genetical genomics (1) study involves the perturbation of thousands of genes at the same time through genetic mechanisms of recombination and segregation to create genome-wide "mosaics" of naturally occurring gene variants. Genetical genomics experiments then correlate gene expression variation with DNA variation for tens of thousands of genes, performing tens of thousands times an analysis similar to traditional QTL analysis of a classical phenotypic trait. The analysis of variance (ANOVA) methods offer a framework well suited for such QTL analyses.

Over the past few years, a large number of mouse recombinant inbred populations (RILs, e.g., the BXD or BXA panels) and tissues have been studied in eQTL screens (2–9). The field is now expanding with the study of outbred mice (10). Many of

K. DiPetrillo (ed.), *Cardiovascular Genomics*, Methods in Molecular Biology 573,
DOI 10.1007/978-1-60761-247-6_16,

these data have been uploaded to the GeneNetwork database (11), which have made this the central repository for mouse and rat eQTL data. While eQTL publications on rats have been scarcer, there have been a few studies, for example, using the BXH/HXB panel of recombinant inbred strains (12, 13).

This chapter provides a computational protocol (**Fig. 16.1**) for eQTL analysis on RIL crosses in mice and rats. The protocol can easily be adapted to suit other genetic populations, such as backcrosses or intercrosses (14).

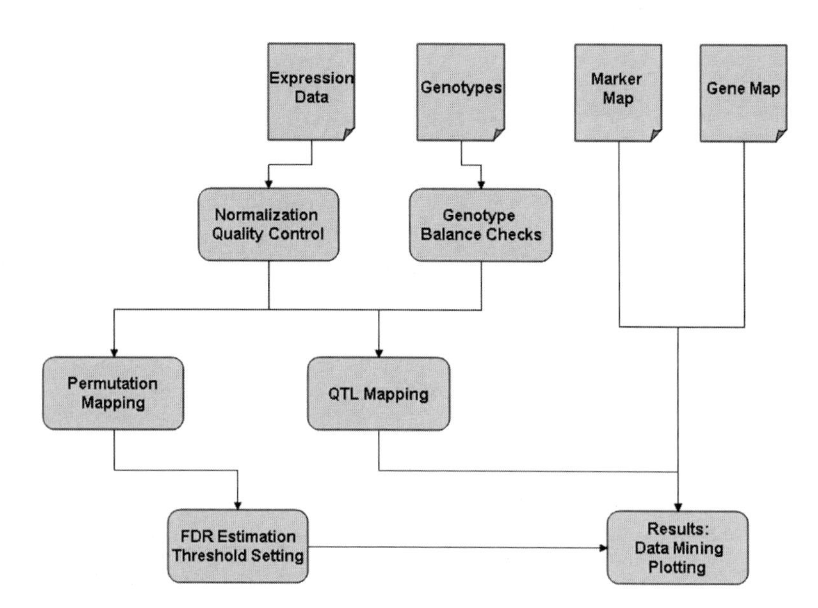

Fig. 16.1. Flowchart of eQTL mapping protocol.

2. Materials

2.1. Hardware and Software Requirements

The protocol requires the following:

- R (www.r-project.org): R is a programming environment for statistical computing and graphics. It is available under the GNU General Public License on Windows, Linux/Unix and Mac systems. R has a command line-based interface and is widely used in the field of biostatistics thanks to the availability of multiple add-on packages designed to address specific biological analyses. All the code lines and functions presented in `courier font` are written in R language. Detailed knowledge of R programming is not required but the interested reader can go to the R tutorial: http://cran.r-project.org/doc/manuals/R-intro.pdf.

- CPU/memory requirements: this protocol is illustrated with a sample dataset of 100 genes, so that the protocol will run well on a regular desktop computer. For a real genome-wide experiment, you are strongly advised to use multiple-core machines (*see* **Note 1** on parallel computation).

2.2. Dataset

The methods we describe here are showcased on Illumina Bead Array data. Illumina is an increasingly popular technology for gene expression profiling and uses arrays containing multiple beads with 50-mer probes attached. Illumina bead arrays have been developed for a number of species including humans, mice, and rats. The protocol described in this chapter is not, however, specific to Illumina data and can be applied to virtually any technology. Some adjustments will need to be made in the particular case of Affymetrix arrays (*see* **Note 2**) due to specificities of this technology (i.e., multiple probes per gene).

For this protocol, we use a small sample dataset of 100 expression traits for efficiency purposes. This dataset was extracted from a survey of hematopoietic stem cells in a population of 24 mouse recombinant inbred strains (BXD). This dataset and an electronic version of the code presented in this chapter are available at the following URL: http://gbic.biol.rug.nl/supplementary/2008/linkageGG.

2.2.1. Genotype Data

The genotype data should be prepared as a tab-delimited file: each column represents one individual, each row a different marker (*see* **Table 16.1**). Values are either 1 for the first parental strain, 2 for the second parental strain, or 1.5 in the case of heterozygote individuals (these should be rare in the case of RILs). The markers are ordered by genomic location as in the marker map file (**Section 2.2.3**).

Table 16.1
Example of genotype data in tab-delimited format. The columns represent the different recombinant inbred lines (here from the BXD cross). The rows are different markers. RILS are homozygous 1 for the B6 allele or homozygous 2 for the DBA2 allele

	BXD6	BXD28	BXD19	BXD15	BXD40	BXD12	BXD31
rs6376963	2	2	2	2	1	2	1
rs6298633	2	2	2	2	1	2	1
D1Mit1	2	2	2	2	1	2	1
rs3654866	2	1	2	2	1	2	1
rs3088964	2	1	2	2	1	2	1

Genetical genomics screens can be very expensive; the costs of sample preparation, microarrays, and genotyping must be multiplied by the population size. Selective genotyping is usually not a realistic option to reduce the costs in this context because the samples with the most informative genotypes depend on which gene is being considered. We therefore assume here that the genotype data are complete as is usually the case in genome-wide eQTL studies on recombinant inbred lines. The interested reader may want to refer to **Note 3** for software to handle missing information or sparse marker maps.

2.2.2. Expression Data

BeadStudio, the standard Illumina software, produces probe data from bead-level intensities. The output of BeadStudio contains several columns per sample. In this chapter, we use the raw bead summary data as output by BeadStudio. From the BeadStudio output files, we extract the AVG_SIGNAL columns per sample. These columns contain raw-averaged bead intensities for each probe. The expression data are stored in a tab-delimited file, where each column refers to one individual and each row to a probe, as shown in **Table 16.2**.

Table 16.2
Example of raw expression data in tab-delimited format. The first column shows the unique probe IDs, the other columns refer to the samples denoted here by their RIL numbers

	BXD6	BXD28	BXD19	BXD15	BXD40
GI_84579826-I	341.668	349.7453	509.0667	495.4675	591.0002
GI_84579830-A	105.3439	113.3545	117.8497	111.6411	109.7728
GI_84579883-I	121.9119	126.5275	126.1814	132.6144	119.5611
GI_84579884-A	138.155	138.7963	158.4077	150.2157	133.7268
GI_84579905-A	189.2942	180.8074	274.7991	367.868	204.4543
GI_84662726-I	148.5721	147.1926	153.2858	145.0625	135.5658
GI_84662775-S	132.148	125.3375	139.3298	136.605	130.0435

Some authors have suggested alternative pre-processing methods for Illumina data (*see* **Note 4** for references).

2.2.3. Marker Map

We also need a genetic map with the genomic positions of the markers in the genotype file. These positions can be specified in centimorgans (cM) or in mega base pairs (as one or both present). The marker map should be a text file in tab-delimited format (*see* **Table 16.3**).

Table 16.3
Example of marker data in tab-delimited format. The first column contains marker IDs, the second column contains chromosome numbers, the third column contains centimorgan positions, and the last column base pair positions

	Marker_Chr	Marker_cM	Marker_Mb
rs6376963	1	0.895	5.008089
rs6298633	1	2.367	6.820241
D1Mit1	1	3.549	11.50072
rs3654866	1	5.797	13.69223
rs3088964	1	6.962	15.19202

2.2.4. Probe and Gene Annotation Data

We finally need a file containing all the relevant probe information, including the genes targeted by the probes and genomic positions of the probes. This file should also be in tab-delimited format (**Table 16.4**). Annotations provided by microarray manufacturers are often not complete or up-to-date. It is sometimes necessary to re-annotate the probes based on a BLAT search of probe to genome sequence. *See* **Note 5** for some tools which enable such re-annotation.

Table 16.4
Probe annotations. The first column contains the probe IDs, the second column contains chromosome numbers, the third column base pair positions and the last columns give gene information

	Probe_Chr	Probe_Mb	Gene_Symbol	Gene_Description	Gene_ID
GI_84579826-I	10	87.87682	*Gnptab*	*N*-Acetylglucosamine-1-phosphate transferase, alpha and beta subunits	432486
GI_84579830-A	12	8.560399	*Slc7a15*	Solute carrier family 7 (cationic amino acid transporter, y+ system), member 15	328059
GI_84579883-I	11	120.7646	*Slc16a3*	Solute carrier family 16 (monocarboxylic acid transporters), member 3	80879

3. Methods

3.1. Experimental Design

When profiling many samples with microarrays, it is often necessary to divide the samples into batches, which may then be profiled at different times or even dates. Attention should be paid to the random assignment of samples to batches in order to minimize the influence of confounding factors. The number of batches should be small (so that it will not take too many degrees of freedom in the analysis) and the batches should preferably be of equal size. Obviously, you should keep track of this batch organization and of any relevant information possibly associated with the profiling process. The eQTL analysis procedure can be adapted to take into account batch effects as described in **Note 6**.

Special considerations apply to sex, treatment, or environmental factors. If sex is not a factor of interest in the study, it is safer to limit the experiment to males or females only. Alternatively, if individuals of different sexes or different treatments or conditions are used, the model used for the eQTL analysis should account for these as additional factors (*see* **Section 3.6.3**). Strategies have been developed to optimize the power of the eQTL study with multiple conditions (e.g., *see* (15)).

The population size is obviously a critical choice. While more is always better in terms of statistical power, you must find the right balance between the costs incurred by microarray screens and the number of degrees of freedom necessary to fit the models you are using in your study. The relationship between population size and power in classical QTL analysis is discussed and illustrated in (16).

3.2. Loading the Data into R

The following commands will import the data into the R workspace:

```
>rawExpr <- as.matrix(read.csv(file="raw_data.txt",row.names=1,
   header=TRUE,sep="\t"))
>genotypes <-as.matrix(read.csv(file="genotypes.txt",row.names=1,
   header=TRUE,sep="\t"))
>markerMap<- as.matrix(read.csv(file="markerMap.txt",row.names=1,
   header=TRUE,sep="\t"))
>geneMap<- as.matrix(read.csv(file="geneMap.txt",row.names=1,
   header=TRUE,sep="\t"))
```

We convert the data to logarithmic scale with the following:

```
>log2expr <- log2(rawExpr);
```

3.3. Useful Checks on the Data

3.3.1. Clustering of the Expression Data

A rapid clustering of the samples can detect major correlation structure such as those caused by batch effects (**Fig. 16.2**, *see* **Note 6** on how to deal with these artifacts).

```
>sample_clustering <- hclust(dist(t(log2expr)));
>plot(sample_clustering);
```

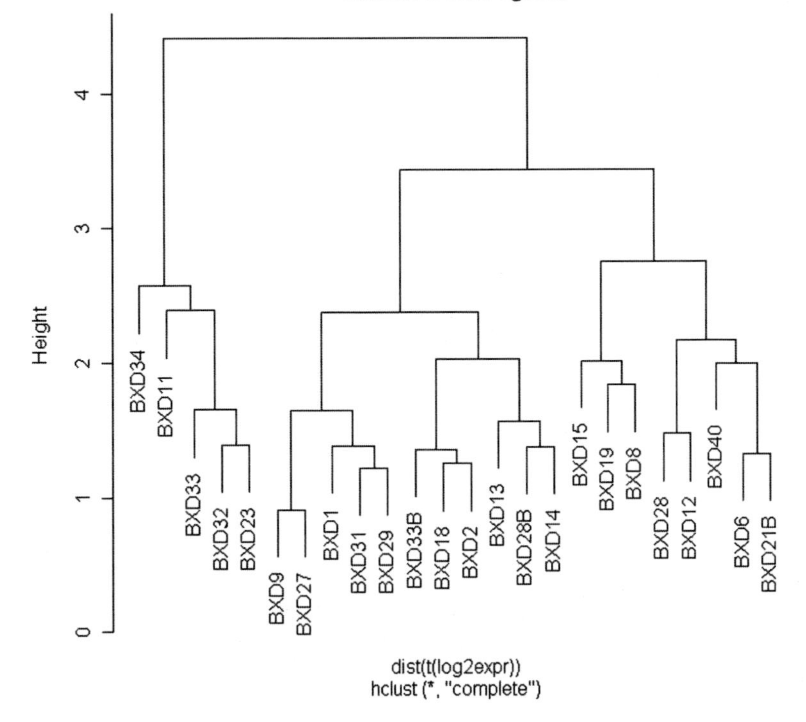

Fig. 16.2. Hierarchical clustering of samples using the raw expression data. From this plot, the samples may appear to be divided into three separate clusters. If these correspond to experimental batches, it would be wise to include them in the model (*see* **Note 6**).

At this stage, it is advisable to go back to the information collected during the wet lab process (*see* **Section 3.1**) to try to match that information with possible clusters.

3.3.2. Genotype Imbalance

It may happen that one of the parental genotypes is very poorly represented at some markers, especially with a small population size. Such imbalance may be caused at random or by segregation distortion, and can lead to an acute sensitivity to outliers; it should therefore be watched carefully. **Fig. 16.3** illustrates the genotype distribution across the markers. The code for plotting the genotypes diagnosis graph is as follows:

```
#The following vector contains all the chromosome lengths
#in Mb for plotting purposes
>chr.lengths<-c(198,182,160,156,152,150,146,133,125,130,
                   122,121,121,124,104,99,96,91,62,166);
>names(chr.lengths)<- c("1","2","3","4","5","6","7","8",
                             "9","10","11","12","13","14",
                             "15","16","17","18","19","X");
#Check of segregation distortion
>plotGenotypeBalance <- function(genotypes,markerMap,chr.lengths)
{
  op <- par()
  mk_pos<-as.numeric(markerMap [ ,"Marker_Mb"] ) +
     diffinv(chr.lengths)[ match(markerMap [ ,"Marker_Chr"] ,names(chr.lengths))] ;
```

```
breaks<- c(0,
      apply(cbind(mk_pos [ 1:length(mk_pos)-1] ,mk_pos [ 2:length(mk_pos)] ),1,mean),
      mk_pos[ length(mk_pos)] )
pos<-rep(mk_pos,ncol(genotypes))
geno_fac<-factor(c(as.numeric(genotypes)))
par(fig=c(0.001, 0.999, 0.28, 1),mai=c(0,0,1,0))
spineplot(pos,geno_fac,breaks=breaks,xaxlabels='',
          border=NA,col=c("white","black","grey"),
          yaxlabels='',main="Genotype balance")
chr_col<-c("GREY","WHITE")[ match(markerMap [ ,"Marker_Chr"] ,names(chr.lengths))
   %%2+1]
par(new=T, fig=c(0.001, 0.999, 0.1, 0.28),mai=c(1,0,0,0))
spineplot(mk_pos,factor(chr_col),breaks=breaks,
          border=NA,xaxlabels="",yaxlabels='',
          ylab="Chr",xlab="Marker Positions")
par(op)
}
>plotGenotypeBalance(genotypes,markerMap,chr.lengths)
```

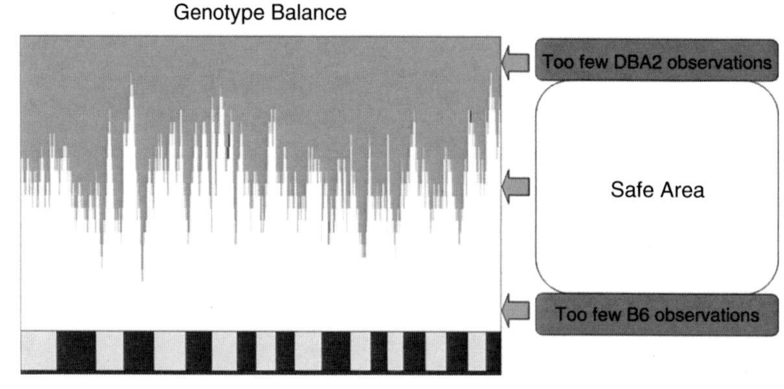

Fig. 16.3. Genotype balance plot. This plot represents the proportions of individuals with either genotype: *white* and *grey* denote the two parental genotypes; heterozygotes appear in *black* in the *middle*.

If this "information content" plot reveals a region with such imbalance, QTLs mapped in this region should be carefully scrutinized, since the minor genotype group will be extremely sensitive to outlier samples. The superimposition of the information content on the QTL profiles can provide additional insight into local variations in the statistical power available to detect eQTLs: the regions with the best power being those where the genotypes are perfectly balanced (50% for both parental genotypes).

3.4. Normalization of the Expression Data

Microarray data of multiple samples need to be normalized (i.e., converted to the same scale) to allow them to be compared across samples. Normalization removes some of the between-array technical variation. A robust, simple, and efficient method is quantile normalization (17), which is widely used and has been shown to be one of the most appropriate methods in the context of eQTL mapping (18). Quantile normalization orders intensities per sample and then replaces the intensity by the mean of the measurement at that rank in all the samples. An implementation of this normalization method is available in the Bioconductor Affy package (19). Our sample dataset has already been normalized, and should therefore not be re-normalized.

The commands to normalize a complete microarray dataset are as follows:

Installing and loading Affy R library:

```
>source("http://bioconductor.org/biocLite.R");
>biocLite("affy");
>library(affy);
```

These are the commands that apply to quantile normalization:

```
>normExpr <- normalize.quantiles(log2expr);
>dimnames(normExpr)<-dimnames(rawExpr);
```

It is advisable to perform similar checks to those described in **Section 3.3.1** on the normalized data to control how the normalization procedure affects the data structure.

```
#Our sample dataset was extracted from a complete dataset which was
   already normalized.
> normExpr<-log2expr;
```

3.5. Mapping

3.5.1. Definition of the Model

The first step of the actual eQTL analysis is to define the relevant model to use. In the simplest case, namely single-marker mapping without batches and without different environments, the model only includes the genotype effect:

$$\mathbf{Y_i = m_i + G_j + e_{ij}}$$

where Y_i is the expression measurement for probe i, m_i is the mean intensity of probe i over all samples, G_j is a factor containing the genotypes at marker j, and e_{ij} is the error term.

3.5.2. Fitting the Model

The following commands are used to first fit the model using the `lm()` function and then retrieve significances (p-values) at each marker along the genome. It is assumed that the order of the columns (samples) in the `normExpr` matrix matches the order of the columns in the `genotypes` matrix.

```
#Single Marker Mapping function
>singleMarkerMapping<-function(traits,genotypes)
{
  qtl_profiles <- NULL;
  for (i in 1:nrow(traits))
  {
    current_profile<-NULL;
    for (j in 1:nrow(genotypes))
    {
      model <- traits [i,] ~ genotypes [j,];
      anova_table<-anova(lm(model));
      current_profile<-c(current_profile,
                                   -log10(anova_table [1, 5]));
    }
    qtl_profiles<-rbind(qtl_profiles,current_profile);
  }
  rownames(qtl_profiles) <- rownames(traits);
```

```
  colnames(qtl_profiles) <- rownames(genotypes);
  qtl_profiles;
}
```

Then applying the single-marker mapping function to the expression values and the genotypes:

```
>qtlProfiles<-singleMarkerMapping(
     traits = normExpr, genotypes = genotypes);
```

Warning: this step can be computationally very intensive (*see* **Note 1**).

3.5.3. Processing and Visualizing the Results

Using this single-marker mapping approach, we obtain *p*-values for linkage for each gene with each of the markers on our genetic map. The *p*-value distribution across the genome for a given gene is termed the QTL profile of that gene and can be plotted as shown in **Fig. 16.4** using the following function:

```
>plotQTLProfile<-
function(qtl_profile,markerMap,chr.lengths)
{
  chrStrips<-seq(0,0,length=sum(chr.lengths))
  for(i in 2*0:as.integer((length(chr.lengths)-1)/2)+1)
  {
    for (j in
        (diffinv(chr.lengths)[ i] :diffinv(chr.lengths)[ i+1] ))
    {
      chrStrips[ j] <-1;
    }
  }
 plot(chrStrips,type=' h' ,col="#ECECEC",xlab=' ' ,
      ylab=' ' ,axes=F,ylim=c(0,1));
 par(new=TRUE);
 marker_x_positions <-
   as.numeric(markerMap [ ,"Marker_Mb"] ) +
   diffinv(chr.lengths)[
        match(markerMap [ ,"Marker_Chr"] ,names(chr.lengths))
                      ];
 plot(
        y=qtl_profile,x=marker_x_positions,
        xlim=c(0,sum(chr.lengths)),
        ylim=c(0,max(max(qtl_profile)+1,6)),
        type=' l' ,xlab="Marker Position", ylab="-log(p)"
        );
}
```

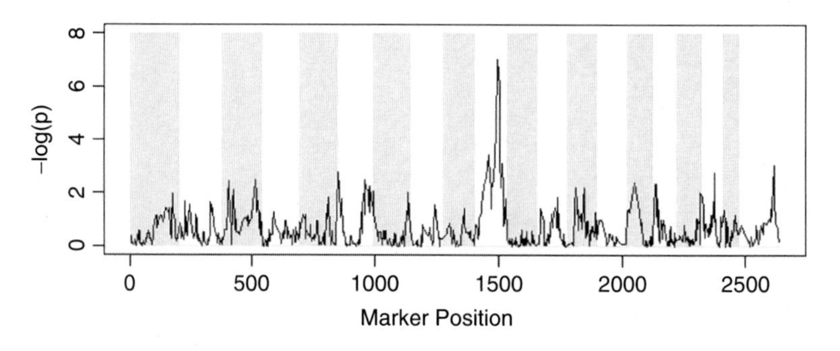

Fig. 16.4. Example of a QTL profile plot. This profile shows a QTL peak for the *Gnptab* gene on chromosome 10.

We use this function to plot the QTL profile of our first probe, which targets the *Gnptab* gene.

```
>plotQTLProfile(qtlProfiles[ 1,] ,markerMap,chr.lengths)
```

Here we set the significance threshold for detection of an eQTL to log10(p-value) >6 (*see* **Section 3.7** on how to set significance thresholds). The following function extracts primary QTL peaks from the QTL profiles:

```
>getQTLMaxPeaks  <-  function(qtl_profiles,threshold)
{
  max_index  <-  function(v)
  {
   which(v  ==  max(v,na.rm=T))  [ 1] ;
  }
  maxQTLs  <-  cbind(
                 rownames(qtl_profiles),
                 colnames(qtl_profiles)[
                 apply(qtl_profiles,1,max_index)] ,
                 apply(qtl_profiles,1,max,na.rm=T)) ;
  maxQTLsThreshold  <-
  matrix(
       maxQTLs [ which(as.numeric(maxQTLs [ ,3] )>=threshold),] ,
       ncol=3) ;
  colnames(maxQTLsThreshold)  <-  c("Probe","Marker","p") ;
  maxQTLsThreshold;
}
>QTLPeaksThresh3  <-
        getQTLMaxPeaks(qtlProfiles,threshold=3)
```

We now have a list of all the significant primary eQTLs, reported in triplets containing the probe, the marker with the smallest linkage p-value, and the largest "minus log p-value" of that linkage. It is sometimes useful to report a confidence interval too. Two approaches are commonly employed: bootstrap and 1-lod-score drop off (20, 21).

We can now generate an eQTL dot plot (**Fig.16.5**), which provides an informative summary of the mapping results.

```
>qtlDotPlot  <-
function(QTLPeaks,markerMap,geneMap,chr.lengths)
{
 chrStrips  <-  seq(0,0,length=sum(chr.lengths))
 for(i  in  2*0:as.integer((length(chr.lengths)-1)/2)+1)
 {
  for  (j  in
  (diffinv(chr.lengths)[ i] :diffinv(chr.lengths)[ i+1] ))
  {
   chrStrips [ j ]  <-  1;
  }
 }
 plot(chrStrips,type=' h' ,col="#ECECEC",xlab='' ,
       ylab='' ,axes=F,ylim=c(0,1)) ;
 par(new=TRUE) ;
 QTL_Positions  <-
  as.numeric(
       markerMap [ QTLPeaks [ ,"Marker"] ,"Marker_Mb"] )  +
  diffinv(chr.lengths)[
    match(
          markerMap [ QTLPeaks [ ,"Marker"] ,"Marker_Chr"] ,
          names(chr.lengths)
                 )] ;
 Gene_Positions  <-
```

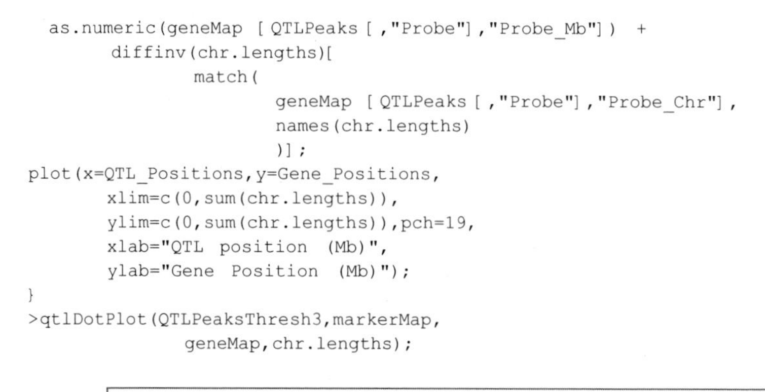

```
    as.numeric(geneMap [ QTLPeaks [ ,"Probe"] ,"Probe_Mb"] )  +
          diffinv(chr.lengths)[
                  match(
                          geneMap [ QTLPeaks [ ,"Probe"] ,"Probe_Chr"] ,
                          names(chr.lengths)
                          )] ;
plot(x=QTL_Positions,y=Gene_Positions,
        xlim=c(0,sum(chr.lengths)),
        ylim=c(0,sum(chr.lengths)),pch=19,
        xlab="QTL  position  (Mb)",
        ylab="Gene  Position  (Mb)");
}
>qtlDotPlot(QTLPeaksThresh3,markerMap,
               geneMap,chr.lengths);
```

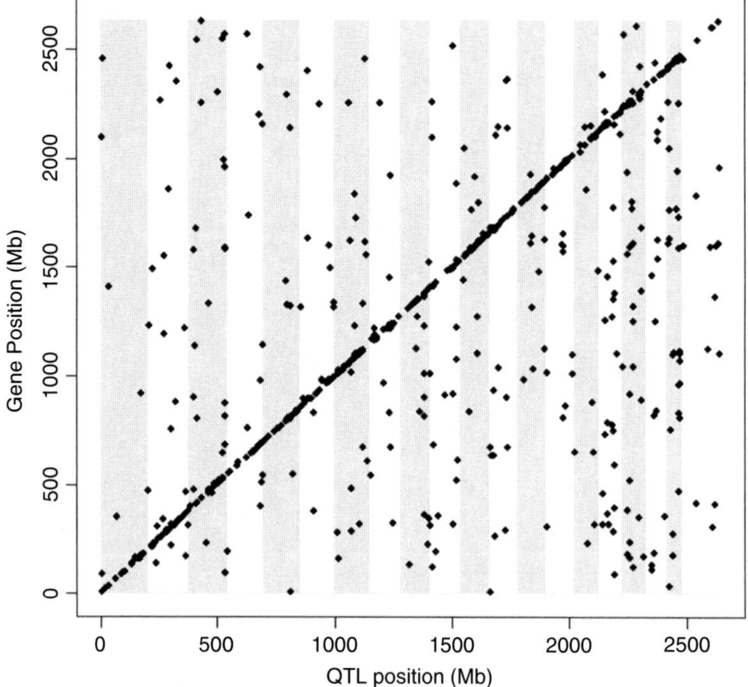

Fig. 16.5. An example of an eQTL dot plot. Each *dot* represents a significant eQTL, with the gene position on the *Y*-axis and the QTL position on the *X*-axis. (This plot was obtained using the complete dataset, not just the 100 probes subset we are using as a sample for this chapter).

Locally acting eQTLs appear on the diagonal and are often over-represented. You can also sometimes see the presence of vertical bands (typically between 0 and 8), which have been suggested to reflect the presence of regulation hotspots: a distant eQTL controlling or regulating many genes (22).

3.6. More Elaborate Models

3.6.1. Multiple QTL Mapping

To potentially improve statistical power for eQTL detection, it can be worthwhile fitting multiple QTL models, for example, by a stepwise procedure: correct for the first (most significant) eQTL effect found, and then map the corrected data to detect a second eQTL. The two-eQTL model is as follows:

$$\mathbf{Y_i = m_i + G_k + G_j + e_{ij}}$$

where G_k is the genotype vector at the first QTL position.

In this code example, we look for secondary eQTLs for the probes, for which a primary QTL has already been identified in **Section 3.5**: we use the list maxQTLPeaksThresh3.

```
>secondaryMarkerMapping <-
function(traits,genotypes,primaryQTLs)
{
 if (paste(rownames(traits),collapse='')!=
       paste(primaryQTLs [,"Probe"],collapse=''))
 {
       print("Error: Traits submitted do not match traits
              with primary eQTLs.");
       return;
 }
 qtl_profiles <- NULL;
 for (i in 1:nrow(traits))
 {
  current_profile <- NULL;
  for (j in 1:nrow(genotypes))
  {
  model <- traits [i,] ~
     genotypes [primaryQTLs [i,"Marker"],]+ genotypes [j,];
  anova_table <- anova(lm(model));
  current_profile <- c(current_profile,
     -log10(anova_table [2,5]));
  }
  qtl_profiles <- rbind(qtl_profiles,current_profile);
  }
  rownames(qtl_profiles) <- rownames(traits);
  colnames(qtl_profiles) <- rownames(genotypes);
  qtl_profiles
 }
 >secondaryQTLProfiles <- secondaryMarkerMapping(
                          normExpr [QTLPeaksThresh3 [,"Probe"],],
                          genotypes,QTLPeaksThresh3);
 >secondaryQTLPeaksThresh3 <-
        getQTLMaxPeaks(secondaryQTLProfiles,threshold=3);
```

Using the function defined in **Section 3.5** to extract QTL peaks, we can now create a list of the significant secondary eQTLs for a given threshold:

```
>secondaryQTLPeaksThresh3 <-
       getQTLMaxPeaks(secondaryQTLProfiles,threshold=3);
```

It is, of course, possible to include three or more QTLs per gene by extending the model. However, you should be cautious because of over-fitting issues. This sequential way of defining the co-factors to include in the model may not be optimal, and there are a number of more advanced strategies which address the problem of model selection (*see* **Note 7**).

3.6.2. Epistasis

In the previous step we have presented how to detect multiple QTLs per gene. However, we have not tested for interaction between the eQTLs (i.e., if the effect of one eQTL is modulated by the effect of the second one). Such complex mechanisms are common in gene regulation and are termed epistasis. Our eQTL analysis model can again be extended to take such epistasis effects into account.

$$\mathbf{Y_i = m_i + G_k + G_j + G_k{}^*G_j + e_{ijk}}$$

The code below tests for epistasis between two eQTLs that we identified for the gene in **Section 3.6.1**:

```
>my_probe<-secondaryQTLPeaksThresh3 [ 1,1] ;
#Probe of the gene we will test for epistasis
>G1 <-genotypes [ QTLPeaksThresh3 [
 which(QTLPeaksThresh3 [ ,"Probe"] ==my_probe), "Marker"] ,] ;
>G2 <- genotypes [ secondaryQTLPeaksThresh3 [ 1, "Marker"] ,] ;
>model_epistasis<- normExpr [ my_probe,] ~ G1 + G2 + G1:G2;
>anova_table <- anova(lm(model_epistasis));
>interaction_p_value <- anova_table [ 3,5] ;
```

In this example, the *p*-value is insignificant and there is no evidence for epistasis.

Interactions also occur between two loci whose main effects (terms G1 and G2 in the model) may not be significant on their own. It can therefore be relevant to screen for interactions for any possible pairs of loci, but this can sometimes be computationally unrealistic (a two-dimensional genome scan leads to a huge multiple-testing problem). For more guidance on strategies for epistasis testing, *see* **Note 7**.

3.6.3. Adding Environments/Treatments

Genetical genomics studies can provide insights into the way different environments or treatments affect the regulation of gene expression. When combining the genetic perturbation naturally present in inbred populations with the effect of different environments, the study of the interaction between those two causes of variation can teach us about the plasticity of eQTLs (15, 23). We can illustrate this with the example of the study of gene expression regulation across several cell types. In the example below, expression profiles were collected from four distinct cell types. The following model can be used:

$$\mathbf{Y_i = m_i + CT + G_j + CT^*G_j + e_{ij}}$$

where CT is the cell type factor. For this example, we need to load new data files:

```
>genotypes4CT <-
      as.matrix(read.csv(file="genotypes4ct.txt",
                        sep="\t",row.names=1));
>expr4CT <-as.matrix(read.csv(file="expr4ct.txt",
                                    sep="\t",row.names=1));
#the cell types are coded as "1", "2", "3"and "4"
>CT.factor <-
    factor(c(rep(1,24),rep(2,25),rep(3,22),rep(4,25)));
```

The mapping function therefore becomes

```
>singleMarkerMappingWithEnv <-
function (traits, genotypes, env.factor)
{
  P1 <- P2 <- P3 <- P4 <- NULL;
  for (i in 1:nrow(traits))
   {
   p1 <- p2 <- p3 <- NULL;
```

```
for (j in 1:nrow(genotypes))
{
  model_environment<- traits [ i,] ~ factor(env.factor)+
       genotypes [ j,] + factor(env.factor):genotypes [ j,];
  anova_table <- anova(lm(model_environment));
  p1 <- c(p1,-log10(anova_table [[ 5]] [ 1]));
  p2 <- c(p2,-log10(anova_table [[ 5]] [ 2]));
  p3 <- c(p3,-log10(anova_table [[ 5]] [ 3]));
}
P1 <- rbind(P1,p1); # Env
P2 <- rbind(P2,p2); #qtl
P3 <- rbind(P3,p3); #qtlxEnv
}
dimnames(P1) <-
       list(rownames(traits),rownames(genotypes));
dimnames(P2) <-
       list(rownames(traits),rownames(genotypes));
dimnames(P3) <-
       list(rownames(traits),rownames(genotypes));
results<-list();
results$Profiles_Environment <- P1;
results$Profiles_QTL <- P2;
results$Profiles_QTLxEnvironment <- P3;
results;
}
```

This function outputs three *p*-values for each trait-marker pair: the first *p*-value indicates the significance level of the environment term (a low *p*-value indicates a clear overall influence of the environment on the trait; this *p*-value is *not* valid if the environment has not been randomly allocated to samples). The second *p*-value is the significance of the main genotype effect at that marker, while the third *p*-value reflects the significance of the genotype by environment interaction term.

```
>results4CT <-singleMarkerMappingWithEnv(
              expr4CT,genotypes4CT,env.factor=CT.factor);
>interactionQTLsThresh3<-getQTLMaxPeaks(
              results4CT$Profiles_QTLxEnvironment,
              threshold=3);
```

A norm of reaction plot (**Fig.16.6**) can show how the eQTL effect is modulated by the environment. It can be obtained using the following function:

```
>plotNormOfReaction<-function(trait,genotype,env.factor)
{
 env.factor <- as.numeric(env.factor);
 plottingColors <- c("black","black","lightgrey","grey");
 yrange <- range(trait) +
        c((range(trait) [ 1] - range(trait) [ 2])/5,
          (range(trait) [ 2] - range(trait) [ 1])/5);
 plot(y=trait,x=env.factor,xlim=range(env.factor),
      ylim=yrange,col=plottingColors [ 2*genotype],
      xaxt=' n' ,pch=19,xlab="Environment",
      ylab="expression for individuals");
 meanGroupValues <-
  matrix(nrow=2,ncol=length(unique(env.factor)));
 for (env in 1:length(unique(env.factor)))
 {
 meanGroupValues [ 1,env] <-
      mean(trait [ intersect(
             which(genotype == 1),
             which(env.factor == env))]);
```

```
      meanGroupValues [2,env] <-
            mean(trait [intersect(
                  which(genotype == 2),
                  which(env.factor == env))]);
      text(env,yrange [1],env,cex=1.5,col="black");
      }
      par(new=T);
      matplot(y=t(meanGroupValues),xlim=range(env.factor),
            ylim=yrange,xlab='',ylab='',xaxt='n',yaxt='n',
            type='l',lty=1,lwd=4,col=c("black","grey"));
}
>plotNormOfReaction(
  expr4CT [interactionQTLsThresh3 [1,"Probe"],],
  genotypes4CT [interactionQTLsThresh3 [1,"Marker"],],
  CT.factor);
```

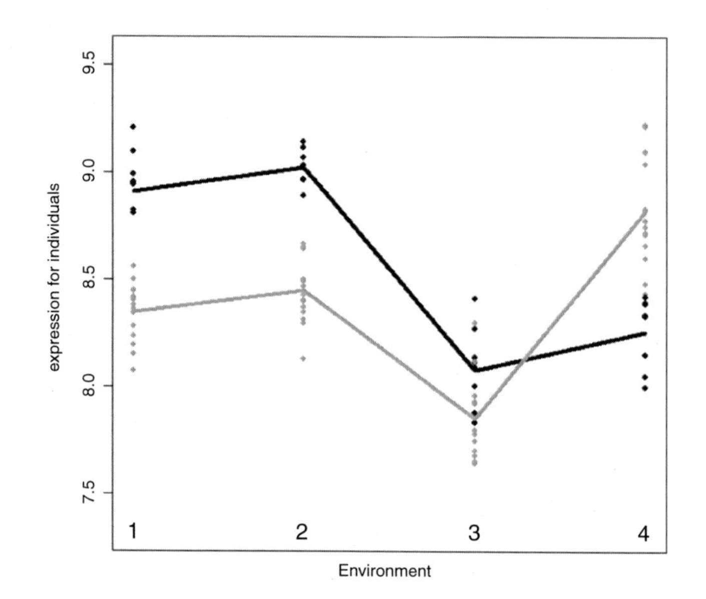

Fig. 16.6. Norm of reaction plot: eQTL by environment interaction, with the cell types on the *X*-axis and the gene expression on the *Y*-axis. Each *dot* is an individual sample measurement (*black* = B6, *grey* = DBA2). The *lines* represent mean values. The effect of the QTL is here reversed in cell type 4 compared with the other three cell types.

3.7. Determining the Significance Threshold

In eQTL analysis, the determination of the threshold for statistical significance is a critical aspect since multiple testing issues arise from both the high number of genes studied and the high number of genomic loci at which linkage is tested. The *p*-values yielded by the ANOVA must be adjusted to take into account these multiple testing issues. Bonferroni correction is somewhat too drastic here since the tests are not independent: firstly, the markers tested are intrinsically linked and thus correlated to their neighbors on each chromosome; and, secondly, large families of genes are known to be co-regulated, so there is also a correlation structure in that dimension.

A more appropriate approach is to estimate a false discovery rate (FDR) based on a permutation strategy (*24*). A carefully designed permutation procedure will make it possible to estimate the null distribution. The principle is to apply the exact same analysis protocol to permuted datasets, calculate the average number of rejected null hypotheses for a certain *p*-value threshold in those permuted datasets, and then derive an FDR estimate, at that *p*-value threshold, as the average number of rejected hypotheses in the permuted datasets divided by the number of rejected hypotheses at the same threshold in the true data.

Different permutation strategies are possible: we advise permuting only the genotypes of the individuals, while conserving trait values (gene expression measurements). This ensures that the permutation procedure does not break the internal correlation structure of the data (within markers and within genes), but that any linkage detected between a marker and gene expression in a permuted dataset is a false-positive (25).

The following function here estimates the number of false-positives obtained with a permuted dataset for a range of *p*-value thresholds in the single-marker mapping case discussed in **Section 3.5**.

```
>estimateFalsePositives <- function(traits,genotypes,threshold_range,
   nperm)
{
  permuteGenotypes <- function(geno)
  {
  geno [ ,sample(1:ncol(geno),ncol(geno),replace=F)];
  }
  Counters <- NULL
  for (i in 1:nperm)
  {
    permGenotypes <- permuteGenotypes(genotypes);
    current_profiles <-
        singleMarkerMapping(traits,permGenotypes);
    current_counters <- NULL;
    for (thresh in threshold_range)
    {
      current_counters <-
               c(
               current_counters,
               length(
               which(apply(current_profiles,1,max)>=thresh)
               ));
    }
    Counters <- rbind(counters,current_counters);
  }
  colnames(counters) <-
        apply(matrix(threshold_range),1,as.character);
  for (j in 1:nrow(counters))
  {
   rownames(counters)[ j] <-
               paste("permutation round",j);
  }
  Counters;
}
>false_positive_estimates <-
       estimateFalsePositives(normExpr,genotypes,
       threshold_range = c(3,4,5,6,7), nperm=10);
#In this example, for efficiency purpose we run only
#10 permutations round. For a reliable estimation,
#100 permutations would be a minimum.
```

We can derive an estimate of the FDR, for example, for a *p*-value threshold of 3:

```
#number of rejected null hypothesis:
positives_thresh3 <- nrow(QTLPeaksThresh3)
#number of false positives
false_positives_thresh3 <-
       mean(false_positive_estimates [,"3"])
#FDR
FDR_thresh3 <- false_positives_thresh3/positives_thresh3;
```

The result here gives a very high FDR (>50%) which means we need to use a more stringent threshold than log$p$ >3.

This code can easily be adapted to estimate the FDRs for other mapping procedures, the principle being that the permuted data should be analyzed with the same model and the same procedure as the real data. The cases of complex models for stratified data or interacting factors require adapted permutation procedures (26).

Another advantage of this permutation procedure is that it allows an unbiased estimation of the significance of the number and size of eQTL hotspots. There is some speculation that some hotspots may be the result of false-positive linkage of groups of correlated genes to random genome positions (with no regulatory connection) (27, 28). Calculating the size and the number of the hotspots obtained with permuted datasets that have retained the correlation between genes is a straightforward manner of testing the significance of hotspots (25).

Some authors have suggested using different thresholds for local and distant eQTLs: detecting local effects does not involve genome-wide testing of loci and can therefore be controlled with relaxed thresholds (29). Finally, it is important to take into account the fact that sex chromosomes have specific properties which require different thresholds for sex and autosomal chromosomes (*see* **Note 8**).

3.8. Interpretation of the Results

A typical eQTL analysis will yield hundreds or thousands of genetic linkages. Extracting meaningful biological information from the results can prove challenging.

Local eQTLs typically offer insights into possible *cis*-regulatory differences between the two alleles. Inspection of the polymorphisms in the regulatory regions of the gene can provide insight into the possible molecular mechanism (e.g., an SNP located in a transcription factor binding site located in the promoter region of the gene). Polymorphisms located within the probe target regions can also create a technically false-positive eQTL (*see* **Note 9**).

A distant eQTL indicates the presence of a distant regulator (e.g., a transcription factor or an miRNA gene) at the QTL location. This regulator may either be locally regulated or contain a non-synonymous polymorphism affecting its function. It is, however, usually difficult to directly pinpoint the regulator because of the relatively poor mapping resolution (a QTL typically spans several Mb and contains tens to hundreds of candidate genes).

Hotspots, which are large groups of genes having co-localizing eQTLs, may reveal the action of master regulators (i.e., genes controlling many others). It is possible to design strategies to reduce the large number of candidate regulators (typically hundreds) that fall into the hotspot QTL region. We present here one small sample hotspot, illustrated in **Fig. 16.7**. Sixteen genes were found to share a QTL. We investigate the possibility of a common regulator located in that QTL region. Different candidates (genes physically located within the QTL) are ranked according to their correlation with each of the hotspot genes (**Table 16.5**). Using Rank Product (30) it is then possible to prioritize the candidates.

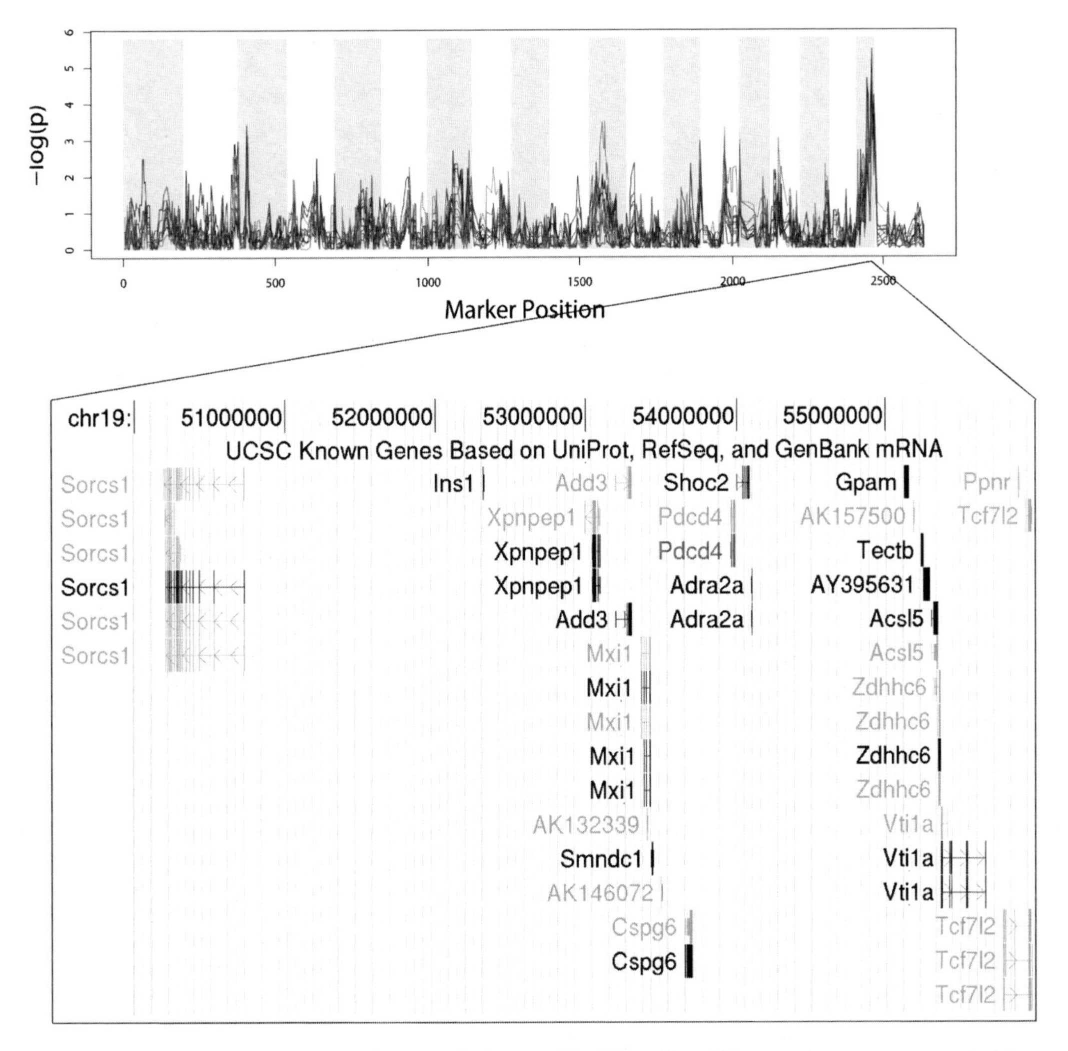

Fig. 16.7. Multiple possible candidate regulators: on the *top panel* the QTL profiles of 16 genes show a common peak. A large number of genes, illustrated by a UCSC genome browser screenshot (*lower panel*), lie within the confidence interval of that eQTL.

Table 16.5
Prioritization of candidate regulators based on Rank Product of correlation with hotspot genes. The genes with the lowest *p*-values are those that correlate best with the hotspot genes and are therefore given top priority. *Mxi1* and *Add3* are the most likely candidates according to this correlation criterion

Candidate regulators	Rank Product of correlation with hotspot genes	*p*-value
Mxi1	2.412555	<0.00001
Add3	2.641117	<0.00001
Smndc1	3.808562	0.0012
Shoc2	3.883223	0.00135
Gpam	4.179971	0.0027
Sorcs1	4.571672	0.00825
5830416P10Rik	5.282127	0.03055
Adra2a	7.849839	0.38
1700001K23Rik	9.409524	0.69665
Pdcd4	10.60708	0.8643
Gucy2g	10.74565	0.878475
Dusp5	11.54332	0.93845
Tcf7l2	11.95649	0.9578
Tectb	13.5352	0.99195
Zdhhc6	14.56703	0.99825
Ins1	15.25569	0.99965
Vti1a	15.5439	0.99975
Rbm20	16.02744	0.99995
Acsl5	16.25449	0.99995
Xpnpep1	18.07143	1

Hotspot elucidation and more generally QTL gene candidate search are data-driven research processes which integrate heterogeneous types of information (31): we have illustrated the use of correlation measurements. Other data types can include Transcription Factor Binding Site (TFBS) modules investigation, gene annotations, and databases of known protein–protein interactions (32). Methods such as the Rank Product can be used to prioritize candidates based on these criteria.

4. Notes

1. *Computational capacity issues*

   Some of the protocol steps (mapping, permutation procedure) can be computationally very intensive. If available, it is advisable to use a multi-core machine or a cluster of computers to perform these steps. The jobs can easily be separated by groups of probes, since every probe is here mapped separately. R/Parallel (33) is a useful R package, which allows R to run iterative tasks in parallel on multiple processors. Another trick that can be used to reduce the computing time is to drop redundant markers (neighboring markers with identical genotypes for all samples) at the start of the analysis.

   Memory issues can also arise when huge matrices are created within R. The amount of memory used by R is, for example, limited to 1 GB in the default windows setup of the program. This memory limit can be extended using the command `memory.limit(size_MB)`. However, the maximum memory cannot exceed the physical memory available in the computer. A possible workaround is to divide the traits into smaller entities and to write intermediary results to files.

2. *Affymetrix-related issues*

   Affymetrix arrays differ from alternative expression profiling technologies by their use of probe sets made of multiple (10–16) probes targeting one gene. While a number of studies have focused on probe set-summarized data to perform eQTL mapping, we suggest using a more subtle approach taking into account probe-level intensities. This approach has been extensively described in (34).

3. *Alternative software solutions*

   R/QTL (35) is an R package which includes many functions for mapping, including an algorithm to infer missing genotype data using Hidden Markov Models. GeneNetwork (www.genenetwork.org (11)) also offers eQTL analysis for user uploaded data, one trait at a time, and genome-wide analysis tools for a number of published datasets.

4. *Alternative Illumina data pre-processing*

   Compared with Affymetrix, for example, Illumina is a relatively new technology and standard analysis guidelines have yet to emerge. While in this chapter we illustrate our eQTL analysis with raw probe-summarized data as output by BeadStudio, there are alternative possibilities which make use of bead-level intensity data. See, for example, work by Dunning et al. (36, 37) for some methods and software.

5. *Probe (re-)annotations*

   The correct annotation of probes is a critical aspect of any microarray analysis. It is especially crucial in the case of eQTL studies since the presence of subtle differences in the probe target sequences (SNPs) between parental lines can produce technical false-positive eQTLs (*see* **Note 9**). Annotation files provided by array manufacturers tend to be incomplete and outdated and do not include genetic variation information across different strains.

   The strategy of probe re-annotation should therefore comprise a systematic BLAT search of each of the probe sequences against the latest genome assembly build, combined with a mining of polymorphism databases. This is obviously a gruesome task for which there are a few software tools available (38, 39, 40).

6. *Batches and hidden factors*

   Microarray data are known to be very sensitive to the effect of batches, which can create artificial correlation. In the context of eQTL mapping, these effects can act as confounding factors and cause multiple spurious genetic linkages, often forming apparent hotspots: if the confounding factor influences many genes (as is the case with microarray batches) and if there is, by chance, correlation of that factor with the genotypes at a certain genomic locus, then all genes will artificially map to that region, misleadingly suggesting the presence of a master regulator (34). If the confounding factor is known, it is possible to correct for its effect by adapting the mapping model. In the single-marker mapping case:

   $$\mathbf{Y_i = m_i + B + G_j + e_{ij}}$$

   where B is the batch factor.

   Caution: if multiple environments are used, it is usually required to account for the batch effects in an environment-specific fashion. In this case, an appropriate model would be

   $$\mathbf{Y_i = m_i + B + T + B^*T + G_j + G_j^*\,T + e_{ij}}$$

   The $B * T$ allows for a more careful batch effect correction: for example, if one gene was only expressed in one of the environments (in our example one tissue), then the batch effect could affect only the gene in that tissue, and should not be corrected in the samples belonging to other environments.

7. *Advanced model selection procedures*

   The selection of relevant co-factors and interaction terms in generalized models, and particularly in the context of QTL mapping, has been widely discussed in the scientific literature. Mapping of multiple QTLs and epistasis testing can be seen as model selection problems. For examples, *see* (41, 42).

8. *The case of sex chromosomes*

   While most of the Y chromosome does not undergo recombination, the recombination rate of the X chromosome is slower than that of the autosomes. This has important consequences on the detection of significant QTLs. For a comprehensive view of these issues, *see* (43).

9. *Probe hybridization artifacts*

   When several probes are available for the same gene, it is not uncommon to observe a difference in the mapping results of those probes: "the probes tell different stories" or statistically there is eQTL by probe interaction (34). This can be explained either by biological mechanisms (alternative splicing) or by technical artifacts. Such technical artifacts may arise when a polymorphism is present in the sequence targeted by a probe (44). If the probe was designed specifically based on the genome sequence of one of the parental strains, it is possible that some polymorphism causes the other genotype mRNA products to have a weaker binding affinity and thus a lower signal. Such effects will yield spurious local eQTL linkages. If the probes have been designed specifically based on the sequence of one of the two parental strains (say, strain A, and not strain B), it is possible to estimate roughly the number of local eQTLs affected by this issue. For example, if 65% of local eQTLs are linked with a higher expression of the gene for the A allele, while for the other 35% local eQTLs the B allele is more highly expressed. This contrasts with the 50–50% expected without hybridization effect. In this case, we would expect 65–35 = 30% of eQTLs to be caused by this hybridization difference rather than by a real differential expression effect.

References

1. Jansen, RC, Nap, JP. (2001) Genetical genomics: the added value from segregation. *Trends Genet* **17**(7), 388–391.
2. Schadt, EE, Monks, SA, Drake, TA, et al. (2003) Genetics of gene expression surveyed in maize, mouse and man. *Nature* **422** (6929), 297–302.
3. Bystrykh, L, Weersing, E, Dontje, B, et al. (2005) Uncovering regulatory pathways that affect hematopoietic stem cell function

using 'genetical genomics'. *Nat Genet* **37**(3), 225–232.

4. Chesler, EJ, Lu, L, Shou, S, et al. (2005) Complex trait analysis of gene expression uncovers polygenic and pleiotropic networks that modulate nervous system function. *Nat Genet* **37**(3), 233–242.
5. Mehrabian, M, Allayee, H, Stockton, J, et al. (2005) Integrating genotypic and expression data in a segregating mouse population to identify 5-lipoxygenase as a susceptibility gene for obesity and bone traits. *Nat Genet* **37**(11), 1224–1233.
6. Lan, H, Chen, M, Flowers, JB, et al. (2006) Combined expression trait correlations and expression quantitative trait locus mapping. *PLoS Genet* **2**(1), e6.
7. Wang, S, Yehya, N, Schadt, EE, et al. (2006) Genetic and genomic analysis of a fat mass trait with complex inheritance reveals marked sex specificity. *PLoS Genet* **2**(2), e15.
8. McClurg, P, Janes, J, Wu, C, et al. (2007) Genomewide association analysis in diverse inbred mice: power and population structure. *Genetics* **176**(1), 675–683.
9. Wu, C, Delano, DL, Mitro, N, et al. (2008) Gene set enrichment in eQTL data identifies novel annotations and pathway regulators. *PLoS Genet* **4**(5), e1000070.
10. Ghazalpour, A, Doss, S, Kang, H, et al. (2008) High-resolution mapping of gene expression using association in an outbred mouse stock. *PLoS Genet* **4**(8), e1000149.
11. Wang, J, Williams, RW, Manly, KF. (2003) WebQTL: web-based complex trait analysis. *Neuroinformatics* **1**(4), 299–308.
12. Hubner, N, Wallace, CA, Zimdahl, H, et al. (2005) Integrated transcriptional profiling and linkage analysis for identification of genes underlying disease. *Nat Genet* **37**(3), 243–253.
13. Petretto, E, Mangion, J, Dickens, NJ, et al. (2006) Heritability and tissue specificity of expression quantitative trait loci. *PLoS Genet* **2**(10), e172.
14. Darvasi, A, Soller, M (1995) Advanced intercross lines, an experimental population for fine genetic mapping. *Genetics* **141**(3), 1199–1207.
15. Li, Y, Breitling, R, Jansen, RC. (2008) Generalizing genetical genomics: getting added value from environmental perturbation. *Trends Genet* **24**.
16. Van Ooijen, JW. (1999) LOD significance thresholds for QTL analysis in experimental populations of diploid species. *Heredity* **83** (**Pt 5**), 613–624.
17. Bolstad, BM, Irizarry, RA, Astrand, M, et al. (2003) A comparison of normalization methods for high density oligonucleotide array data based on variance and bias. *Bioinformatics (Oxford, England)* **19**(2), 185–193.
18. Williams, RBH, Cotsapas, CJ, Cowley, MJ, et al. (2006) Normalization procedures and detection of linkage signal in genetical-genomics experiments. *Nat Genet* **38**(8), 855–856.
19. Gautier, L, Cope, L, Bolstad, BM, et al. (2004) affy – analysis of Affymetrix GeneChip data at the probe level. *Bioinformatics (Oxford, England)* **20**(3):307–315.
20. Visscher, PM, Thompson, R, Haley, CS. (1996) Confidence intervals in QTL mapping by bootstrapping. *Genetics* **143**(2), 1013–1020.
21. Lander, ES, Botstein, D. (1989) Mapping Mendelian factors underlying quantitative traits using RFLP linkage maps. *Genetics* **121**(1):185–199.
22. Darvasi, A. (2003) Genomics: gene expression meets genetics. *Nature* **422**(6929), 269–270.
23. Li, Y, Alvarez, OA, Gutteling, EW, et al. (2006) Mapping determinants of gene expression plasticity by genetical genomics in C. elegans. *PLoS Genet* **2**(12), e222.
24. Churchill, GA, Doerge, RW. (1994) Empirical threshold values for quantitative trait mapping. *Genetics* **138**(3), 963–971.
25. Breitling, R, Li, Y, Tesson, BM, et al. (2008) Genetical genomics: spotlight on QTL hotspots. *PLoS Genet* **4**(10), e1000232.
26. Churchill, GA, Doerge, RW. (2008) Naive application of permutation testing leads to inflated type I error rates. *Genetics* **178**(1), 609–610.
27. de Koning, DJ, Haley CS. (2005) Genetical genomics in humans and model organisms. *Trends Genet* **21**(7), 377–381.
28. Perez-Enciso, M. (2004) In silico study of transcriptome genetic variation in outbred populations. *Genetics* **166**(1), 547–554.
29. Doss, S, Schadt, EE, Drake, TA, et al. (2005) Cis-acting expression quantitative trait loci in mice. *Genome Res* **15**(5), 681–691.
30. Breitling, R, Armengaud, P, Amtmann, A, et al. (2004) Rank products: a simple, yet powerful, new method to detect differentially

regulated genes in replicated microarray experiments. *FEBS Lett* 573(1–3), 83–92.

31. Stylianou, IM, Affourtit, JP, Shockley, KR, et al. (2008) Applying gene expression, proteomics and single-nucleotide polymorphism analysis for complex trait gene identification. *Genetics* **178**(3), 1795–1805.
32. Zhu, J, Zhang, B, Smith, EN, et al. (2008) Integrating large-scale functional genomic data to dissect the complexity of yeast regulatory networks. *Nat Genet* **40**(7), 854–861.
33. Vera G, Jansen RC, Suppi RL. (2008) R/parallel–speeding up bioinformatics analysis with R. *BMC Bioinformatics* **9**, 390.
34. Alberts, R, Terpstra, P, Bystrykh, LV, et al. (2005) A statistical multiprobe model for analyzing cis and trans genes in genetical genomics experiments with short-oligonucleotide arrays. *Genetics* **171**(3), 1437–1439.
35. Broman, KW, Wu, H, Sen, S, et al. (2003) R/qtl: QTL mapping in experimental crosses. *Bioinformatics (Oxford, England)* **19**(7), 889–890.
36. Dunning, MJ, Smith, ML, Ritchie, ME, et al. (2007) beadarray: R classes and methods for Illumina bead-based data. *Bioinformatics (Oxford, England)* **23**(16), 2183–2184.
37. Dunning, MJ, Barbosa-Morais, NL, Lynch, AG, et al. (2008) Statistical issues in the analysis of Illumina data. *BMC Bioinformatics* **9**, 85.
38. Verdugo RA, Medrano JF. (2006) Comparison of gene coverage of mouse oligonucleotide microarray platforms. *BMC Genomics* **7**, 58.
39. Alberts R, Vera G, Jansen RC. (2008) affyGG: computational protocols for genetical genomics with Affymetrix arrays. *Bioinformatics (Oxford, England)* **24**(3), 433–434.
40. Alberts, R, Terpstra, P, Hardonk, M, et al. (2007) A verification protocol for the probe sequences of Affymetrix genome arrays reveals high probe accuracy for studies in mouse, human and rat. *BMC Bioinformatics* **8**, 132.
41. Jansen, R. (2007) Quantitative trait loci in inbred lines. In: Handbook of statistical Genetics. Balding, D, Bishop, M, Cannings, C, (eds), Vol. 1, 3rd edn. Wiley, New York, 616–617.
42. Aylor, DL, Zeng, ZB. (2008) From classical genetics to quantitative genetics to systems biology: modeling epistasis. *PLoS Genet* **4**(3), e1000029.
43. Broman, KW, Sen, S, Owens, SE, et al. (2006) The X chromosome in quantitative trait locus mapping. *Genetics* **174**(4), 2151–2158.
44. Alberts, R, Terpstra, P, Li, Y, et al. (2007) Sequence polymorphisms cause many false cis eQTLs. *PLoS ONE* **2**(7), e622.

Chapter 17

eQTL Analysis in Humans

Lude Franke and Ritsert C. Jansen

Abstract

Improving human health is a major aim of medical research, but it requires that variation between individuals be taken into account since each person carries a different combination of gene variants and is exposed to different environmental conditions, which can cause differences in susceptibility to diseases. With the advent of molecular markers in the 1980s, it became possible to genotype individuals (i.e., to detect the presence or absence of local DNA sequence variants at each of hundreds of genome positions). This DNA sequence variation could then be related to disease susceptibility by using pedigree data. Such linkage analyses proved to be difficult for more complex diseases. Recently, with the decreasing costs of genotyping, analyses of large natural populations of unrelated individuals became possible and resulted in the association of many genes (and genetic variants in these genes) with complex diseases. Unfortunately, for a considerable proportion of these genes and their proteins, it is not yet clear what their downstream effects are. Studying the expression of these genes and proteins can help to uncover the effects of these variants on the expression of these and other genes, proteins, metabolites, and phenotypes. In this chapter, we focus on the high-throughput and genome-wide measurement of gene expression in a natural population of unrelated humans, and on the subsequent association of variation in expression to "expression quantitative trait loci" (eQTLs) on DNA using oligonucleotide arrays with hundreds of thousands of single-nucleotide polymorphism (SNP) markers that capture most of the human genetic variation well. This strategy has been successfully applied to several diseases such as celiac disease (Hunt et al. 2008, *Nat Genet* 40, 395–402) and asthma (Moffatt et al. 2007, *Nature* 448, 470–473): associated genetic variants have been identified that affect levels of gene expression in *cis* or in *trans*, providing insight into the biological pathways affected by these diseases.

Key words: Quantitative trait locus, eQTL, SNP, genetical genomics, human, oligonucleotide array, parametric test, non-parametric test.

1. Introduction

Recent "genetical genomics" studies of human gene expression (1–6) mostly rely on association analysis in natural populations of unrelated individuals. One reason has been that today's oligonucleotide arrays now offer the large number of markers (>100,000)

K. DiPetrillo (ed.), *Cardiovascular Genomics*, Methods in Molecular Biology 573,
DOI 10.1007/978-1-60761-247-6_17, © Humana Press, a part of Springer Science+Business Media, LLC 2009

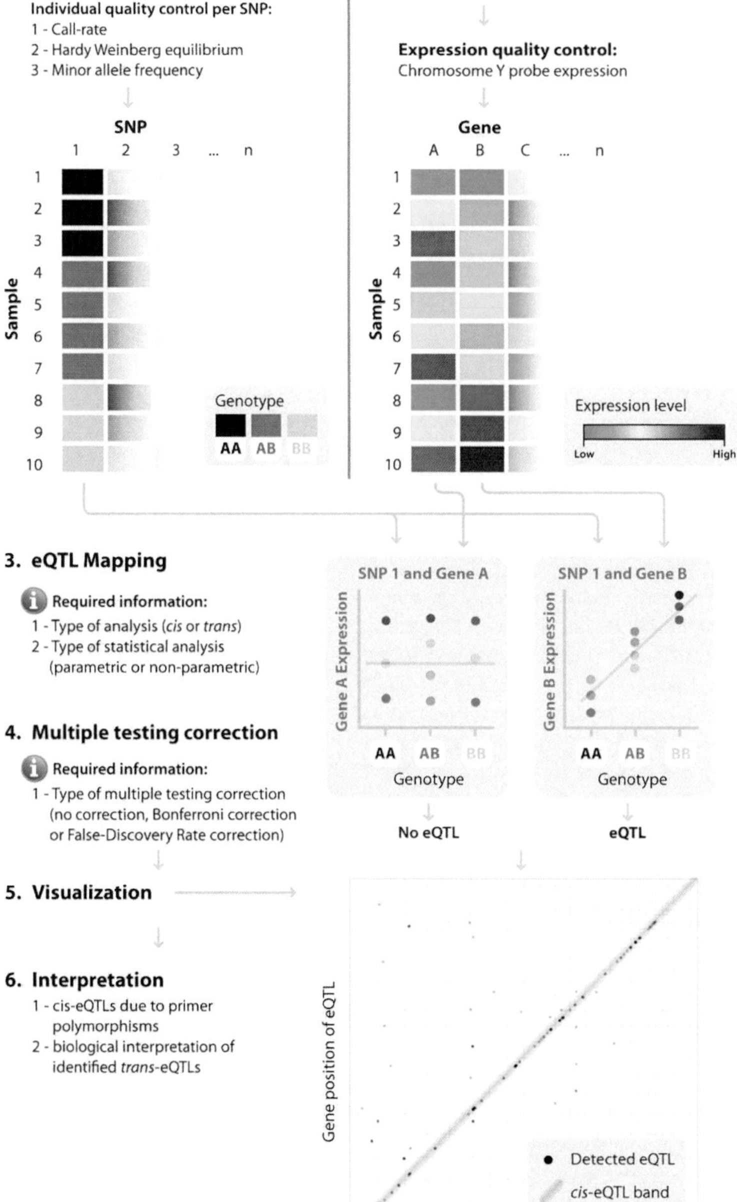

Fig. 17.1. Workflow for association analysis of genetical genomics data.

needed for association analyses. Numerous cohorts have already been genotyped using these arrays for other reasons, which limits the costs of a genetical genomics study considerably. Another reason is that the genotyping and the expression arrays are now cheap enough to afford to study the large number of individuals required for association analyses. *See* **Note 1** for a comparison of association analysis in natural populations to linkage analysis in pedigrees.

In this chapter, we provide a step-by-step protocol for association analysis of genetical genomics data (**Fig. 17.1**). We illustrate the protocol using Illumina-based genotype and RNA expression data freely available from the HapMap (*7*)/ GENEVAR project (*5, 6*). The population consists of 90 Caucasian, 90 Yoruban, 45 Chinese, and 45 Japanese individuals. The expression data from these individuals have been derived from EBV-transformed B cells. Throughout this chapter we confine the analysis to a sample set of 47 Caucasian individuals.

Box 1. Differences Between Linkage and Association Analysis

Generally, genetic studies can be performed either through linkage or through association analysis. Linkage analysis can detect rare variants and is insensitive to allelic heterogeneity, but generally results in the detection of linked loci that can be large (usually between 50 and 500 genes reside within such loci). Association analysis generally has higher statistical power to detect common variants and typically results in associated loci that contain only a few genes, but has little power to detect rare variants, especially when allelic heterogeneity is present. Another aspect to consider is cost as genome-wide linkage analysis is usually performed using fewer than 1,000 markers, while genome-wide association analysis requires genotyping of hundreds of thousands of markers in order to capture most of the genetic variation.

2. Materials

2.1. Hardware and Software

The software for the association analysis of genetical genomics data ("associationGG") can be freely downloaded from http://gbic.biol.rug.nl/supplementary/2008/associationGG/data/associati

ongg.zip and is written in Java. Please ensure that Java 1.5 or higher has been installed (freely available from http://java.sun.com). The example analyses require less than 1 GB of free hard drive space and 1 GB of available system memory. We assume a basic knowledge of using the command line interface in Windows, Mac OSX, Unix, or Linux.

2.2. Required Types of Genotype Data

The data consist of (I) SNP information, (II) genotype data at these SNPs in the human samples, and (III) additional information about these human samples (**Fig. 17.1**).

2.3. SNP Information

The physical position on the genome of SNPs should be provided in the following tab-delimited format:

Chromosome	Chromosomal position	SNP name
1	89456346	rs123545
3	45494732	rs689873

The SNP information file should be named *SNP Information.txt.*

Manufacturers of genome-wide oligonucleotide DNA arrays (such as Illumina and Affymetrix) provide mapping information for these SNPs that is continuously being updated and reflects the most current genome assembly. Please ensure that you have downloaded this information and formatted the data in the tab-delimited format described here.

2.4. Genotype Data

Genotype data should be in the following tab-delimited format:

SNP name	Sample ID	Allele1	Allele2
rs123545	Sample1	A	C
rs689873	Sample1	C	G
rs123545	Sample2	C	C
rs689873	Sample2	G	G

Illumina's BeadStation software can produce genotype data in this format through its "Report Wizard" (this can be found in the "Analysis" menu → "Reports" → "Report Wizard ..."). You should generate a "full" tab-delimited report and export the "SNP Name," "Sample ID," "Allele1," and "Allele2" columns.

The genotype information file should be named "*GenotypeData.txt.*"

2.5. Sample Information

Additional sample data should be in the following tab-delimited format:

Sample ID	Sex
Sample1	Female
Sample2	Male

The "Sample ID" reflects the samples that have been included. "Sex" reflects the sex status of each individual. The phenotype information file should be named *PhenotypeInformation.txt*.

2.6. Required Probe Information and Expression Data

Expression data (*see* **Fig. 17.2**) should be in the following tab-delimited format:

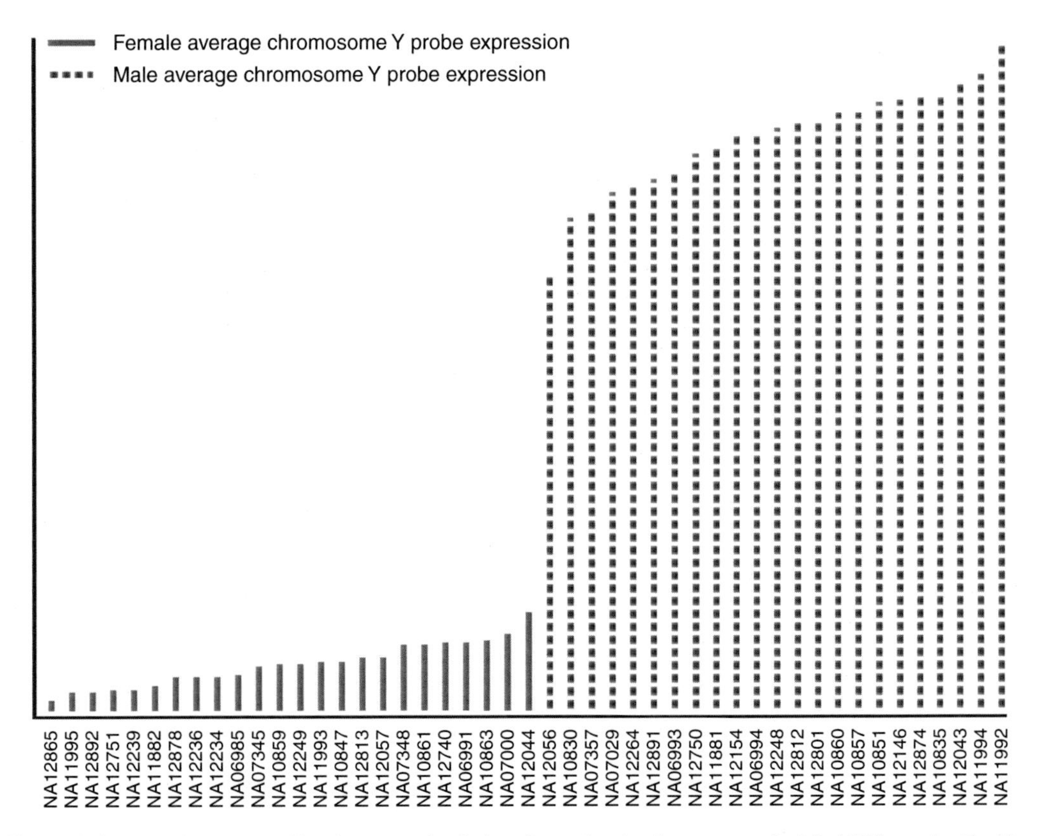

Fig. 17.2. Average chromosome Y probe expression in females and males for our example data (47 Caucasian HapMap samples, expression data from Epstein–Barr Virus-transformed B cells). As expected, males show a higher average measured expression for probes that map to chromosome Y than females.

Probe identifier	Chromosome	Chromosome start (bp)	Chromosome end (bp)	HGNC name	Sample1	Sample2
GI_14149786	1	900147	900196	PLEKHN1	6.1414	6.1264
GI_4826773	1	939723	939772	ISG15	12.3446	13.7679

HGNC – official human gene name

The "probe identifier" reflects a unique identifier for each probe, while "chromosome," "chromosome start," and "chromosome end" describe the physical location to which each probe maps on the human genome. For oligonucleotide RNA arrays, decent probe mapping information is usually available from the manufacturers, although these tend to be of varying quality (8). Remove probes that cross-hybridize to multiple genomic positions or that are not perfectly complementary to either the human transcriptome or human genome. Ensure that the genome assembly used for mapping probes is identical to the genome assembly used to map the SNPs.

If a probe maps within a gene, the "HGNC name" describes the official human gene name. Subsequent columns contain the individual measured expression levels for each sample. Two samples (*sample1* and *sample2*) are shown here.

This expression information file should be named *ExpressionData.txt.*

2.7. Example Genotype and Expression Data

We have provided example genotype and expression data for 47 HapMap samples, which were genotyped on Illumina HumanHap300 oligonucleotide arrays. Expression data for these samples were derived from EBV-transformed B cells and assayed using Illumina HumanRef-6 v1 Expression BeadChips. The example data can be downloaded from http://gbic.biol.rug.nl/supplementary/2008/associationGG/data/exampledata.zip. First unzip this file (on Unix or Linux use the command "unzip exampledata.zip," on Windows XP, Windows Vista, and Mac OSX double click on "exampledata.zip"). This will generate a directory ("exampledata/") containing five files:

Genotype data:	GenotypeData.txt
Phenotype information:	PhenotypeInformation.txt
SNP information:	SNPInformation.txt
Expression data:	ExpressionData.txt
associationGG instruction file:	associationGGInstructions.txt

2.8. Instructions Required for eQTL Mapping

The software program *associationGG* is used to perform the actual genetical genomics analysis and can be freely downloaded from http://gbic.biol.rug.nl/supplementary/2008/associationGG/data/associationgg.zip. First unzip this file (on Unix or Linux use the command "unzip associationgg.zip," on Windows XP, Windows Vista, and Mac OSX double click on "associationgg.zip"). This will generate a directory ("associationGG/"). Move the "exampledata/" directory into the "associationGG/" directory. All programs and data are now in place.

An instruction file "associationGGInstructions.txt" should be present in the following tab-delimited format (**Fig. 17.3**):

Dataset	HapMap	exampledata/
SNP_Call_Rate_Threshold	0.95	
SNP_HWE_Threshold	0.001	
SNP_MAF_Threshold	0.05	
RegulationType	Cis	250,000
CorrelationMeasure	Spearman	
MultipleTestingCorrection	FDR	0.05
OutputIndividualPlots	0.000001	exampledata/eqtls/

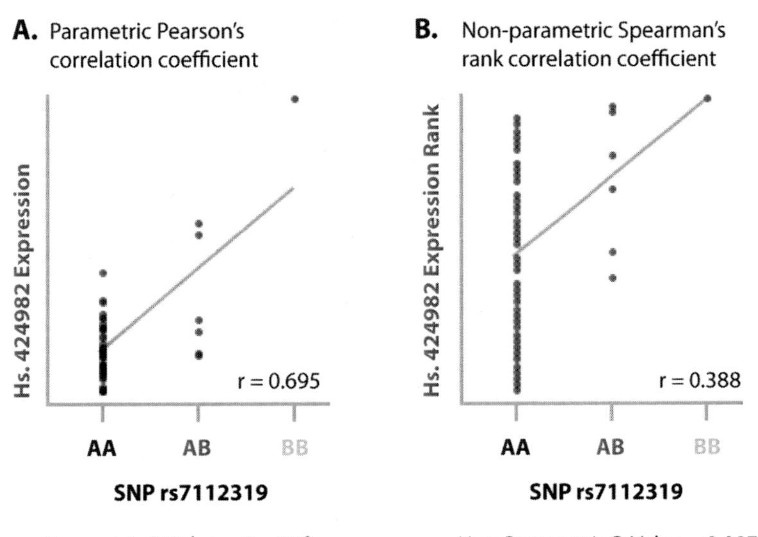

Fig. 17.3. Differences between parametric and non-parametric testing. A putative *cis*-eQTL (SNP rs7112319 and probe Hs. 424982) has a very high (parametric) Pearson product-moment correlation coefficient, due to a potential outlier (**Fig. 17.3A**). The corresponding *P*-value is highly significant (*P*-value = 6×10^{-8}). This significance diminishes when permuting the data (empiric *P*-value = 2×10^{-4}). A non-parametric Spearman's rank correlation coefficient test results in an even lower

The first row specifies the name of the file containing the data ("HapMap") and the directory where to find this file ("exampledata/"). Rows two to eight specify the user's choices for the pre-processing and processing steps in *associationGG*. These options will be discussed in Section 3, some additional optional instructions are described in **Note 7**. The example data come with an instruction file.

3. Methods

Once genotype and expression data have been converted to formats as described above, *associationGG* can be invoked. Start the command-line and change the directory to the directory where *associationGG* has been extracted. Start *associationGG* with the following command:

```
java -Xmx512m -jar associationGG.jar
exampledata/associationGGSettings.txt
```

associationGG will then use the settings defined in "*exampledata/ssociationGGSettings.txt*" to start the eQTL mapping workflow, as described in **Fig. 17.1**.

3.1. Pre-processing of Genotype Data

The first step in the eQTL mapping workflow is pre-processing of the genotype data (**Fig. 17.1**), as thorough quality control (QC) on genotypes is essential to avoid false-positive and false-negative findings (detection of false eQTLs or non-detection of true eQTLs, respectively). Three QC filtering criteria are used:

1. Call-rate: If the genotype for a particular SNP cannot be established for a considerable number of samples, this suggests that the SNP assay is of low quality. Generally, a call-rate threshold of at least 95% is used to include an SNP for subsequent analyses. The user can set the "*SNP_Call_Rate_Threshold*" parameter in the instruction file or use the default setting (0.95).
2. Hardy–Weinberg Equilibrium (HWE): SNPs tend to show deviations from HWE when genotypes have been incorrectly assigned, although there are also other explanations (SNPs mapping to copy number variation (CNV) regions or SNPs

Fig. 17.3 (continued) significance (*P*-value = 0.007, **Fig. 17.3B**). Although each of these *P*-values are nominally significant (*P*-value < 0.05), in a genetical genomics analysis many SNP-probe pairs are tested, making multiple testing correction essential. Consequently, after the multiple-testing correction, this SNP–probe pair will not remain significant when using permutations or a non-parametric test, whereas it will remain significant when using the original parametric *P*-value.

that map to sex chromosomes). An exact HWE test can determine whether there are deviations present, and an exact HWE *P*-value threshold of 0.001 is usually used to filter out these SNPs. The user can set the *SNP_HWE_Threshold* in the instruction file or use the default setting (0.001).

3. Minor Allele Frequency (MAF): If nearly all samples have the same genotype for an SNP, and, as such, a low MAF, it will not be possible to identify eQTLs due to lack of statistical power. An MAF below 5% is normally used to filter out these SNPs. The user can set *SNP_MAF_Threshold* in the instruction file or use the default setting (0.05).

3.2. Normalizing Expression Data

It is essential to normalize the expression data across arrays to make intensity levels between different samples comparable (**Fig. 17.2**). This has to do with technical variation during sample preparation, such as inducing differences in RNA amounts or concentrations between different samples. While a higher RNA concentration generally amounts to a higher measured raw intensity for each probe, this relationship is neither linear nor the same for each probe. Apart from this, other factors (such as RNA quality) can also influence the relationship between the measured raw intensity and the actual RNA expression.

A frequently used generic normalization procedure is "quantile normalization" (*9*), which orders measurements for each sample according to their raw intensity values and then replaces each value by the mean of measurements at that rank for all the samples. An implementation of this normalization method is available in current versions of Illumina's BeadStation, through the Bioconductor "*Affy*" package of the R programming language, and also in *associationGG*. *See* **Note 4** for details on two-channel oligonucleotide arrays.

To normalize the expression data in *associationGG*, you should add a *NormalizeExpressionData* parameter to the instruction file and set it to "Quantile." As the example expression data have already been quantile normalized, it is not necessary to normalize the data by setting this parameter for this dataset.

3.3. Expression Analysis of Probes Mapping to Chromosome Y

Once expression data have been normalized, a quality check will be performed on probes that map to the non-pseudoautosomal region of chromosome Y (**Fig. 17.2**). Males show clear gene expression of these probes, whereas females should show hardly any expression. If these chromosome Y probes have been assessed and included in the dataset, *associationGG* will provide summary statistics for each of the samples, permitting you to assess whether males indeed have higher expression of these probes than females. This is helpful to identify individuals for whom genotype data and expression data have been accidently mixed up in the lab.

A chromosome Y probe expression plot for the example data is shown in **Fig. 17.2**. As expected, all males show an average expression that is markedly higher than females.

3.4. Regulation Type

DNA sequence variation can have *cis*- and *trans*-regulatory effects on the expression variation of a local gene or one or more distant genes, respectively. The test for a potential *cis*-regulatory effect involves an association analysis of only those SNP–gene pairs that map on the same chromosome in each other's vicinity, while the test for a potential *trans*-regulatory effect involves a genome-wide scan for association between an SNP and all genes, resulting in considerably more statistical tests (**Fig. 17.3**). Different thresholds therefore apply to the different regulation tests. The user can set *RegulationType* to *cis* or to *trans* in the instruction file. (If the user wants to confine the analysis to the detection of *cis*-eQTLs, the maximum allowed distance between SNP and probe mid-point can be altered, or the default setting (250 kb) can be used.)

Trans-effects indicate that there is a biological relationship between the locus to which the SNP maps and the gene affected by this SNP. Correct genomic mapping of SNPs and probes is essential to find out whether an eQTLs is *trans* or *cis*. If these are incorrect, a *cis*-eQTL might be deemed a *trans*-eQTL and this could lead to a lot of unnecessary literature and practical research aiming to biologically link the locus to which the SNP maps with the gene to which the probe maps.

3.5. Correlation Measure

In a parametric analysis, the gene expression can be regressed on SNP genotypes, for example, by assuming additive allele effects. The square of the variance explained is equal to the Pearson product-moment correlation between the expression variation and the number of SNP alleles (0, 1, 2 times allele B). To perform this parametric analysis, you can set the *CorrelationMeasure* to *Pearson* in the instruction file. However, if for instance you only have one sample that has the BB genotype and its expression level is an extreme outlier, the correlation is likely to become highly significant (**Fig. 17.3A**).

To prevent these false-positives in a parametric analysis, it is essential to perform a permutation-based analysis to get an accurate, empiric *P*-value. *associationGG* will automatically perform this permutation-based correction when instructed by using the option *CorrelationMeasurePearson*. You can also choose to use a non-parametric analysis that is more robust with outliers (**Fig. 17.3B**). Such an approach does not use the expression levels for a particular probe, but uses a rank of the expression levels for this probe and then calculates a correlation coefficient through a Spearman's rank correlation coefficient analysis. By default, *associationGG* performs a non-parametric analysis (**Fig. 17.3**) because no permutations are necessary to establish accurate *P*-values, resulting in substantially fewer computations (*see* **Note 2**). *See* **Note 6** for other factors that might result in false-positive eQTLs.

3.6. Multiple Testing Correction

Once genotype and expression data have been pre-processed, expression data have been normalized, QC has been applied, and the type of analysis (*cis* or *trans*, parametric or non-parametric) has been defined, the actual eQTL mapping will be conducted (**Fig. 17.3**). In eQTL mapping, tens of thousands of gene expression traits are analyzed for association with hundreds of thousands of SNPs. As such, a correction for multiple testing (**Fig. 17.4**) is needed (although the burden of multiple testing correction is

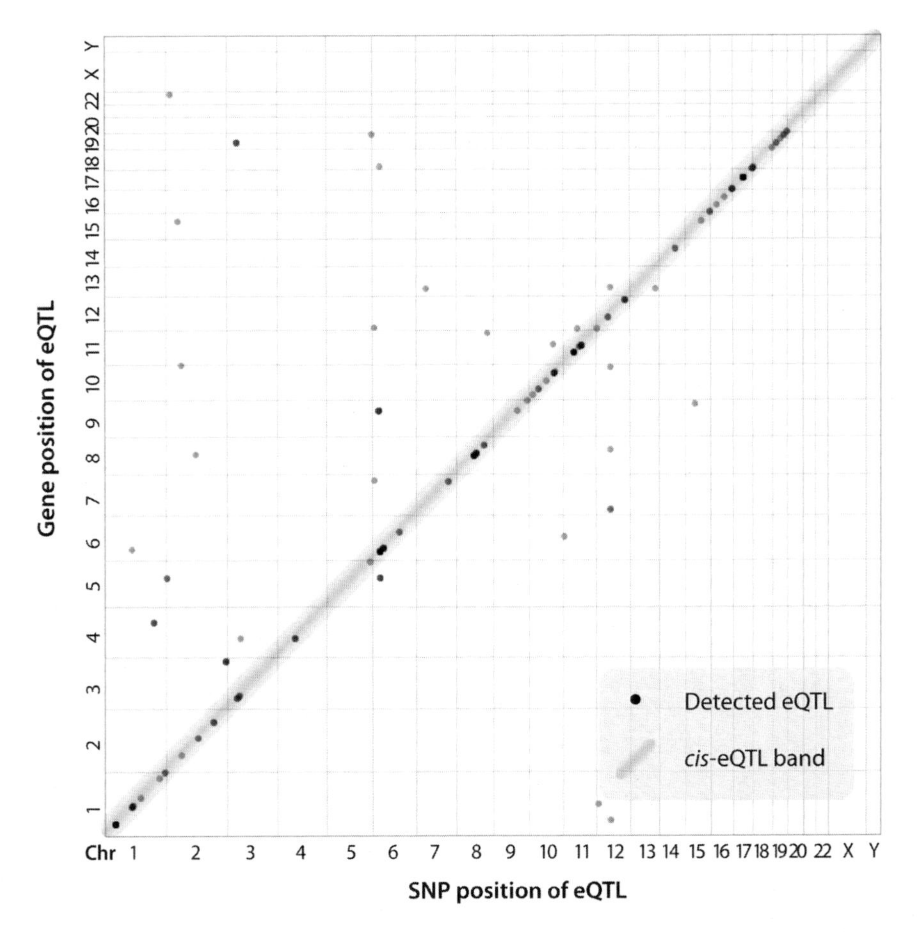

Fig. 17.4. eQTL dot plot of 185 identified *cis*- and *trans*-eQTLs (using a false discovery rate (FDR) of 0.05 and 100 permutations). Each dot represents an eQTL that has been detected in the provided dataset (nominal Spearman's correlation *P*-value $< 1.52 \times 10^{-9}$, corresponding to an FDR of 0.05). The SNP position of each SNP is depicted on the *X*-axis. The probe mid-point position is depicted on the *Y*-axis. eQTLs that fall on the diagonal reflect *cis*-eQTLs, whereas the others reflect *trans*-eQTLs.

somewhat smaller in the analysis for *cis*-eQTLs than for *trans*-eQTLs). There are various methods to correct for multiple testing. The most conservative way to do this is by using a Bonferroni correction. When using this procedure, the corresponding *P*-value for each detected eQTL is multiplied by the total number of tests that have been performed. However, this is overly conservative, as SNPs are frequently correlated due to linkage disequilibrium. In addition, many genes have correlated expression levels.

Another important consideration is that we expect to identify multiple true-positive eQTLs. Even when SNPs and gene expression levels are not correlated, applying a Bonferroni correction is over-conservative: For the most significant eQTL it makes sense to multiply its *P*-value with the total number of tests. However, for the next most significant eQTL, we should not multiply its significance by the total number of tests, but rather by the total number of tests minus one, as the number of tests has now been reduced by one (we consider the previous test independent and are allowed to do so, since in that previous test we have corrected for all tests that have been performed).

A false discovery rate (FDR) analysis properly accounts for this. An FDR-based correction for multiple testing effectively controls for the number of false-positives among the identified eQTLs. For instance, when using an FDR of 0.05, 5% of the list of most significant eQTLs eventually produced will be false-positives. Through permuting the sample labels of the expression matrix (keeping the correlation structure within the genotype and within the expression data), we can control both the FDR and correct for the fact that there is correlation between genotypes and between expression probes.

You can set the *MultipleTestingCorrection* in the instruction file to *FDR* and specify the FDR level or use the default setting (0.05).

3.7. Visualizing QTL Mapping Results

Once a list of significant eQTLs has been established using multiple testing correction, *associationGG* summarizes all identified *cis*- and *trans*-eQTLs by generating an eQTL dot plot (**Fig. 17.5**). An

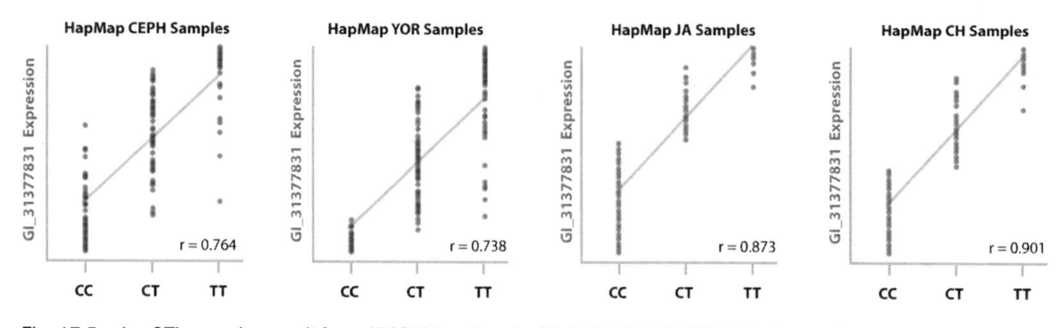

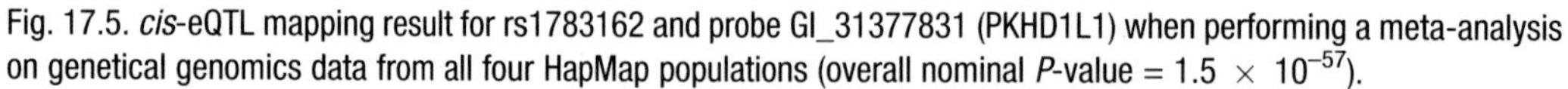
Fig. 17.5. *cis*-eQTL mapping result for rs1783162 and probe GI_31377831 (PKHD1L1) when performing a meta-analysis on genetical genomics data from all four HapMap populations (overall nominal *P*-value = 1.5×10^{-57}).

eQTL dot plot is given for a non-parametric *cis-* and *trans-*analysis using an FDR of 0.05 (**Fig. 17.4**) on the 47 HapMap samples; 185 eQTLs have been detected, of which the majority (119/185) are *cis-*eQTLs (probe mid-point distance to SNP $<$ 250 kb).

3.8. Interpreting eQTL Mapping Results

Once eQTLs have been identified, it is important to biologically interpret the findings. However, the interpretation of eQTL mapping results depends highly on the intended goal. When you have found genetic association for a certain set of SNPs, you could consider confining the analysis to only those SNPs and initially performing only a *cis-*eQTL analysis to have the highest statistical power to detect a potential functional *cis-*effect for any of the associated SNPs. This is a strategy that has been utilized by various groups (10, 11).

The results observed by these groups also serve as another QC method to assess whether eQTL mapping has been performed correctly: as some of these *cis-*eQTLs have been detected in nearly all genetical genomics datasets, they are expected to be present in the genetical genomics dataset under investigation (1, 12, 13). However, a word of caution is justified for these strong *cis-*eQTLs; a considerable proportion of these reflect false-positive findings due to secondary polymorphisms that are present in the sequence to which the probe hybridizes and which are in strong linkage disequilibrium (LD) with the *cis-*SNP (14). When interpreting the results, it is important to know for which cis-eQTLs this is the case. *See* **Note 3** for additional information on how this can be assessed.

When you want to investigate potential downstream effects of genetic variants, a *trans-*eQTL analysis is appropriate. *Trans-*effects suggest a biological relationship between genes or microRNAs to which the SNP maps and other genes affected in expression. The most simple biological explanation is that you are dealing with a non-synonymous SNP in a gene that codes for a transcription factor, directly affecting expression of the *trans-*regulated gene. However, this relationship can be more complicated, involving multiple indirect steps. To gain insight into these relationships, you could assess whether the gene to which the SNP maps has a biological function in common with the *trans-*regulated gene, such as through assessing over-representation of Gene Ontology terms or through using a human biological network (15) to see whether genes cluster in each other's vicinity.

When genetical genomics data are used with the aim to identify previously unknown pathways, post hoc analyses of these *trans-*eQTLs are important. However, as *trans-*eQTLs are generally difficult to detect in small datasets, there are alternative strategies to identify previously unknown pathways and biological relationships. A frequently used method is to assess whether genes are co-expressed. If this is observed for gene X and gene Y, it is likely that genes X and Y are biologically related. These co-expressed genes are often controlled by another upstream regulatory gene. However, when either gene X or gene Y is *cis*-regulated, co-expression between these genes will become more difficult to detect. On such occasions, it is worthwhile to perform a conditioned co-expression analysis, where *cis*-genotypic effects are removed from the data. In this way, biological relationships can be detected that would otherwise have gone unnoticed (12).

Through genetical genomics, it has been established that the expression of many genes is under genetic control. With the availability of larger genetical genomics datasets and using meta-analysis (*see* **Note 5**), it is likely that it will become evident that even more genes are under genetic control. Such analyses will also help to detect new biological relationships through identification of *trans*-eQTLs. By integrating additional biological information (such as known biological interactions and networks (15)) through extending *associationGG* (*see* **Note 8**), our fundamental insight into the biological consequences of genetic variants associated with disease is likely to expand.

4. Notes

1. *Experimental design considerations*

   An important consideration for the detection of eQTLs is the tissue from which expression data are derived. If a gene is not expressed at all in a tissue, it will be impossible to identify an eQTL for this gene. As such, if you want to conduct a genetical genomics study in a disease-specific setting, you should first assess whether the genes you are interested in are sufficiently expressed in those tissues that are relevant for the disease under investigation

   Another consideration is sample size. It has been established that if the sample size is limited (< 200 samples), detection of *trans*-eQTLs is usually difficult (16). As such, sample sizes should be considerably larger if you want to use genetical genomics to identify *trans*-eQTLs and be able to establish biological links between SNP variants and genes that map to different genomic positions.

   Results also depend on the genotyping and expression platforms being used. While most current genome-wide genotyping platforms cover most human genetic variation well, platforms that assess expression can vary considerably. There are arrays that assess only a single exon for a subset of all known genes, while some arrays assess hundreds of thousands of different exons. It has been observed that two different probes within a single gene constituted a *cis*-eQTL with the same SNP, but with opposite allelic effects (3, 12). This suggests that genetic variation can have different, opposite effects on expression of different exons of the same gene.

2. *Computational capacity issues*

   In contrast to linkage analysis, association analysis is generally straightforward, but the number of statistical tests performed is usually much higher. *associationGG* has been implemented in such a way that a typical full analysis (of both *cis*- and *trans*-eQTLs) can be performed within a reasonable time. A full non-parametric *trans*-eQTL analysis on the example data, using ten permutations to estimate the FDR on the example dataset and one MacBook Pro notebook (Intel Core Two Duo 2.33 GHz, 3 GB internal memory), took approximately 2 days. Running more permutations will result in a more accurate estimate of the FDR, although this will not greatly influence the number of reported significant eQTLs.

3. *Probe hybridization artifacts*

   Sometimes significant *cis*-eQTLs will be identified that are due to the presence of untyped SNPs within the sequence targeted by the probes that are in strong LD with the genotyped *cis*-SNP. These SNPs can affect the hybridization of the probes and, as a result, suggest that the expression is strongly altered when this is entirely due to a technical hybridization artifact (14). To control this you should use dbSNP and HapMap to assess whether SNPs are known to map within these probe regions, permitting you to flag these probes. If there is also LD information between these SNPs and the *cis*-eQTL SNP in HapMap, you can determine whether these are, as expected, in strong LD. If this is the case, it is likely that *cis*-eQTL can be discarded as having a real expression effect on the probe.

4. *Two-channel oligonucleotide arrays:* We have not discussed two-channel oligonucleotide arrays, where the expression intensity of a particular gene is determined by taking the ratio between the measured gene intensity for a sample of interest and the measured gene intensity for a pool of all samples. Although the genetical genomics analyses for these types of oligonucleotide arrays are similar, normalization of the expression data is very different compared to normalization of single-spot intensity arrays from Illumina or Affymetrix, for example.

5. *Further analyses: meta-analysis of genetical genomics datasets*

   For complex diseases, meta-analyses of different datasets investigating the same diseases may identify genetic variants with small effects that would not have been identified in a single dataset with a smaller sample size. This is likely true for genetical genomics as well; when sample sizes increase, it is likely that more eQTLs will be identified. As various genetical genomics datasets have recently become publicly available (5, 17), meta-analysis has now become a feasible strategy to increase the detection of eQTLs (12).

associationGG provides functionality for performing a meta-analysis on different genetical genomics datasets. The independent genetical genomics datasets first need to be converted into the formats that have been described earlier. Then *associationGG* needs to be instructed to use multiple datasets. To do this, a *Dataset* entry needs to be present within "*associationGGInstructions.txt*" for each genetical genomics dataset and the directory where the genetical genomics datasets reside has to be indicated. Subsequent analyses are identical to a genetical genomics analysis of only one dataset. However, the eQTL mapping and multiple testing correction now takes all genetical genomics datasets into account and weighs evidence for individual QTLs through a weighted *Z*-test (18). For each individual genetical genomics dataset, the *P*-value of a potential eQTL is *Z*-transformed. A weighted *Z*-score is then calculated to combine the evidence for each eQTL:

$$z_w = \sum z_i \times \sqrt{N_i / N_{tot}}$$

where z_w is the weighted *Z*-score from the meta-analysis, z_i is the individual *Z*-score for an eQTL from dataset i, N_i is the number of genotyped samples (that have passed QC) for dataset i, and N_{tot} is the total number of genotyped samples (that have passed QC). Finally, the weighted *Z*-score is transformed back into an overall *P*-value.

When *associationGG* has been instructed to generate individual eQTL plots, a plot is generated for each dataset, permitting visual assessment of each of the identified eQTLs (**Fig. 17.5**).

However, meta-analysis can be complicated when genetical genomics datasets have been generated on different platforms (e.g., genotypes for one dataset generated on the Illumina platform, whereas genotypes for the other dataset generated on the Affymetrix platform). In such a situation, only a small subset of all SNPs will have been genotyped on both platforms, greatly diminishing the number of eQTLs for which data from both datasets are available. To overcome this, genotypes that are present on one platform but not on the other can often be successfully imputed (19, 20).

Another word of caution is due when performing a meta-analysis on genetical genomics datasets that have been generated on different types of Illumina expression arrays. Illumina has used probes with identical probe identifiers for different chips, but sometimes these probes have different oligonucleotide sequences and they consequently map to different genomic positions. If you are performing an analysis that uses different Illumina expression arrays, it is strongly recommended that you only jointly analyze eQTLs if the probes have identical oligonucleotide sequences.

6. *Batches and hidden factors*

   Batch effects can introduce false-positive associations. *associationGG* has been designed to carry out a meta-analysis on different genetical genomics studies. If batch effects are present, you can split the datasets into several different ones and perform a meta-analysis on these. Although this will result in some loss of statistical power (due to a weighted Z-method to integrate *P*-values from the different datasets (18)), this permits effective control of batch effects.

7. *Optional* associationGG *parameters*

   Apart from performing meta-analysis on genetical genomics datasets, some other optional parameters are available for *associationGG* that can be defined in *associationGGInstructions.txt*:

   1. To confine the analysis to a subset of SNPs, you can make a file that contains only the subset of SNPs that need to be analyzed. To do so, define SNPSubset and name the file that contains this subset of SNPs (e.g., SNPSubset*SNPSubset.txt*).
   2. To confine the analysis to a subset of expression probes, you can make a file that contains the subset of probes that need to be analyzed. To do so, define ProbeSubset and name the file that contains this subset of probes (e.g., ProbeSubset*ProbeSubset.txt*).
   3. Individual plots for the detected eQTLs can be generated by associationGG. To do so, define OutputIndividualPlots. Define a nominal P-value threshold when generating a plot for an eQTL and defining an output directory (e.g., OutputIndividualPlots *0.000001 exampledata/eqtls/*).

8. *Altering* associationGG

   The Java source code for *associationGG* is publicly available, permitting researchers to alter it to suit specific needs. *associationGG* comes with extensive documentation, enabling developers to change it relatively easily. The implementation of *associationGG* is currently optimized to perform a meta-analysis of multiple genetical genomics datasets within a reasonable time.

References

1. Dixon, AL, Liang, L, Moffatt, MF, et al. (2007) A genome-wide association study of global gene expression. *Nat Genet* **39,** 1202–1207.
2. Goring, HH, Curran, JE, Johnson, MP, et al. (2007) Discovery of expression QTLs using large-scale transcriptional profiling in human lymphocytes. *Nat Genet* **39,** 1208–1216.
3. Kwan, T, Benovoy, D, Dias, C, et al. (2008) Genome-wide analysis of transcript isoform variation in humans. *Nat Genet* **451,** 359–362.
4. Morley, M, Molony, CM, Weber, TM, et al. (2004) Genetic analysis of genome-wide variation in human gene expression. *Nature* **430,** 743–747.
5. Stranger, BE, Forrest, MS, Dunning, M, et al. (2007) Relative impact of nucleotide and copy number variation on gene expression phenotypes. *Science* (New York, NY) **315,** 848–853.
6. Stranger, BE, Nica, AC, Forrest, MS, et al. (2007) Population genomics of human gene expression. Nat Genet **39,** 1217–1224.
7. Frazer, KA, Ballinger, DG, Cox, DR, et al. (2007) A second generation human haplotype map of over 3.1 million SNPs. Nature **449,** 851–861.
8. Dai, M, Wang, P, Boyd, AD, et al. (2005) Evolving gene/transcript definitions significantly alter the interpretation of GeneChip data. *Nucleic Acids Res* **33,** e175.
9. Bolstad, BM, Irizarry, RA, Astrand, M, et al. (2003) A comparison of normalization methods for high density oligonucleotide array data based on variance and bias. *Bioinformatics* (Oxford, England) **19,** 185–193.
10. Hunt, KA, Zhernakova, A, Turner, G, et al. (2008) Newly identified genetic risk variants for celiac disease related to the immune response. *Nat Genet* **40,** 395–402.
11. Moffatt, MF, Kabesch, M, Liang, L, et al. (2007) Genetic variants regulating ORMDL3 expression contribute to the risk of childhood asthma. *Nature* **448,** 470–473.
12. Heap, GA, Trynka, G, Jansen, RC, et al. (2009) Complex nature of SNP genotype effects on gene expression in primary human leucocytes. *BMC Med Genomics* 7, 2:1.
13. Stranger, BE, Forrest, MS, Clark, AG, et al. (2005) Genome-wide associations of gene expression variation in humans. *PLoS Genet* **1,** e78.
14. Alberts, R, Terpstra, P, Li, Y, et al. (2007) Sequence polymorphisms cause many false cis eQTLs. PLoS ONE **2,** e622.
15. Franke, L, van Bakel, H, Fokkens, L. et al. (2006) Reconstruction of a functional human gene network, with an application for prioritizing positional candidate genes. *Am J Hum Genet* **78,** 1011–1025.
16. Breitling, R, Li, Y, Tesson, BM, et al. (2008) Genetical genomics: spotlight on QTL Hotspots. *PLoS Genet* **10,** e1000232.
17. Myers, AJ, Gibbs, JR, Webster, JA, et al. (2007) A survey of genetic human cortical gene expression. *Nat Genet* **39,** 1494–1499.
18. Whitlock, MC. (2005) Combining probability from independent tests: the weighted Z-method is superior to Fisher's approach. *J Evol Biol* **18,** 1368–1373.
19. Marchini, J, Howie, B, Myers, S, et al. (2007) A new multipoint method for genome-wide association studies by imputation of genotypes. *Nat Genet* **39,** 906–913.
20. Purcell, S, Neale, B, Todd-Brown, K, et al. (2007) PLINK: a tool set for whole-genome association and population-based linkage analyses. *Am J Hum Genet* **81,** 559–575.

Chapter 18

A Systematic Strategy for the Discovery of Candidate Genes Responsible for Phenotypic Variation

Paul Fisher, Harry Noyes, Stephen Kemp, Robert Stevens, and Andrew Brass

Abstract

It is increasingly common to combine genome-wide expression data with quantitative trait mapping data to aid in the search for sequence polymorphisms responsible for phenotypic variation. By joining these complex but different data types at the level of the biological pathway, we can take advantage of existing biological knowledge to systematically identify possible mechanisms of genotype–phenotype interaction. With the development of web services and workflows, this process can be made rapid and systematic. Our methodology was applied to a use case of resistance to African trypanosomiasis in mice. Workflows developed in this investigation, including a guide to loading and executing them with example data, are available at http://www.myexperiment.org/users/43/workflows.

Key words: Genotype, phenotype, QTL, microarray, workflows, web services.

1. Introduction

Discovering associations between genotype and phenotype plays a crucial role in understanding the various biological processes which contribute to overall cellular, tissue, and organism responses, particularly when under a disease state (1, 2). Large-scale sequencing projects have provided researchers with genomes from an increasing number of organisms (3). This has enabled the development of a range of high-throughput technologies, notably genome-wide expression analysis. In combination with systems for identifying and screening genome sequence variants, it is now possible to perform very high-throughput experiments, aimed at understanding genome function (4, 5).

K. DiPetrillo (ed.), *Cardiovascular Genomics,* Methods in Molecular Biology 573,
DOI 10.1007/978-1-60761-247-6_18, © Humana Press, a part of Springer Science+Business Media, LLC 2009

The challenge now lies in the analysis of this biological data, which is vital to understanding the role genes play in organism responses to disease and other phenotypic variation.

One of the first, and classic, examples of correlating genotype and phenotype was the discovery of the Huntington allele (6). This enabled predictive tests for "age of onset" and "severity of disease" to be established. Since then, researchers have discovered single-gene lesions for a large number of simple Mendelian traits. It has proved much more difficult, however, to discover genes underlying genetically complex traits that are the product of multiple genes or biological processes (7). There are a number of ways in which this problem may be reduced, including utilising quantitative trait loci (QTL) and microarray gene expression data in genotype–phenotype studies.

The application of microarrays varies between individual experiments, from investigating the expression of genes in specific stages of a cell cycle to measuring gene expression levels in relation to abnormal cellular responses or disease states (8). By identifying which genes are differentially expressed in a disease state, we obtain information as to which genes may play a role in the expression of the observed phenotype. Many genes, however, will be expressed but not associated with the phenotype. In order to enrich the set of expressed genes with those that are most likely to be contributing to the phenotype, we can combine the gene expression data with QTL information.

QTL represent regions of chromosome, typically spanning several centimorgans (cM), which provide some quantitative evidence that a given phenotype is directly influenced by genes residing within that chromosomal region (8, 9). One of the goals of QTL mapping is to determine which genes residing in the region influence a given phenotype. Integrating QTL information with that of gene expression studies enables us to investigate the expression of genes at the level of chromosomal regions (8) and link candidate genes (significantly expressed in a microarray study) to a QTL region known to be involved in an expressed phenotype.

A further measure to determine the precise biological factors influencing a given phenotype is to investigate phenotypic variation at the level of biological pathways. The expression of genes within their biological pathways contributes to the expression of an observed phenotype. By investigating links between genotype and phenotype at the level of biological pathways, it is possible to obtain a broader view of the processes which may contribute to the expression of the phenotype (10). Using pathways in this manner provides a means of noise and dimensionality reduction. By cross-correlating pathway lists from both genes in the QTL and from the microarray study, we can begin to see which pathways are directly influencing the phenotype (**Fig. 18.1**) (4). Additionally the explicit identification of responding pathways naturally leads to hypotheses that can be tested in the laboratory.

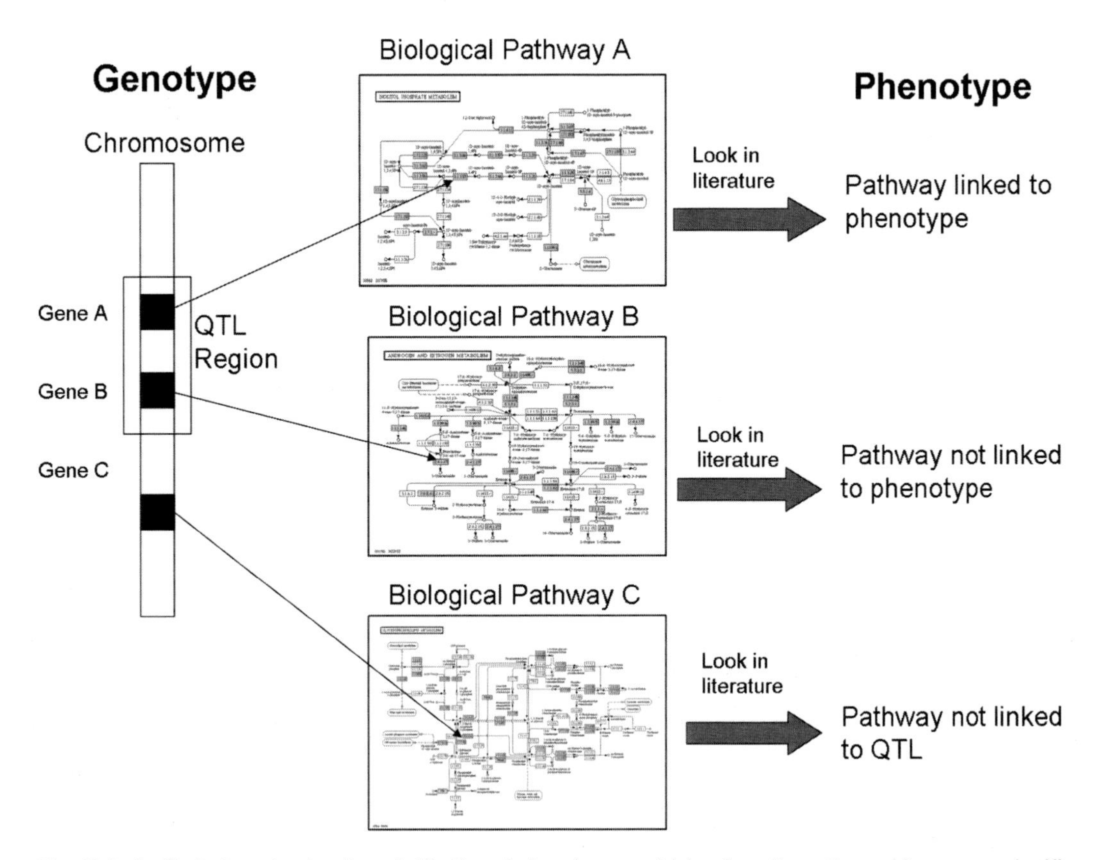

Fig. 18.1. An illustration showing the prioritisation of phenotype candidates from the pathway-driven approach. All pathways are differentially expressed in the microarray data. Those pathways that contain genes from the QTL region are assigned a higher priority (*pathways A and B*) than those with no link to the QTL region (*pathway C*). Higher priority pathways are then ranked according to their involvement in the phenotype, based on literature evidence. Abbreviations: CHR – chromosome; QTL – quantitative trait loci.

We discuss the correlation of genotype and phenotype for the identification of candidate genes responsible for phenotypic variation. We will explain how microarray gene expression and quantitative trait loci (QTL) data may be used in combination with a pathway-driven approach to investigating genotype–phenotype correlations; discuss the issues facing the manual analysis of microarray and QTL data; briefly outline web service and workflow technology; describe a systematic workflow-based approach to the analysis of microarray and QTL data; and finally showcase our methods by applying a use case from the Wellcome Trust Pathogen–Host African Trypanosomiasis project.

The process of linking genotype and phenotype using this method currently requires the use of numerous, and often specialised, resources to begin to piece the fragmented biological information into a hypothesis. Such large-scale projects, however, typically involving the analysis of large amounts of data, rely heavily on computer applications and distributed services to perform rudimentary and often repetitive tasks (11). The issues surrounding these types of analyses are described in full in the next section of this chapter.

2. Issues with Current Analysis Techniques

In conducting microarray gene expression and QTL mapping experiments, we quickly begin to accumulate data in vast quantities (12–14); these datasets regularly contain tens of thousands of genes. Researchers can quickly become overwhelmed with a wealth of data which require analysis.

The identification of candidate genes is a huge task to undertake, requiring large amounts of time, human resource, and repetitive searches through numerous software tools. With genomic databases in continuous flux, chromosomal regions may be reassembled, and the task of clarifying the candidate genes, notably in QTL regions, becomes a problem. With tens to hundreds of genes lying within each QTL region, it is vital that the identification, analysis, and functional testing of Quantitative Trait genes (QTg) is carried out systematically, without bias introduced from prior assumptions (7). The sheer scale of data being generated by such high-throughput experiments has led some investigators to follow a hypothesis-driven approach (15), where the selection of candidate genes is based on some prior knowledge or assumption (4). Although these techniques for candidate gene identification are able to detect some QTg, they also run the risk of overlooking genes that have less obvious associations with the phenotype, namely genes which have an indirect involvement in the phenotypes expression (16).

A further complication is that ad hoc methods for candidate gene identification are inherently difficult to replicate, which is compounded by poor documentation of experimental methods used to generate and capture the data from such investigations (17). An example is the widespread use of "link integration" (18) in bioinformatics. This process of hyper-linking through any number of data resources further exacerbates the problem of capturing the methods used for obtaining in silico results.

In order to investigate whether such problems were present in current bioinformatics analysis techniques, we conducted a review of the literature surrounding combined QTL and microarray analyses (**Table 18.1**) (4). Results from this review enabled us to identify specific issues facing the manual analysis of microarray and QTL data, including the selection of candidate genes and pathways. These issues are as follows:

Table 18.1
The literature surrounding combined QTL and microarray analyses on which a comprehensive survey of data analysis techniques was conducted

Reference	Phenotype
H.C. Liu (2001)	Marek's disease
M.F. de Buhr (2006)	Inflammatory bowel disease
A.M. Rodriguez de Ledesma	Open-field activity
R. DeCook (2006)	Shoot development
M.L. Wayne (2002)	Ovariole number
I.M. Stylianou	Obesity
I.A. Eaves (2002)	Type 1 diabetes
D.R. Prows (2003)	Nickel-induced acute lung injury
T. Okuda (2001)	Hypersensitive rats
D.N. Cook (2004)	Endotoxin-induced airway disease
N.R. Street (2006)	Drought response
C.Q. Lai (2007)	Life span
L.G. Carr (2007)	Alcohol preference

1. Premature filtering of datasets to reduce the amount of information requiring investigation.
2. Predominantly hypothesis-driven investigations instead of a complementary hypothesis and data-driven analysis.
3. User bias introduced on datasets and results, leading to narrow-focused, expert-driven hypotheses.
4. Implicit data analysis methodologies.
5. Constant flux of data resources and software interfaces hinder reproducing searches and re-analysis of data.
6. Error proliferation from any of the listed issues.

3. The Systematic Pathway-Driven Approach

Web services are now being used to offer programmatic access to a large number of bioinformatics resources (19), experiments previously conducted manually can now be replaced by automated experiments, capable of processing a far greater volume of data. This enables us to process the data in a much more efficient, reliable, un-biased, and explicit manner.

3.1. The Taverna Workflow Workbench

Numerous bioinformatics tools have been developed, including Taverna (20), Kepler (21), and Talisman (22), which are capable of capturing and representing workflows. For the purpose of implementing our systematic, pathway-driven approach, we chose to use the Taverna workflow workbench (version 1.7.1). Although this workbench offers many features for workflow construction, the workflows discussed in this chapter may be constructed using any of the well-established workflow workbenches currently available. The Taverna workbench may be downloaded from http://taverna.sf.net.

The Taverna workflow workbench was established to provide a means of accessing web services from a vast array of resources, enabling bioinformaticians to conduct in silico experiments efficiently. **Figure 18.2** shows the user interface of Taverna, including the three distinct panes of

1. Advanced Model Explorer (AME) – used to connect third party services to one another.
2. Available Services – enables the researcher to add and connect services.
3. Workflow Diagram – used to visualise the workflow currently being constructed/run.

3.2. Automating QTL and Microarray Analyses

We constructed three prototype workflows to provide gene annotations for microarray data, annotate genes known to be located in a given QTL region, and identify common pathway annotations between microarray and QTL data. In order to construct the workflows we:

1. Identified and mapped the process of gene annotation for both QTL and microarray data.
2. Determined which genes reside in a QTL region of choice by identifying the physical boundaries of each QTL with either the left and right flank or peak markers of the QTL (*see* **Note 1**).
3. Converted the marker positions to physical base pair positions using a well-established genome annotation tool, the Ensembl database (23).

Each gene in the QTL was then annotated with its associated biological pathways. The same process of pathway annotation was also carried out for the genes found to be differentially expressed in the microarray study. These two sets of pathway data enabled us to obtain a subset of common pathways that contain genes within the QTL region that are differentially expressed in the microarray study (**Fig. 18.3**).

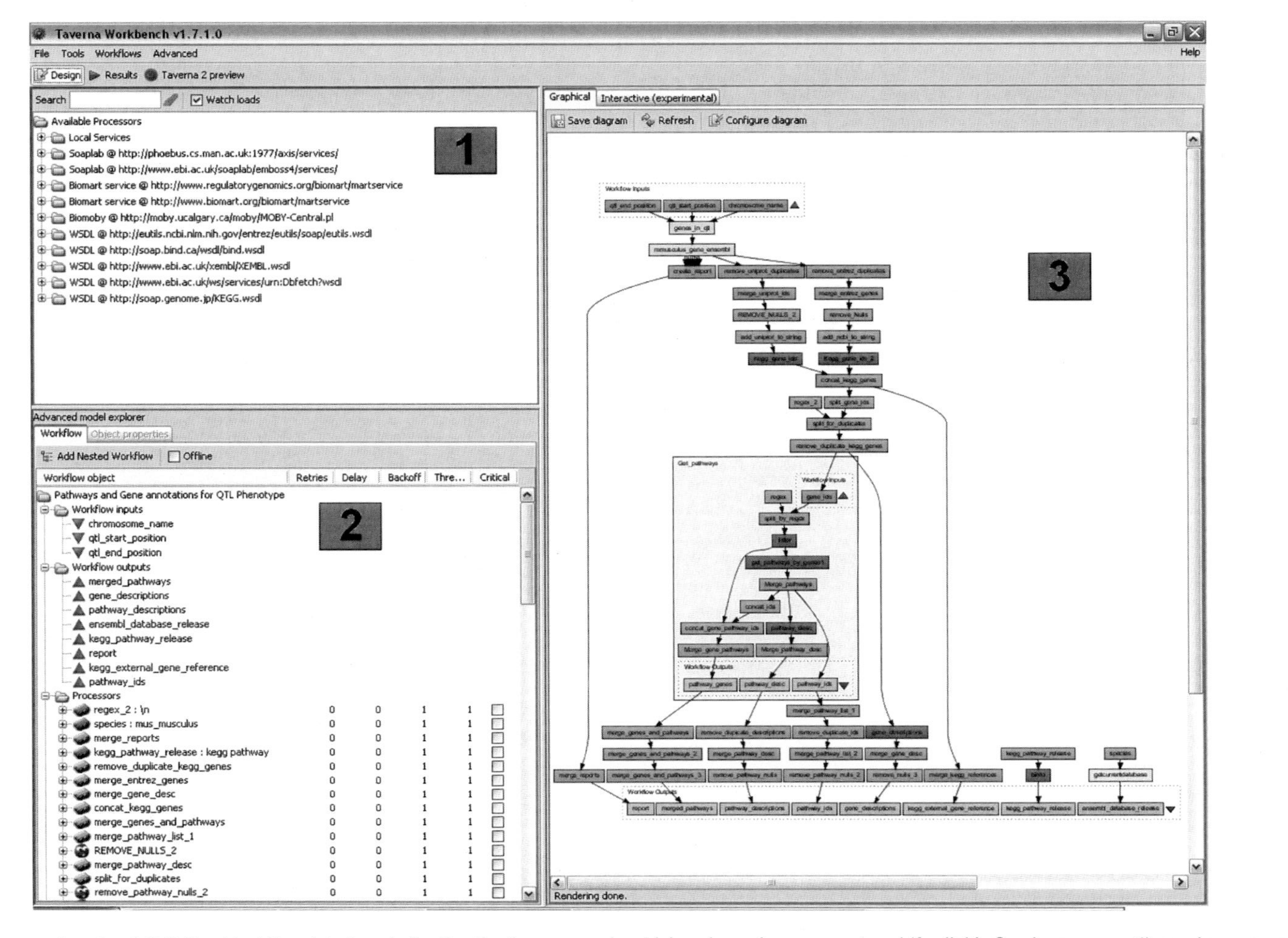

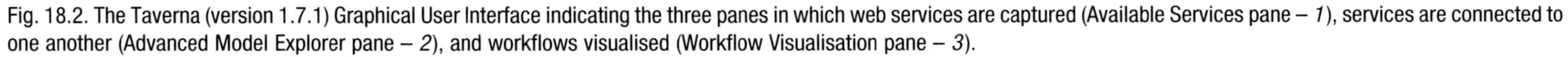
Fig. 18.2. The Taverna (version 1.7.1) Graphical User Interface indicating the three panes in which web services are captured (Available Services pane – *1*), services are connected to one another (Advanced Model Explorer pane – *2*), and workflows visualised (Workflow Visualisation pane – *3*).

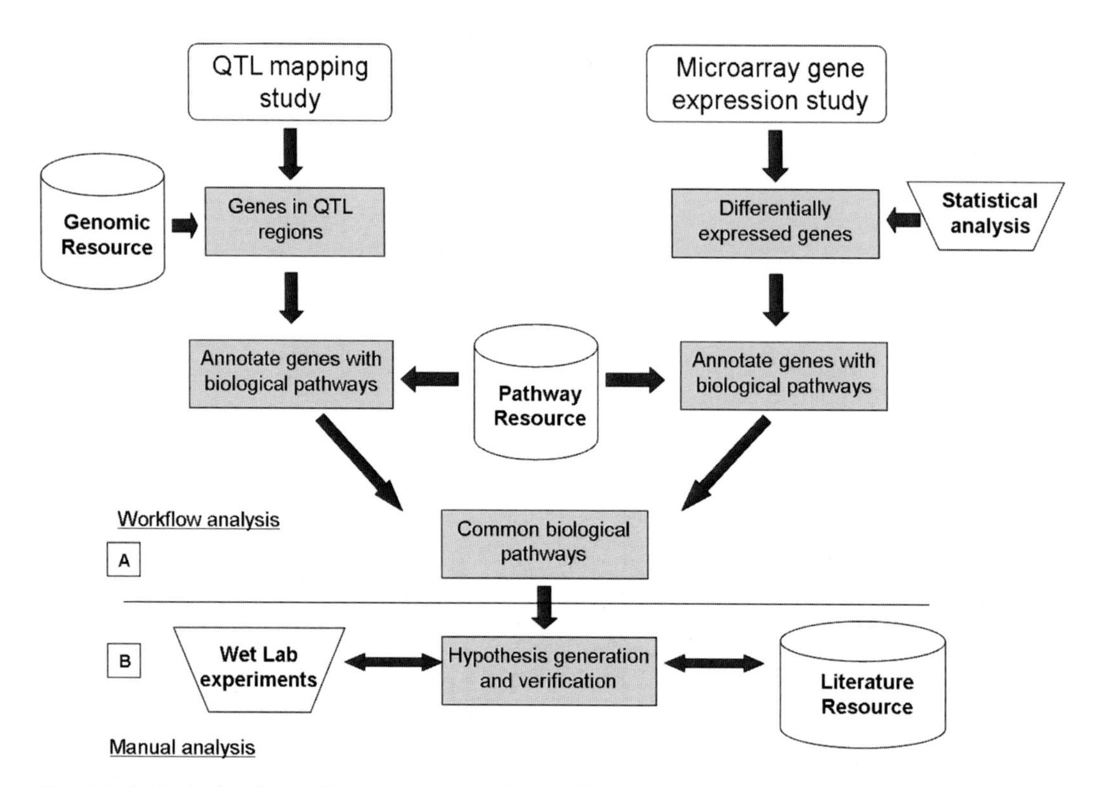

Fig. 18.3. An illustration showing the process of annotating candidate genes from microarray and QTL investigations with their biological pathways. The pathways gathered from both studies are compared and those common to both are extracted as the candidate pathways for further analysis. These pathways represent a set of hypotheses, in that the candidates are the hypothetical processes which may contribute to the expression of the observed phenotype. Subsequent verification is required for each pathway by wet lab investigation and literature searches. This approach is separated into two sections, distinguished by the *dividing line* between the selection of common pathways and the generation of hypotheses. The section labelled *A* represents the workflow side of the investigation, whilst the section labelled *B* represents verification of the hypotheses through wet lab experimentation and literature mining.

3.3. Constructing Workflows Using Taverna

The first workflow constructed, *qtl_pathway*, was implemented to identify genes within a QTL region, and subsequently map them to pathways. We chose to utilise pathways obtained from the KEGG pathway database (24) for both QTL and microarray data.

The first stage of construction required a BioMart[1] processor to be added to the Advanced Model Explorer (AME) pane. The BioMart processor, in the “Available Services” pane of Taverna, was added to the workflow by locating the BioMart plug-in. The “Ensembl 49 Genes (Sanger)” database and then the species (*Mus musculus* is named mmusculus_gene_ensembl) were added to the workflow model. This processor required configuration, including the parameters to be taken as input, and the attributes to be returned as output. In order to capture the genes within the QTL

[1] http://www.biomart.org/

region of choice, three input parameters were required: the chromosome on which the QTL resides, start position of QTL in base pairs, and finally end position of QTL in base pairs. The attributes that returned from BioMart were chosen based on the need for cross-referencing to KEGG gene identifiers, including Entrez gene (25) identifier and UniProt/SwissProt (26) accession number.

Once this service had been correctly configured, the next phase of construction involved adding the remaining services to the workflow model and connecting input and output ports to one another. In order to feed the outputs of one service directly into the input of another, a series of shim[2] services (written in Java) were required. These were necessary to transform the results of one service into a format that met the needs of the next (i.e. changing the format of results from tab-delimited into comma separated). The complete workflow, including all services, is given in **Fig. 18.4**.

The second workflow, *probeset_to_pathway*, provided annotation of microarray probeset identifiers. We chose to use probeset identifiers to allow the workflow to be used directly after gene expression data normalisation. The construction of this second workflow is built upon the same gene annotation mechanisms as the first workflow, *qtl_pathway*. An additional BioMart service was required to convert the input list of probeset identifiers into Ensembl gene identifiers.

To obtain the pathways which both intersect the QTL region and are present in the gene expression data, we used a third workflow, named *common_pathways*. This workflow compared both sets of KEGG pathway annotations, returning one version of a common pathway and a separate set of unique pathways, both listed as KEGG pathway descriptions.

3.4. Loading and Executing Workflows in Taverna

This section describes the pre-requisites required before running Taverna; downloading and installing Taverna; loading a workflow; executing a workflow; and examining the experimental results.

3.4.1. Pre-requisites

The Java 1.5 run-time environment is required to run Taverna (http://java.sun.com/javase/downloads/index.jsp). Those users who are installing Taverna on Linux will also have to install GraphViz onto the system. This is available at http://www.graphviz.org/.

3.4.2. Downloading and Installing Taverna

The Taverna workflow workbench is available at http://taverna.sourceforge.net. On the Taverna download page, users are able to specify the operating system (Mac OS, Windows, Linux), and then download the relevant version of Taverna.

[2] http://www.iscb.org/ism2004/posters/duncan.hullATcs.man.ac.uk_445.html

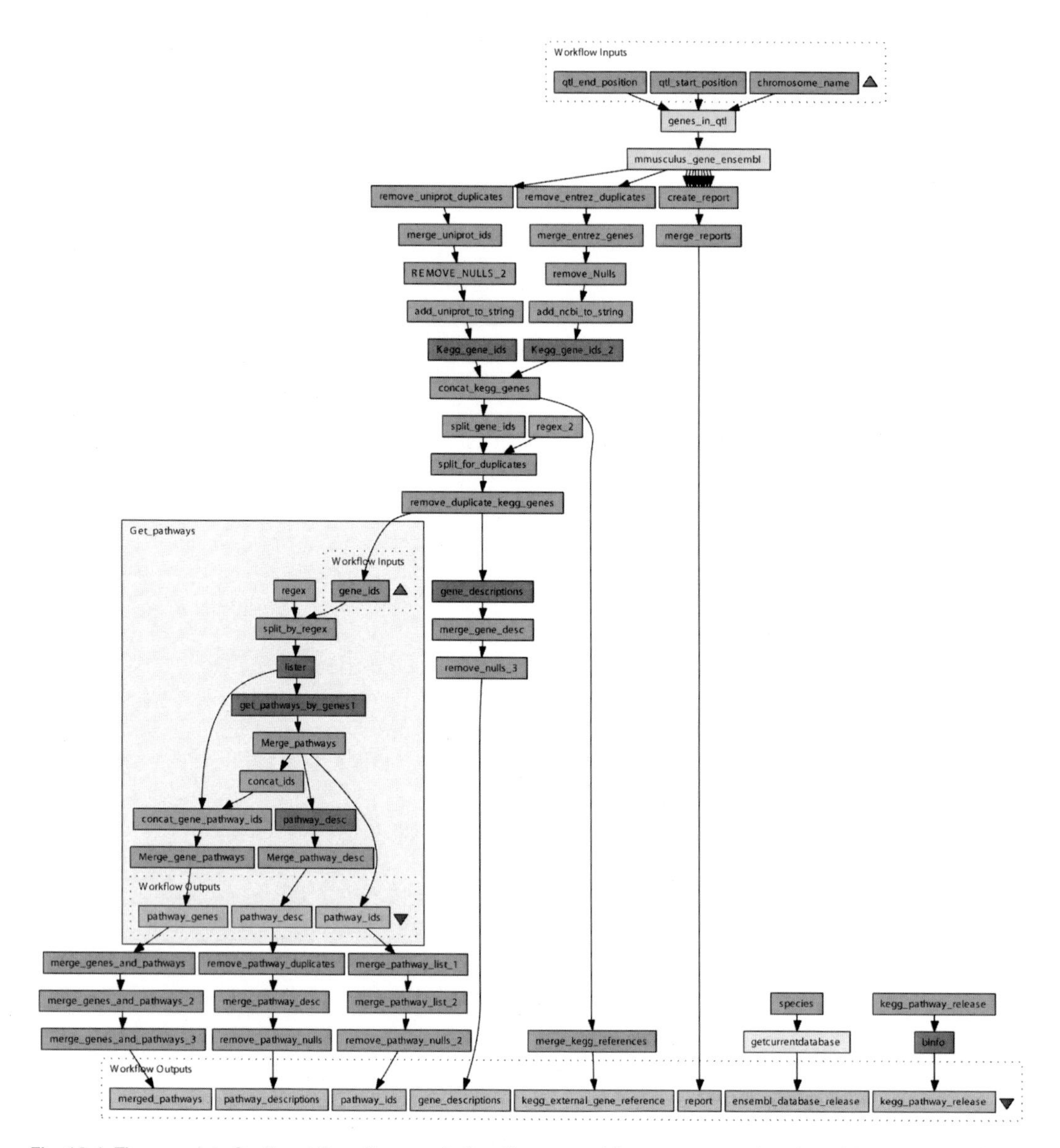

Fig. 18.4. The complete Scufl workflow diagram of qtl_pathway, providing gene annotations for a QTL region identified in the *Mus musculus* Ensembl database. Input parameters are specified at the top of the diagram (named workflow inputs), whilst the results are gathered at the *bottom* of the diagram (named workflow outputs). A nested workflow was added to the workflow (shown in the *inner box*) to iterate over individual KEGG gene identifiers, returning one or more KEGG pathways per KEGG gene id.

3.4.3. Running the Taverna Workbench

After installing Taverna, locate the installation folder and start Taverna by clicking on the relevant "runme" file (".bat" for Windows, ".sh" for Mac OS and Linux). For the first time, Taverna will need to update its components by downloading the files from the central repository across the Internet. Each component will be shown loading in a progress bar. Once this has completed the Taverna workbench will open.

3.4.4. Workflow Re-use in Taverna

In order to execute a workflow in the workbench, users need to firstly load a workflow (*see* **Note 3**). This can be done by either

- Locating the workflow and downloading it to the client machine or
- Opening the workflow directly from the web.

To use all of the workflows discussed in this chapter, users must navigate to the URLs provided below:

1. QTL gene annotation – http://www.myexperiment.org/workflows/16
2. Microarray gene annotation – http://www.myexperiment.org/workflows/19
3. Common pathways – http://www.myexperiment.org/workflows/13

If the user wishes to open the file without downloading it, the URL including the workflow version can be copied onto the client clipboard. The user must choose to open a workflow from a given location, navigating to the "Open Workflow from Location" option in the Taverna menu system. The URL can then be pasted into the resulting dialog box. An example of the workflow with a version number is as follows: http://www.myexperiment.org/workflows/13/download?version = 2. This will open version 2 of the common pathways workflow.

Once opened, the workflows may be executed within the workbench by choosing the "Run Workflow" option from the menu. Relevant data values can then be entered for each input parameter required. In order to execute the workflow *qtl_pathway*, three different input parameters are required: chromosome_name; start_position; end_position. Each of these requires the user to enter data into each of the input ports. This is achieved by clicking on one input and then clicking on the "New Data" option in the pop-up menu system. Relevant data values can then be entered in place of the "Some input data goes here" text. Examples of how the data values are to be entered are given below:

- chromosome_name = 17
- start_position = 28394586
- end_position = 38278830

The workflow may then be executed by pressing the "Run Workflow" button at the bottom of the pop-up box (*See* **Note 2**).

The *probeset_to_pathway* workflow requires a single list of AffyMetrix probeset identifiers. These identifiers should be provided in a list of values, each on a separate new line. The outputs from both the QTL and microarray workflows are returned as a set of text files, representing a flat file database. This is so that IDs for a given pathway can be traced back through the workflow provenance, identifying Ensembl or AffyMetrix probeset IDs from which they were derived. The final workflow, *common_pathways*, requires lists of KEGG pathway identifiers from both the QTL and the microarray workflows. These identifiers can be found in the file out named "pathway_ids.text". These lists should be loaded directly into the workflow by choosing "Load input from location" in the menu of the "Run Workflow" pop-up box. An example input for this workflow is "path:mmu04060". Further examples of workflow outputs can be found in the supplementary material in the paper by Fisher et al. (4).

3.4.5. Viewing Results in Taverna

Upon executing the workflow, the Taverna workbench will change views from "Design" to "Results". The varying colours indicate whether a service is currently being executed (green), awaiting execution (grey), or completed (purple). Once completed, individual results are provided for users to investigate and save as required.

4. A Use Case of African Trypanosomiasis

To illustrate the use of our methodology, we will describe the identification of a number of candidate genes thought to be involved in resistance to African Trypanosomiasis.

African trypanosomiasis, or sleeping sickness, is caused by a number of parasites of the *Trypanosoma* genus. Human sleeping sickness is caused by subspecies of *Trypanosoma brucei*, whilst sleeping sickness in cattle is caused by *Trypanosoma congolense* and *Trypanosoma vivax*. The infection from these parasites is a major restriction on cattle production in the subSaharan region of Africa (27). With no available vaccine, and with heavy and sustained expenditure on trypanocidal and vector control measures, trypanosomiasis is estimated to cost in excess of 4 billion US dollars each year in direct costs and lost production (28).

4.1. QTL Controlling Susceptibility to Infection with T. congolense

Some breeds of African cattle, such as N'Dama, are able to tolerate trypanosome infections and remain productive (27, 28). This trait is presumed to have arisen through natural selection, with cattle continually coming into contact with the parasites. Mouse strains also differ in their resistance to *T. congolense* infection; C57BL/6 (B6) mice survive significantly longer than A/J (A) or BALB/c (C) strains (29, 30). Due to the difficulty of keeping and studying cattle populations for identifying resistance and susceptibility genetic factors, these mouse strains are often used as a model for the modes of resistance in cattle.

The differences in resistance in mice have been mapped to three QTL controlling survival after infection (31). These loci have been designated *Trypanosoma* infection response (Tir), and numbered Tir1, Tir2, and Tir3. The Tir1 QTL was shown to exhibit the largest effect on survival (31) and, so, was chosen as the proof-of-concept target for our methodology. Tir1 is located on mouse chromosome 17 in a particularly gene-rich area. It was hypothesised that this QTL may contain a number of QTg.

In order to determine the genes that lie within the Tir1 QTL region, the positions of flanking markers used in the original mapping studies were identified in mouse Ensembl release 40 (NCBI build 36). These were identified as D17Mit29 and D17Mit11 on chromosome 17 (30), at 28,394,586 and 38,278,830 bp, respectively. The position of D17Mit11 was estimated based on close proximity to the gene *Crisp2* (Mouse Genome Identifier – MGI:98815).

4.2. Gene Expression Analysis of Mice Infected with T. Congolense

A microarray gene expression study, previously conducted by researchers of the Wellcome Trust Pathogen–Host Project (*see* Acknowledgements), was used to survey the mouse genome for genes that were differentially expressed between susceptible (A/J and BALB/c) and resistant (C57BL/6) strains. Microarray data used in this investigation are available in ArrayExpress (E-MEXP-1190). Microarray data were first analysed with DChip (32) to determine any outliers. All hybridisations that passed the DChip analysis were normalised using RMA (33). In addition, PCA analysis was performed to identify any hybridisations that passed the DChip quality-control, but could still be classified as outliers.

Analysis of microarray liver sample data from C57BL/6 and A/J mice at days 3 and 7 post-infection, identified 981 and 1,331 probesets that were differentially expressed on the basis of a corrected t-test with a p-value < 0.01 and a $\log_2$ fold change > 0.5. We chose to focus on the early time points of days 3 and 7 for this investigation since the mouse strains used showed a strong gene expression response to infection at the early time points, compared to that of the later time points. The later time points from the microarray study were found to be the result of secondary effects of infection. Permissive criteria were used at this stage of data analysis in order to reduce the incidence of false-negative results (which could result in missing one of the true QTg). As a result of permissive criteria being employed, 2,312 probesets were chosen for further analysis and annotation with their biological pathways.

5. Results and Discussion

The application of these workflows to the lists of genes from both the QTL and microarray data identified a number of biological pathways. Those biological pathways that were found to contain a high proportion of genes, indicating differential expression following trypanosome infection, were prioritised for further analysis. From the original list of 344 genes initially identified from the chosen Tir1 QTL region, 32 candidate QT genes were subsequently used in further analyses (4).

One such pathway chosen for further analysis was the MapK signalling pathway. Within this pathway, *Daxx* was shown to represent the strongest signal of differential expression at early time points. It was found, through examining the scientific literature and re-sequencing the gene in the laboratory, that this gene was a primary candidate QTg (4). This pathway provides an example of the pathway labelled A in **Fig. 18.1**, with the candidate gene being directly related to the QTL region and known, through literature, to be involved in the phenotype. The remaining candidate genes identified from these workflows are currently undergoing further investigation to establish their role in the trypanosomiasis resistance phenotype. Details of gene sequencing methods, carried out for validating potential candidate QT genes, can be found in the paper by Fisher et al. (4).

5.1. Workflows and the Systematic Approach

By automatically passing data from one service to the next through workflows and web services, we are able to process a much greater volume of data than can be achieved using manual data analysis techniques. This enables us to systematically analyse all the results we obtain without the need to prematurely filter the data. In the study by Fisher et al. (4), it was found that researchers of the Wellcome Trust Pathogen–Host Project had failed to identify *Daxx* as a candidate gene for trypanosomiasis resistance. This was the result of manually analyzing the microarray and QTL data, with data analyses directed by a hypothesis-driven approach. Although the use of a hypothesis-driven approach to data analysis is essential for the construction of a scientifically sound investigation, the use of a data-driven approach must also be considered. This would allow the analyses of the experimental data to evolve in parallel to a given hypothesis, regardless of any previous assumptions (16).

By implementing the manually conducted experiments in the form of workflows, we have shown that an automated systematic approach reduces, if not eliminates, the biases associated with manual data analyses. The workflows also provide an explicit methodology for recording the processes involved and provide a framework for the sharing, re-use, and re-purposing of our experimental methods. The workflows developed in this project are freely available for re-use and have been integrated into the myExperiment (34) project which supports scientific collaboration, discovery, and workflow re-use.

The generality of these workflows allows them to be re-used for the analysis of QTL and microarray data in investigations other than that of response to trypanosomiasis. The QTL gene annotation workflow may be utilised in projects that use the mouse model organism, yet have no gene expression data to back up findings. And similarly, the microarray gene annotation workflow may be used in studies with no quantitative evidence for the observed phenotype.

6. Notes

1. If a given marker has not been mapped in the database, the nearest gene may be used as a temporary marker. Likewise, if only a peak marker is available, a suitable genetic distance may be used to map the outer boundaries of the QTL, provided that the exact reasons for the choices are well documented and noted in the experimental protocol.
2. If Taverna produces an "OutOfMemory" error whilst running a data intensive workflow, the memory allocation for the Java Virtual Machine (JVM) may be increased. This is done by editing the "runme" in the Taverna home directory. By altering the numbers in the line "set ARGS = – Xmx1024m" to the size of available RAM of the computer will fix the problem. Here the JVM is permitted to use 1 GB of RAM.
3. If a workflow is unable to be loaded into Taverna, the user must ensure that all the services are currently functional. This can be done by navigating to the URL of the web services contained within the workflow. The user can copy and paste each URL of the individual services into a standard web browser. The result should show an XML-formatted document. If the service is not available, the user must replace the service with one that performs the same functionality or contact the service provider for further details. Alternatively, Taverna offers a user mailing list to which user based queries can be posted to the Taverna user community.

Acknowledgements

The authors would like to acknowledge the assistance of the myGrid consortium, software developers, and its associated researchers. We would also like to thank the researchers of the Wellcome Trust Host–Pathogen Project (GR066764MA). This work is supported by the UK e-Science EPSRC GR/R67743.

References

1. Hedeler, C, Paton, N, Behnke, J, et al. (2006) A classification of tasks for the systematic study of immune response using functional genomics data. *Parasitology* **132**(Pt 2): 157–167.
2. Mitchell, J, McCray, A, Bodenreider, O. (2003) From phenotype to genotype: issues in navigating the available information resources. *Methods Inf Med* **42**(5): 557–563.
3. Liolios, K, Tavernarakis, N, Hugenholtz, P, et al. (2006) The Genomes on Line Database (GOLD) v.2: a monitor of genome projects worldwide. *Nucleic Acids Res* **34**: D332–D334.
4. Fisher, P, Hedeler, C, Wolstencroft, K, et al. (2007) A systematic strategy for large-scale analysis of genotype phenotype correlations: identification of candidate genes involved in African trypanosomiasis. *Nucleic Acids Res* **35**(16): 5625–5633.
5. Köhler, J, Baumbach, J, Taubert, J, et al. (2006) Graph-based analysis and visualization of experimental results with ONDEX. *Bioinformatics* **22**(11): 1383–1390.
6. Macdonald, M, Ambrose, C, Duyao, M, et al. (1993) A novel gene containing a trinucleotide repeat that is expanded and unstable on Huntington's disease chromosomes. The Huntington's Disease Collaborative Research Group. *Cell* **72**(6): 971–983.
7. Glazier, A, Nadeau, J, Aitman, T, (2002) Finding genes that underlie complex traits. *Science* **298**(5602): 2345–2349.
8. Brown, A, Olver, W, Donnelly, C, et al. (2005) Searching QTL by gene expression: analysis of diabesity. *BMC Genet* **6**(1): 12–20.
9. Doerge, R. (2002) Mapping and analysis of quantitative trait loci in experimental populations. *Nat Rev Genet* **3**(1): 43–52.
10. Schadt, E. (2006) Novel integrative genomics strategies to identify genes for complex traits. *Anim Genet* **37**(1): 18–23.
11. Stevens, R, Tipney, H, Wroe, C, et al. (2004) Exploring Williams-Beuren syndrome using myGrid. *Bioinformatics* **20**(1): 303–310.
12. Hitzemann, R, Malmanger, B, Reed, C, et al. (2003) A strategy for the integration of QTL, gene expression, and sequence analyses. *Mamm Genome* **14**(11): 733–747.
13. Flint, J, Valdar, W, Shifman, S, et al. (2005) Strategies for mapping and cloning quantitative trait genes in rodents. *Nat Rev Genet* **6**(4): 271–286.
14. Dharmadi, Y, Gonzalez, R. (2004) DNA microarrays: experimental issues, data analysis, and application to bacterial systems. *Biotechnol Prog* **20**(5): 1309–1324.
15. Kell, D. (2002) Genotype-phenotype mapping: genes as computer programs. *Trends Genet* **18**(11): 555–559.
16. Kell, D, Oliver, S. (2004) Here is the evidence, now what is the hypothesis? The complementary roles of inductive and hypothesis-driven science in the post-genomic era. *Bioessays* **26**(1): 99–105.
17. Illuminating the black box. *Nature* 2006, **442**(7098): 1.
18. Stein, L. (2003) Integrating biological databases. *Nat Rev Genet* **4**(5): 337–345.
19. Stein, L. (2002) Creating a bioinformatics nation. *Nature* **417**(6885): 119–120.
20. Oinn, T, Addis, M, Ferris, J, et al. Taverna: a tool for the composition and enactment of bioinformatics workflows. *Bioinformatics* **20**(17): 3045–3054.
21. Tomfohr J, Lu J, Kepler T. (2005) Pathway level analysis of gene expression using singular value decomposition. *BMC Bioinformatics* **6**: 225.
22. Oinn, T (2003) Talisman-rapid application development for the grid. *Bioinformatics* **19**(1): i212–i214.
23. Birney, E, Andrews, D, Caccamo, M, et al. (2006) Ensembl 2006. *Nucleic Acids Res* **34**: D556–D561.
24. Kanehisa, M, Goto, S. (2000) KEGG: Kyoto encyclopedia of genes and genomes. *Nucleic Acids Res* **28**(1): 27–30.
25. Maglott, D, Ostell, J, Pruitt, K, et al. (2007) Entrez Gene: gene-centered information at NCBI. *Nucleic Acids Res* **35**: D26–D31.
26. Bairoch, A, Apweiler, R, Wu, C, et al. (2005) The Universal Protein Resource (UniProt). *Nucleic Acids Res* **33**: D115–D119.
27. Hill, E, O'Gorman, G, Agaba, M, et al. (2005) Understanding bovine trypanosomiasis and trypano tolerance: the promise of functional genomics. *Vet Immunol Immunopathol* **105**(3–4): 247–258.
28. Hanotte, O, Ronin, Y, Agaba, M, et al. (2003) Mapping of quantitative trait loci controlling trypanotolerance in a cross of tolerant West African N'Dama and

susceptible East African Boran cattle. *Proc Natl Acad Sci USA* **100**(13): 7443–7448.

29. Iraqi, F, Clapcott, S, Kumari, P, et al. (2000) Fine mapping of trypanosomiasis resistance loci in murine advanced intercross lines. *Mamm Genome* **11**(8): 645–648.
30. Koudandé, O, van Arendonk, J, Iraqi, F. (2005) Marker-assisted introgression of trypanotolerance QTL in mice. *Mamm Genome* **16**(2): 112–119.
31. Kemp, S, Iraqi, F, Darvasi, A, et al. (1997) Localization of genes controlling resistance to trypanosomiasis in mice. *Nature Genetics* **16**(2): 194–196.
32. Li, C, Wong, W. (2001) Model-based analysis of oligonucleotide arrays: model validation, design issues and standard error application. *Genome Biol* **2**(8): research0032.0031-research0032.0011.
33. Irizarry, R, Hobbs, B, Collin, F, et al. (2003) Exploration, normalization, and summaries of high density oligonucleotide array probe level data. *Biostatistics* **4**(2): 249–264.
34. myExperiment: http://myexperiment.org/.

Index

K. DiPetrillo (ed.), *Cardiovascular Genomics*, Methods in Molecular Biology 573,
DOI 10.1007//978-1-60761-247-6, © Humana Press, a part of Springer Science+Business Media, LLC 2009

F

G

H

I

J

K

L

M

T

V

W

Z